I0888026

Innovations in Machine Learning and IoT for Water Management

Abhishek Kumar
University of Castilla-La Mancha, Spain

Arun Lal Srivastav
Chitkara University, India

Ashutosh Kumar Dubey
University of Castilla-La Mancha, Spain

Vishal Dutt
AVN Innovations Pvt. Ltd., India

Narayan Vyas
AVN Innovations Pvt. Ltd., India

IGI Global
PUBLISHER of TIMELY KNOWLEDGE

A volume in the Advances in Civil and Industrial Engineering (ACIE) Book Series

Published in the United States of America by
IGI Global
Engineering Science Reference (an imprint of IGI Global)
701 E. Chocolate Avenue
Hershey PA, USA 17033
Tel: 717-533-8845
Fax: 717-533-8661
E-mail: cust@igi-global.com
Web site: http://www.igi-global.com

Library of Congress Cataloging-in-Publication Data

Names: Kumar, Abhishek, 1989- editor. | Srivastav, Arun Lal, 1984- editor. | Dubey, Ashutosh Kumar (Computer scientist) editor. | Dutt, Vishal, Dr. editor. | Vyas, Narayan, 1998- editor.
Title: Innovations in machine learning and IoT for water management / edited by Abhishek Kumar, Arun Srivastav, Ashutosh Dubey, Vishal Dutt, Narayan Vyas
Description: Hershey, PA : Information Science Reference, 2024 | Includes bibliographical references and index | Summary: "This book investigates the revolutionary potential of ML and IoT technologies in addressing various difficulties associated with water management"-- Provided by publisher
Identifiers: LCCN 2023049101 (print) | LCCN 2023049102 (ebook) | ISBN 9798369311943 (hardcover) | ISBN 9798369311950 (ebook)
Subjects: LCSH: Water-supply--Management--Data processing. | Water quality management--Data processing. | Internet of Things. | Machine learning.
Classification: LCC TD353 .I496 2024 (print) | LCC TD353 (ebook) | DDC 333.9100285/631--dc23/eng/20231108
LC record available at https://lccn.loc.gov/2023049101
LC ebook record available at https://lccn.loc.gov/2023049102

This book is published in the IGI Global book series Advances in Civil and Industrial Engineering (ACIE) (ISSN: 2326-6139; eISSN: 2326-6155)

British Cataloguing in Publication Data
A Cataloguing in Publication record for this book is available from the British Library.

All work contributed to this book is new, previously-unpublished material. The views expressed in this book are those of the authors, but not necessarily of the publisher.

For electronic access to this publication, please contact: eresources@igi-global.com.

Advances in Civil and Industrial Engineering (ACIE) Book Series

Ioan Constantin Dima
University Valahia of Târgovişte, Romania

ISSN:2326-6139
EISSN:2326-6155

Mission

Private and public sector infrastructures begin to age, or require change in the face of developing technologies, the fields of civil and industrial engineering have become increasingly important as a method to mitigate and manage these changes. As governments and the public at large begin to grapple with climate change and growing populations, civil engineering has become more interdisciplinary and the need for publications that discuss the rapid changes and advancements in the field have become more in-demand. Additionally, private corporations and companies are facing similar changes and challenges, with the pressure for new and innovative methods being placed on those involved in industrial engineering.

The **Advances in Civil and Industrial Engineering (ACIE) Book Series** aims to present research and methodology that will provide solutions and discussions to meet such needs. The latest methodologies, applications, tools, and analysis will be published through the books included in **ACIE** in order to keep the available research in civil and industrial engineering as current and timely as possible.

Coverage

- Production Planning and Control
- Coastal Engineering
- Operations Research
- Urban Engineering
- Ergonomics
- Transportation Engineering
- Earthquake engineering
- Quality Engineering
- Optimization Techniques
- Construction Engineering

IGI Global is currently accepting manuscripts for publication within this series. To submit a proposal for a volume in this series, please contact our Acquisition Editors at Acquisitions@igi-global.com or visit: http://www.igi-global.com/publish/.

The Advances in Civil and Industrial Engineering (ACIE) Book Series (ISSN 2326-6139) is published by IGI Global, 701 E. Chocolate Avenue, Hershey, PA 17033-1240, USA, www.igi-global.com. This series is composed of titles available for purchase individually; each title is edited to be contextually exclusive from any other title within the series. For pricing and ordering information please visit http://www.igi-global.com/book-series/advances-civil-industrial-engineering/73673. Postmaster: Send all address changes to above address.

Titles in this Series

For a list of additional titles in this series, please visit: www.igi-global.com/book-series

Intelligent Engineering Applications and Applied Sciences for Sustainability
Brojo Kishore Mishra (NIST Institute of Science and Technology (Autonomous), India)
Engineering Science Reference • copyright 2023 • 542pp • H/C (ISBN: 9798369300442) • US $270.00 (our price)

Global Science's Cooperation Opportunities, Challenges, and Good Practices
Mohamed Moussaoui (National School of Applied Sciences of Tangier (ENSAT), Abdelmalek Essaadi University, Morocco)
Engineering Science Reference • copyright 2023 • 357pp • H/C (ISBN: 9781668478745) • US $245.00 (our price)

Handbook of Research on Inclusive and Innovative Architecture and the Built Environment
Ng Foong Peng (Taylor's University, Malaysia) and Ungku Norani Sonet (Taylor's University, Malaysia)
Engineering Science Reference • copyright 2023 • 526pp • H/C (ISBN: 9781668482537) • US $315.00 (our price)

Artificial Intelligence and Machine Learning Techniques for Civil Engineering
Vagelis Plevris (Qatar University, Qatar) Afaq Ahmad (University of Engineering and Technology, Taxila, Pakistan) and Nikos D. Lagaros (National Technical University of Athens, Greece)
Engineering Science Reference • copyright 2023 • 385pp • H/C (ISBN: 9781668456439) • US $250.00 (our price)

Geoinformatics in Support of Urban Politics and the Development of Civil Engineering
Sérgio António Neves Lousada (University of Madeira, Portugal)
Engineering Science Reference • copyright 2023 • 263pp • H/C (ISBN: 9781668464496) • US $250.00 (our price)

Impact of Digital Twins in Smart Cities Development
Ingrid Vasiliu-Feltes (University of Miami, USA)
Engineering Science Reference • copyright 2023 • 314pp • H/C (ISBN: 9781668438336) • US $260.00 (our price)

AI Techniques for Renewable Source Integration and Battery Charging Methods in Electric Vehicle Applications
S. Angalaeswari (Vellore Institute of Technology, India) T. Deepa (Vellore Institute of Technology, India) and L. Ashok Kumar (PSG College of Technology, India)
Engineering Science Reference • copyright 2023 • 288pp • H/C (ISBN: 9781668488164) • US $260.00 (our price)

Urban Life and the Ambient in Smart Cities, Learning Cities, and Future Cities
H. Patricia McKenna (AmbientEase, Canada)
Engineering Science Reference • copyright 2023 • 261pp • H/C (ISBN: 9781668440964) • US $215.00 (our price)

IGI Global
PUBLISHER of TIMELY KNOWLEDGE

701 East Chocolate Avenue, Hershey, PA 17033, USA
Tel: 717-533-8845 x100 • Fax: 717-533-8661
E-Mail: cust@igi-global.com • www.igi-global.com

Table of Contents

Detailed Table of Contents

Water is unambiguously susceptible to contamination, as it is able to dissolve a broader spectrum of substances than any other liquid on Earth. Increasing population and urbanization have been imposed to monitor water quality and wastewater management in the current global scenario. Conventional water quality monitoring involves water sampling, testing, and investigation, which are usually performed manually and are not dependable. Rapid economic prosperity generates a larger quantity of wastewater enriched with a broad range of pollutants that pose serious threats to the environment and human health. Advancements in artificial intelligence and machine learning approaches have shown breakthrough potential toward large dataset capture and analysis of large datasets to attain complex large-scale water quality monitoring and wastewater management systems. The current chapter summarizes prospects and potentials of AI technologies for the amelioration of water quality monitoring and wastewater management to establish an integrated sustainable biocomputation platform in the near future.

Safe water is becoming a scarce resource, due to the combined effects of increased population, pollution, and climate changes. Due to the vast increase in global industrial output, rural to urban drift and the over-utilization of land, and high use of fertilizers in farms and sea resources, the quality of water available to people has deteriorated greatly. Around 40% of deaths are caused due to contaminated water in the world. Hence, there is a necessity to ensure supply of purified drinking water for the people both in cities and villages. Smart water quality monitoring systems have gained significant attention due to their ability to enhance water management practices and safeguard water resources. These systems integrate advanced technologies such as IoT sensors, data analytics, and machine learning algorithms to continuously monitor and assess water quality parameters in real-time. Smart water quality monitoring harnesses cutting-edge technologies, including internet of things (IoT), sensors, and data analytics, to revolutionize traditional water quality assessment methods.

The water management industry has undergone a sea change since the advent of machine learning (ML) and internet of things (IoT) technology. In this chapter, the utilization of ML and IoT applications for assisting with the fundamentals of water management data gathering and preprocessing will be explored. In order to make educated decisions toward water sustainability, sensors and gadgets connected to the IoT have improved monitoring and evaluation of water resources. In the initial paragraphs, the primary focus of the chapter is introduced: the importance of data collection in water management and the challenges of using traditional data collection techniques. However, before the data acquired from these sensors can be used for analysis and modeling, it must frequently undergo some form of preprocessing. Important

Preface

Water is a precious resource, intricately woven into the fabric of life and indispensable for a myriad of economic activities. Yet, the challenges facing effective water management are formidable, with burgeoning population needs and the specter of climate change looming large. In the crucible of these challenges, recent strides in Artificial Intelligence (AI) and Machine Learning (ML) have emerged as promising beacons capable of reshaping the landscape of water resource management.

This edited reference book meticulously explores the confluence of AI, ML, and the Internet of Things (IoT) in water management. It provides a panoramic view of cutting-edge developments and delves into their practical applications across diverse industries. From the nuanced analysis of satellite imagery for water resource monitoring to the implementation of AI-driven predictive analytics for optimal water usage, the book traverses the intricate terrain where technology meets the fluid dynamics of water management.

The narrative unfolds, shedding light on how IoT facilitates real-time data collection and intelligent decision-making and enhances the overall efficiency of water systems. It delves into the revolutionary potential of ML and IoT technologies in overcoming the multifaceted challenges associated with water management. The comprehensive exploration spans topics such as satellite imagery analysis, aquifer monitoring, hydropower generation, predictive analytics, water quality monitoring, microclimate prediction, weather forecasting, leak detection, wastewater treatment, smart irrigation, demand response systems, water resource planning, flood monitoring, and blockchain-enabled water trading.

This compilation is poised to serve as a valuable resource for researchers, professionals, and decision-makers seeking to harness the power of cutting-edge technologies for the betterment of water management practices. The book addresses a broad audience, including those in water management, environmental science, data science, and IoT. Additionally, engineers, technology enthusiasts, and individuals keen on leveraging AI and IoT to confront the challenges of water resource planning, monitoring, and conservation will find pertinent insights within these pages.

CHAPTER OVERVIEW

The chapters that follow in this volume—ranging from an introduction to ML and IoT for water management to explorations of specific applications such as smart water quality monitoring, automated demand response systems, and AI-driven wastewater treatment—are crafted to impart knowledge and stimulate discourse among professionals and academics alike. We hope this compilation serves as a catalyst for further advancements in the intersection of AI, ML, and IoT, propelling us toward a more sustainable and resilient water future.

Chapter 1: Introduction to ML and IoT for Water Management

Authored by Nalluri Poojitha, B. Kuber, Ambati Vanshika

This inaugural chapter lays the foundation for the book by addressing the critical aspects of planning, regulating, and sustainable utilization of water resources. As the global water shortage jeopardizes ecosystems and human needs, the chapter explores the intersection of Machine Learning (ML) and the Internet of Things (IoT). The authors investigate how this synergy optimizes resource allocation, enhances predictions for water distribution, improves quality surveillance, and identifies leaks. Drawing insights from literature and real case studies, the chapter highlights the transformative potential of ML-IoT fusion in revolutionizing water management practices, providing a glimpse into a more secure and sustainable water future.

Chapter 2: A Comprehensive Exploration of Machine Learning and IoT Applications for Transforming Water Management

Authored by Mandeep Kaur, Rajni Aron, Heena Wadhwa, Righa Tandon, Htet Ne Oo, Ramandeep Sandhu

Addressing pressing issues of water scarcity and environmental concerns, this chapter delves into the innovative realm where the Internet of Things (IoT) and machine learning (ML) intersect to revolutionize water management. The authors scrutinize the effective applications of ML in water resource analysis, emphasizing its role in forecasting water demand and optimizing resource allocation. The chapter concludes by underlining the revolutionary potential of ML and IoT in modernizing water management practices on a global scale, offering a comprehensive exploration of their transformative impact.

Chapter 3: Artificial Intelligence for Water Resource Planning and Management

Authored by Richa Saxena, Vaishnavi Srivastava, Dipti Bharti, Rahul Singh, Amit Kumar, Abhilekha Sharma

In an era marked by growing water scarcity, this chapter underscores the integration of Artificial Intelligence (AI) as a pivotal approach for revolutionizing water resource planning and management. Focused on sustaining ecosystems, human livelihoods, and economic growth, the authors elucidate how AI, encompassing machine learning, data analytics, and optimization, enhances the precision of water availability projections. The chapter emphasizes the role of AI in real-time decision-making, offering valuable insights for preventing water-related emergencies and fostering international collaboration towards a more resilient and water-secure future.

Chapter 4: Forecasting Weather and Water Management Through Machine Learning

Authored by Inzimam Hassan, Zeeshan Lone, Swati, Aya Gamal

This chapter positions machine learning as a powerful tool for data-driven decision-making, particularly in the realms of weather forecasting and water management. Examining the potential of machine learning in optimizing water distribution, anticipating floods and droughts, and predicting water demand, the authors provide a comprehensive review of recent developments in machine learning for weather forecasting and water management. The chapter not only explores the capabilities but also addresses the challenges and presents illustrative examples of effective applications in these domains.

Chapter 5: Optimizing Water Resources With IoT and ML – A Water Management System

Authored by Rakhi Chauhan, Neera Batra, Sonali Goyal, Amandeep Kaur

Addressing the rising demand for water resources and the challenges posed by climate change, this chapter investigates the installation of a Water Management System (WMS) using Machine Learning (ML) and the Internet of Things (IoT). By deploying IoT-enabled sensors in various water infrastructure sites, the proposed WMS gathers real-time data on critical variables, subsequently optimized by sophisticated ML algorithms. Serving as an illustrative example, the chapter demonstrates how ML and IoT have the potential to reshape current approaches to water management, offering solutions for efficient resource utilization.

Chapter 6: Utilizing Machine Learning for Enhanced Weather Forecasting and Sustainable Water Resource Management

Authored by Risha Dhargalkar, Viosha Cruz, Abdullah Alzahrani

This chapter navigates the intersection of nature's influence on human life, focusing on weather forecasting and water management challenges. Highlighting the limitations of traditional weather forecasting models, the authors showcase the revolutionary impact of machine learning and artificial intelligence. Offering an overview of essential weather forecasting attributes and various machine-learning algorithms, the chapter provides insights into forecasting weather conditions, emphasizing the crucial applications in water resource management for automation, control, and efficient distribution of this vital resource.

Chapter 7: AI-Based Smart Water Quality Monitoring and Wastewater Management – An Integrated Bio-Computational Approach

Authored by Dipankar Ghosh, Sayan Adhikary, Srijaa Sau

This chapter delves into the susceptibility of water to contamination, exacerbated by population growth and urbanization. Focusing on water quality monitoring and wastewater management, the authors advocate for advancements in artificial intelligence and machine learning. The chapter outlines the breakthrough potential of AI technologies in large-scale water quality monitoring and wastewater management systems. Emphasizing the integration of bio-computational approaches, the authors envision an integrated sustainable platform for water quality monitoring in the near future.

Chapter 8: Revolutionizing Water Quality Monitoring – The Smart Tech Frontier

Authored by Ambati Vanshika, B. Kuber, Nalluri Poojitha

As safe water becomes increasingly scarce due to population growth, pollution, and climate changes, this chapter explores the frontier of smart water quality monitoring systems. The authors highlight the integration of advanced technologies, including IoT sensors, data analytics, and machine learning algorithms. Positioned as a solution to deteriorating water quality, the Smart Water Quality Monitoring systems continuously assess water quality parameters in real-time, harnessing cutting-edge technologies to revolutionize traditional water quality assessment methods.

Chapter 9: Data-Driven Aquatics – The Future of Water Management With IoT and Machine Learning

Authored by Dankan Gowda V., Anil Sharma, Rama Chaithanya Tanguturi, K. D. V. Prasad, Vasifa Kotwal

This chapter unveils the transformative impact of Machine Learning (ML) and Internet of Things (IoT) on the fundamentals of water management data gathering and preprocessing. Highlighting the importance of data collection and the challenges of traditional techniques, the authors explore the role of sensors and IoT in monitoring and evaluating water resources. The chapter delves into data preparation tasks, including cleansing, outlier identification, and data fusion, enhancing the reliability of future ML algorithms for water sustainability decisions.

Chapter 10: Detection of Ephemeral Sand Rivers Flow Using Hybrid Sandpiper Optimization-Based CNN Model

Authored by Arunadevi Thirumalraj, Anusuya V. S., Manjunatha B.

Focusing on the significance of ephemeral sand rivers as a major water supply in Southern Africa, this chapter investigates the use of satellite optical data for identifying unexpected surface flows. Employing a Hybrid Sandpiper Optimization-Based CNN Model, the authors achieve a remarkable accuracy rate of 99.77%. This chapter showcases the potential of advanced technologies in water resource assessment, emphasizing the importance of accurate surface water identification in challenging conditions.

Chapter 11: Design of IoT-Based Automatic Rain-Gauge Radar System for Rainfall Intensity Monitoring

Authored by Ahmad Budi Setiawan, Danny Ruhiyat, Aries Syamsuddin, Djoko Waluyo, Ardison Ardison

In response to global climate change, this chapter introduces an IoT-based automatic rain-gauge radar system designed to monitor rainfall intensity. Highlighting the environmental impacts of climate change, the authors present a prototype IoT device that detects and records rainfall, providing data through a website and mobile application. The chapter emphasizes the importance of accurate rainfall information for various purposes in the community, outlining the potential impacts on wireless signal transmissions.

Chapter 12: Empowering Safety by Embracing IoT for Leak Detection Excellence

Authored by Neha Bhati, Ronak Duggar, Abeer Saber

This chapter explores the transformative potential of the Internet of Things (IoT) in enhancing safety, particularly in the realm of leak detection. By replacing manual inspections with IoT-driven devices, the authors advocate for real-time monitoring, immediate alerts, and precise leak location capabilities. Beyond technological improvements, the chapter positions IoT-enabled leak detection as a paradigm shift towards safer and more environmentally sustainable production practices.

Chapter 13: Using Augmented Reality (AR) and the Internet of Things (IoT) to Improve Water Management Maintenance and Training

Authored by Muskan Sharma, Yash Mahajan, Abeer Saber

The final chapter explores the convergence of Augmented Reality (AR) and the Internet of Things (IoT) in the context of water management. Highlighting their synergistic potential, the authors delve into how AR overlays real-time data and interactive guidance onto the physical environment, streamlining maintenance, troubleshooting, and training processes. With IoT providing real-time data and remote monitoring, the integration of AR and IoT emerges as a powerful toolkit for addressing water management challenges, promising increased reliability and sustainability in a digitally augmented world.

IN SUMMARY

In concluding this edited reference book on *Innovations in Machine Learning and IoT for Water Management*, we reflect on our esteemed authors' collective expertise and insights. This collaborative effort has woven together a rich tapestry of knowledge to address the urgent global challenges facing water resources. The fusion of Machine Learning (ML) and the Internet of Things (IoT) emerges as a transformative force, propelling water management into a new era of efficiency, sustainability, and resilience.

As the editor(s), we are delighted to witness the diverse perspectives that each chapter brings to the forefront. Each contribution adds a unique layer to our understanding, from the foundational exploration of ML and IoT in water management to the intricacies of weather forecasting, water quality monitoring, and leak detection excellence. The chapters not only dissect current affairs but also project a future where technology becomes an indispensable ally in securing our most vital resource - water.

The comprehensive overviews of ML and IoT applications in water management presented in this book serve as a testament to the collaborative efforts of researchers, scholars, and practitioners in this field. The meticulous analyses, supported by real-world case studies, highlight the potential for these technologies to revolutionize the domain. The applications explored are practical and visionary, from forecasting weather patterns to optimizing water distribution and monitoring water quality in real-time.

This book aims not only to inform but also to inspire. It calls upon water management professionals, researchers, and policymakers to embrace the opportunities presented by ML and IoT. The chapters collectively advocate for a paradigm shift, urging stakeholders to integrate cutting-edge technologies into their practices for a more secure and sustainable water future.

In the face of growing water scarcity, environmental concerns, and the need for efficient resource allocation, the insights shared in this book serve as a beacon of hope. The journey through the chapters underscores the imperative for interdisciplinary collaboration and the seamless integration of technology into water management practices. As editors, we are confident that this compilation will spark further exploration, research, and innovation in the dynamic intersection of ML and the IoT for water management.

Our sincere gratitude goes to all the authors who have contributed their expertise, time, and dedication to making this book a comprehensive and valuable resource. We hope the knowledge encapsulated within these pages propels us toward a future where water, a precious and finite resource, is managed with the precision and care technology can afford.

May this edited reference book serve as a catalyst for positive change, prompting a collective commitment to harness the power of ML and the IoT in pursuing sustainable water management practices worldwide.

Abhishek Kumar
University of Castilla-La Mancha, Spain

Arun Lal Srivastav
Chitkara University, India

Ashutosh Kumar Dubey
University of Castilla-La Mancha, Spain

Vishal Dutt
AVN Innovations Pvt. Ltd., India

Narayan Vyas
AVN Innovations Pvt. Ltd., India

Chapter 1
Introduction to ML and IoT for Water Management

Nalluri Poojitha
Sri Padmavati Mahila Visvavidyalayam, India

B. Ramya Kuber
Sri Padmavati Mahila Visvavidyalayam, India

Ambati Vanshika
Sri Padmavati Mahila Visvavidyalayam, India

ABSTRACT

Planning, regulating, and sustainably using water supplies to meet various demands are all critical to managing water resources. A growing water shortage puts ecosystem health and mortal needs at risk, prompting creative solutions. The main objective is to investigate how ML, along with IoT synergies, optimise resource allocation, improve predictions for water distribution, improve quality surveillance, and identify leaks, ultimately promoting sustainability as well as informed decision-making. The study highlighted the significant potential of the ML-IoT fusion in transforming water management practises. This is accomplished by undertaking a thorough analysis of the current literature, which is supported by actual case studies. This investigation will ultimately shed light on the bright future of water resource management, wherein insights based on data as well as cutting-edge technology will pave the way for a more secure and sustainable water supply.

1. INTRODUCTION

Water management has been transformed by machine learning (ML), which provides data-driven solutions to handle the escalating problems in this crucial area. To enhance the management, distribution, and quality control of water resources, machine learning (ML) algorithms examine large datasets from several sources, such as sensors along with satellite data records (Sun & Scanlon, 2019). They enable distribution network optimization for increased efficiency, leak detection, and predictive modeling of

DOI: 10.4018/979-8-3693-1194-3.ch001

water demand. ML is essential for identifying toxins, keeping track of contaminants, as well as ensuring environmental requirements are followed (Kshirsagar et al., 2022). Moreover, it supports early warning infrastructures for floods, droughts, and floods, supporting authorities in making anticipatory choices to protect water resources and maintain sustainable management.

1.1 Background and Context

With the rising demand for clean water and the prospect of water shortages, efficient water use is a crucial worldwide concern. By allowing real-time data collection, analysis, and optimization of resources, "leveraging the power of machine learning (ML) and the Internet of things (IoT)" may transform water management (Fu et al., 2022). Water management includes planning, implementation, and regulation practices to sustainably use and distribute water resources. It is important to maintain ecosystems as it helps in an effective supply for many requirements, including drinking, agriculture, and industry. It combines human needs with environmental preservation, preserving the future, and is essential due to the developing water shortage (Fox, 2019).

The convergence of "ML and the Internet of Things (IoT)" has developed into an uprising force in today's society, especially in the crucial water management field. The responsible operation of water resources has evolved into a more urgent issue as the world's population continues to increase (Lowe et al., 2022). Combining the IoT's network of interlinked sensors and devices with ML, which can handle enormous volumes of data and extract insightful knowledge, offers a potent remedy for these issues (Fox, 2019).

Figure 1. Water resource management

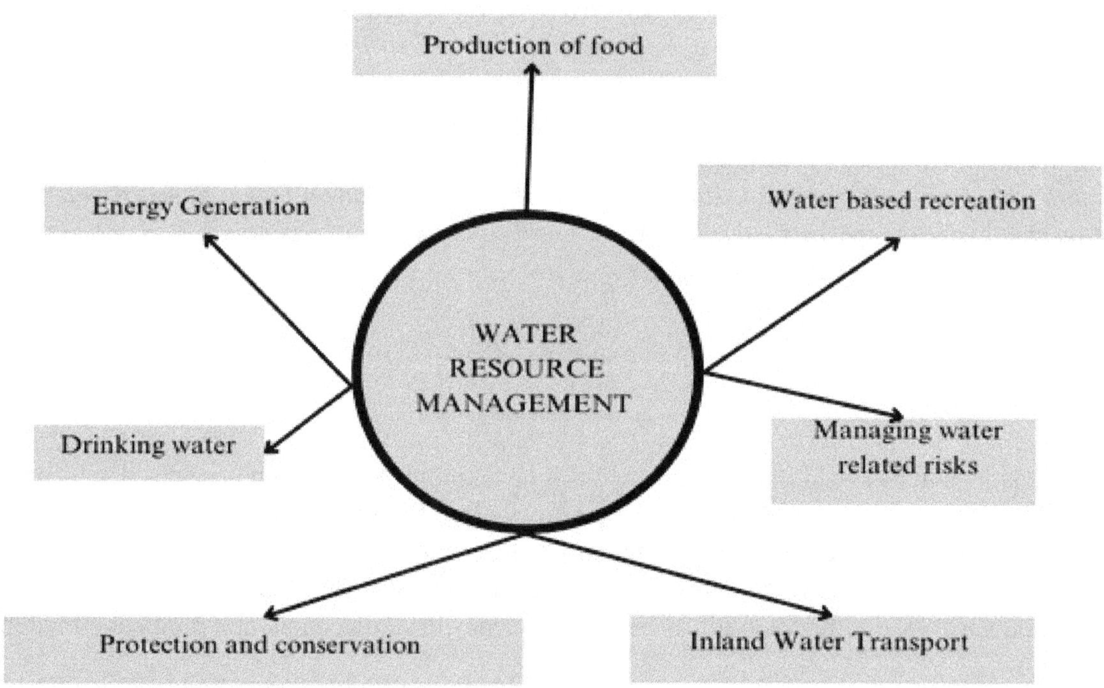

This change in water management has the potential to not only reduce problems with water scarcity but also support environmental sustainability as well as economic development (Kumar et al., 2020). In this investigation of ML along with IoT for water management, this research will dig into the fundamental ideas, uses, and advantages of this ground-breaking combination, illuminating how it is changing how we handle one of the planet's most essential resources (Bedi et al., 2020).

1.2 Significance of ML and IoT Fusion in Water Management

The fusion of ML and the IoT in water management optimizes resource use. ML analyses IoT-generated data to enhance predictive insights for effective water distribution, quality monitoring, and leak detection. As a result, managing water resources becomes more sustainable, waste is reduced, and informed decisions are made (Lowe et al., 2022).

The value of the ML and IoT fusion lies in its capacity to address global water-related issues comprehensively. It makes it possible to allocate resources and create policies based on data, which helps governments and other organizations prioritize their water management projects. Further, by reducing energy use associated with wasteful practices and the ecological consequences of water extraction and transport, it supports environmental sustainability (Bedi et al., 2020). Fundamentally, the incorporation of ML and IoT into water management heralds a paradigm shift in favor of a more equitable, robust, and long-term strategy for safeguarding one of the most precious resources (Fox, 2019).

1.3 Objectives and Scope of Chapter

The goal of this proposal is to offer readers a solid grasp of how ML and IoT technologies are employed in water management by swiftly covering the "Introduction to ML and IoT for Water Management" chapter. It highlights their contributions to data-driven decision-making, resource efficiency, and sustainable water resource usage.

The chapter further clarifies the broader picture of the sustainable use of water resources. It examines the effects of ML and IoT technologies on the economy and the environment for water management. The chapter looks at their ability to reduce water shortages, enhance water quality, and promote environmental protection to emphasize their crucial role in fostering sustainable development.

The scope of this chapter is investigating the various purposes of ML and IoT in water management, with a focus on decision-making based on data, efficient use of resources, and sustainability. While emphasizing its role in alleviating water scarcity & boosting environmental preservation, it also investigates economic as well as environmental ramifications.

2. MACHINE LEARNING: A FOUNDATION FOR DATA-DRIVEN DECISION-MAKING

In the context of water management, it is an essential resource that explores the intersection between contemporary technology and significant decision-making processes. Never before has there been a greater need for data-driven solutions, particularly now when resources are few and the environment is suffering. Overcoming the formidable obstacles requires comprehending and anticipating complex hydrological cycles. Machine learning makes it possible to do in-depth data analysis, trend forecasting,

Figure 2. Main aspects of machine learning in decision-making

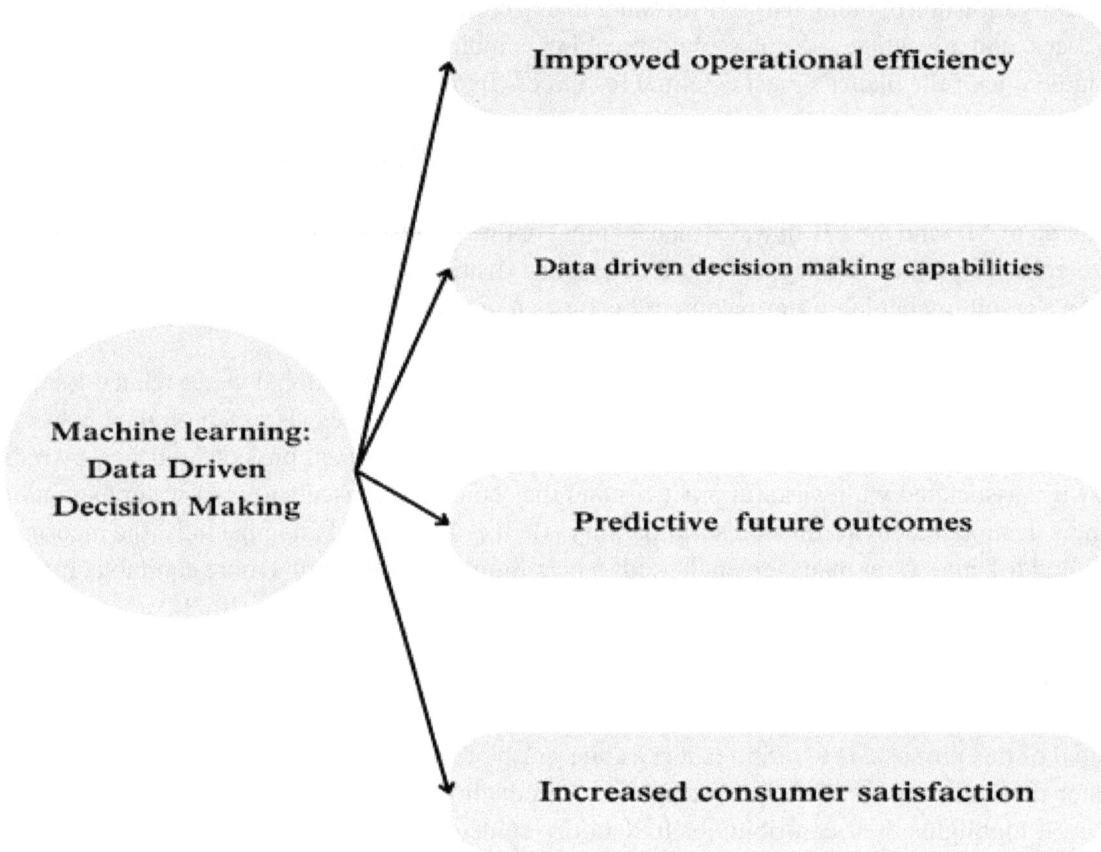

resource allocation optimization, as well as even real-time identification of anomalies or approaching crises (Bhardwaj et al., 2022). The following image explains the main aspects of machine learning in decision-making:

2.1 Understanding Machine Learning in Water Management

Traditional ML techniques, like "multi-layer perceptron (MLP) neural networks," are constrained in how much raw data they can handle and require domain knowledge to digest data before learning. This issue is resolved by ML, which enables automated feature extraction from various representational levels, ranging from the most basic to the most complex (Gupta et al., 2022). The impact of ML on water management is astounding statistically. A World Bank study claims that ML-driven solutions have significantly increased the efficiency of allocating water resources in some areas by 20% (World Bank, 2023). In addition, research presented in the journal "Water Resources Research" demonstrated how

ML algorithms were able to predict water quality metrics with an astounding 95% accuracy, enabling the early identification of contamination episodes (World Bank, 2023). By minimizing water waste and protecting this priceless resource, this data-driven strategy has not only revolutionized decision-making processes but also produced considerable environmental advantages.

"Support vector regression" (SVR), "partial least squares" (PLS), and "deep neural network" (DNN) models all had morc accuracy than traditional models based on real-time data and satellite data when compared with the results generated by experimental ML experiments. A focus neural network built on "convolutional neural networks" (CNN) can also distinguish between clean and contaminated water utilizing water image data (Zhu et al., 2022).

2.2 ML Techniques for Data Analysis and Interpretation

The Environment and Water Management (EWM) field might be significantly impacted by big data and ML technology (Sun & Scanlon, 2019). Water quality and demand forecasting are aided by algorithms like "Support Vector Machines," "Random Forests," and "Decision Trees." Furthermore, "K-Means" and other clustering techniques help to discover use trends. Deep Learning techniques and neural networks offer insights into hydrological predictions. These methods improve the ability to sustainably make well-informed decisions for managing water resources.

The efficiency of various ML techniques is supported by statistics. According to a thorough investigation by the United Nations, the application of Support Vector Machines in water quality prediction achieved an astounding 90% accuracy rate, boosting the ability to safeguard water quality and public health (HESS, 2023). In addition to these developments, clustering methods like "K-Means" have made it possible for EWM experts to identify complex water usage trends. The Earth Observation program of the "European Space Agency" used K-Means clustering in collaboration with ML experts to identify previously unrecognized trends in water consumption across agricultural regions (ESA, 2023). In response to concerns about food security, this finding has enhanced irrigation methods and resulted in a considerable 20% increase in agricultural production (HESS, 2023).

Due to its capacity to analyze enormous volumes of data from several sources, AI has become a potent tool for improving water management. AI can offer insightful information about how water management procedures are carried out by combining data from weather patterns, satellite photography, and sensor networks (Nova, 2023).

According to the research of Elbeltagi et al. (2023), statistical measures typically provide findings that are worse for each model during the testing stage than they do during the training stage. Nevertheless, all of the models meet the requirements to be used for the currently analyzed stations due to their appropriate dimensions within every performance matrix and their consistent statistical accuracy during the testing and training stages. The mathematical results of the "LR, AR, RSS, RF, REPTree, and M5P models" that have been applied to all of the "VPD data sets", based on the study's conclusions, serve as the basis for the forecasting of the VPD (Elbeltagi et al., 2023).

Consequently, the VPD technique may be employed for water management coupled with AI and IoT algorithms. These model inputs were chosen due to their high performance and superior capacity for learning complicated as well as highly nonlinear relationships, including the "VPD time series". The results show that all of the hybrid structures can reasonably predict VPD using them (Elbeltagi et al., 2023).

Figure 3. Flowchart of VPD estimation methodology in water management

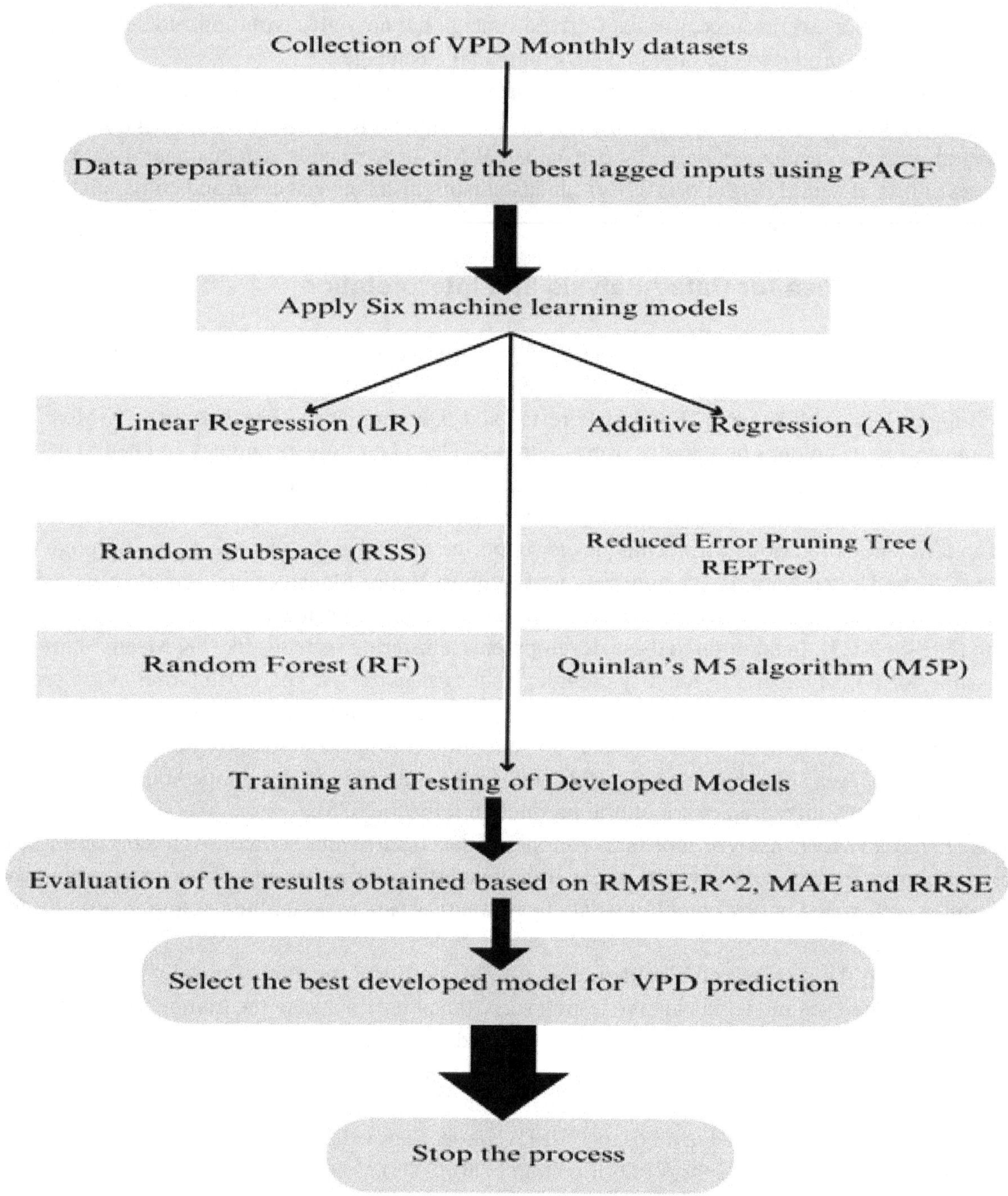

2.3 Role of Machine Learning in Water Quality Assessment

Machine learning plays a crucial part in water quality assessment by quickly and effectively analyzing complicated data from various sources, including sensors, satellites, and environmental databases (Nova, 2023). It improves the accuracy of contamination detection, forecasts pollution trends, and supports prompt decision-making. ML models allow scientists to decipher complex patterns, resulting in safer and more "sustainable management of water resources" (Bedi et al., 2020).

In terms of water quality assessment, this algorithm is preferred as SMOTE only functions with "continuous-valued features," as initially suggested. It was designed to address issues caused by unbalanced data by producing artificial minority class cases (Bedi et al., 2020). The instances of minority classes that were fabricated might provide the model with more information on minority classes (Bedi et al., 2020).

3. THE INTERNET OF THINGS: ENABLING REAL-TIME WATER MONITORING

The management of water is one of the major areas where the Internet of Things has changed how people live. An important tool for resolving these issues in an era marked by growing worries about water scarcity and environmental sustainability is real-time water monitoring using IoT technology (Salam & Salam, 2020). With the help of this ground-breaking approach, it is now feasible to continually monitor

Figure 4. Flowchart of use of IoT modules in water management

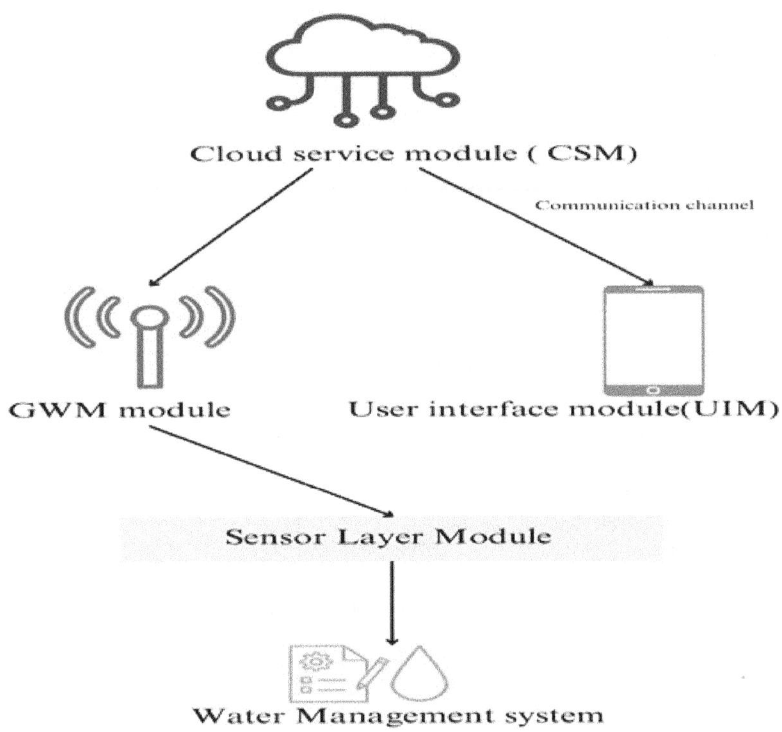

water quality, consumption patterns, and infrastructure performance, providing decision-makers with access to vital data. Due to enhanced efficiency, less waste, and optimal resource allocation, IoT-driven water management helps preserve this scarce resource while also cutting costs and having a less harmful impact on the environment (Geetha & Gouthami, 2016). To preserve the sustainability of the world in this circumstance, it is essential to understand the function that IoT plays in water management.

3.1 Exploring the Internet of Things (IoT) and Its Applications

Resource efficiency and conservation have changed due to IoT adoption in water management. Through a network of connected devices and sensors, the IoT enables real-time monitoring of water quality, consumption, and distribution systems (Salam & Salam, 2020). It is possible to design a building that can handle the water network efficiently in a setting of limited resources by taking into thought the requirements for the development of an affordable and "sustainable water management system" (Narendran et al., 2017). The different layered architecture of IoT system is shown below:

Figure 5. IoT system layered architecture

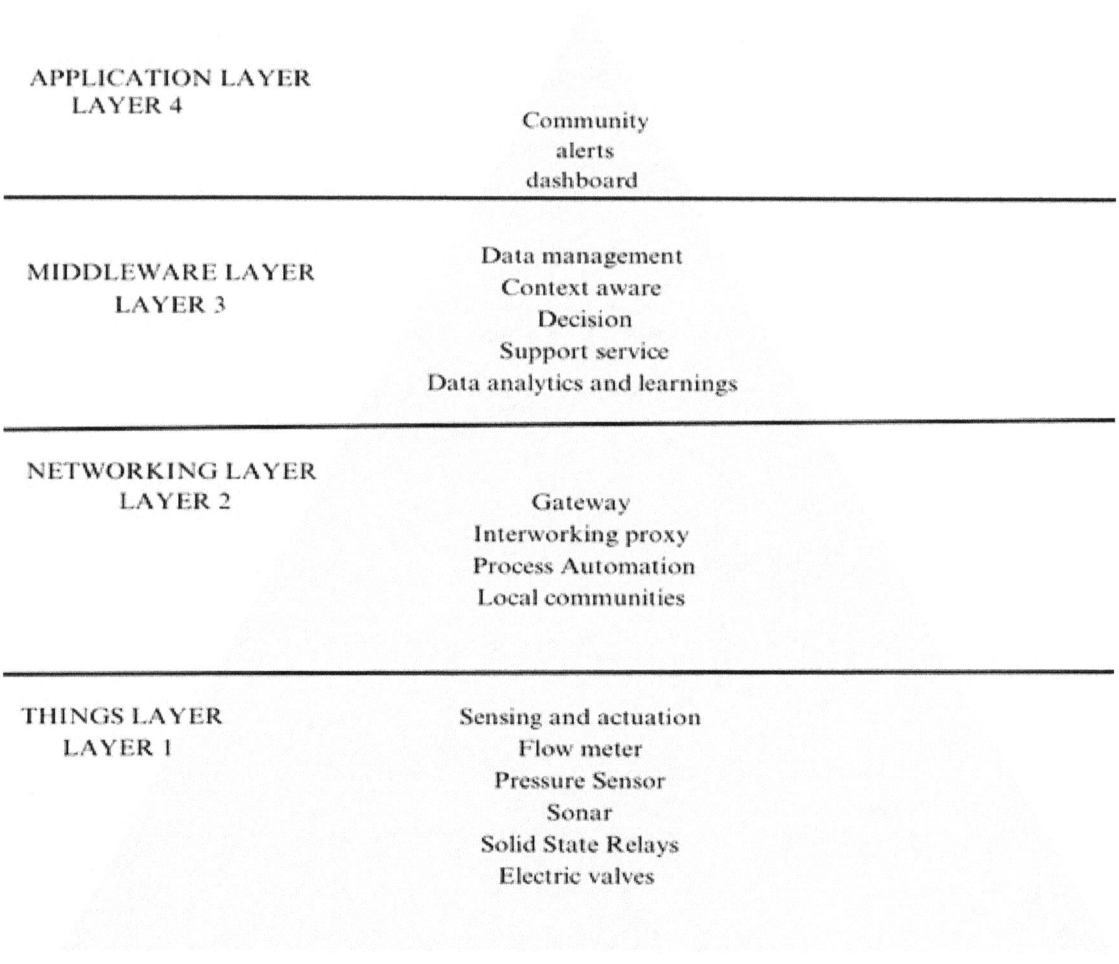

New technologies may be created, integrated into the system, and used to anticipate resource availability, the water cycle, and the relationship between land use and water quantity. As a result, it is possible to build the inventory along with indicators of the bottom-line conditions. In addition, contaminants can spread from watersheds along with other water bodies, including estuaries, rivers, and other coastal areas. Additionally, systems for in-water "HAB observations", such as environmental factors that favor HAB, may be created in real-time, which will help with swift species identification as well as prompt mitigation measures (Salam & Salam, 2020).

3.2 IoT Sensors and Connectivity for Water Parameter Monitoring

The sensor consists of various wireless sensing devices created especially for water monitoring, which may be either quality or quantity, depending on a specific water feature that must be monitored. It has nodes in charge of sensing that are coupled to a microcontroller to enable an interface for water monitoring (Kamaruidzaman & Rahmat, 2020). To measure water level, ultrasonic and liquid level sensors are frequently employed. At the same time, testing essential variables like temperature, turbidity, and pH level of water will determine the quality of the water (Kamaruidzaman & Rahmat, 2020). Moreover, the sensing rate is used to record the water quality to prevent an initially unstable network. The following image explains the basic framework of IoT in Water Parameter Monitoring that can be developed further.

Figure 6. Basic framework of IoT in water parameter monitoring

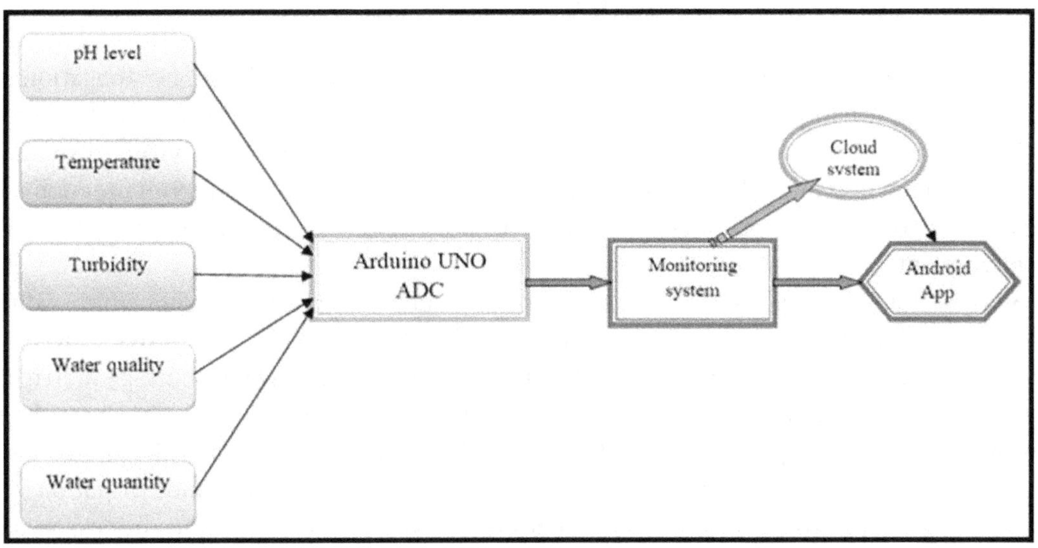

In addition, the IoT has enabled the placement of sensors across the water facilities, which are presently continuously collecting immediate information on its condition. The "IoT cloud mix" has made it easier to address sensor network wireless storage problems. The information given by different water quality monitoring systems during transit, harvesting and retention, IoT sensor machinery, and cloud servers are used to calculate an overall water parameter computation threshold (Chiu et al., 2022). The study suggests a machine learning-based way to evaluate and analyze water quality by forecasting its

Table 1. Data from automated sensors used to forecast turbidity

Parameters	Description
Measuring ID	Beach name and measurement location unique ID record
Measuring timestamp	Date and time the measurements are supposed to be made
Turbidity	Nephelometric Turbidity Units (NTU) measure the turbidity of water. Acceptable, Permissible, and Rejected categories.
Transducer depth	Depth of transducer in meters
Flow speed	Speed of water in meters/seconds
Flow height	Height of water in meters
Water temperature	The temperature of the water is to be measured in Celsius.
Battery life	Battery voltage in mV
Device uptime	Functionality of the device

turbidity based on automated sensor data. Data from automated sensors used to forecast turbidity are shown in Table 1. The optimal turbidity for drinking water is less than 1 NTU, while the maximum acceptable level is 5 NTU (Bhardwaj et al., 2022).

3.3 Case Studies: Real-Life Examples of IoT in Water Management

AgriSensfor Water Management of Irrigated Crop

According to the case study by Roy et al. (2020), AgriSens, an IoT-based variable irrigation planning system, utilizes the IoT to provide real-time and dynamic water treatment during the different growth phases throughout a crop's life cycles. The AgriSens' user-friendly interface, visual display, internet device, and online portal allow farmers to get field information. In terms of energy use, delivery ratio, data validation, and failure, it operates effectively under a variety of climatic situations and with dynamic watering treatments.

AgriSens for Water Management of Irrigated Crops is a state-of-the-art technology that revolutionizes water management in agriculture by fusing "remote sensing, the IoT, and ML. It provides a comprehensive strategy to deal with the problems posed by resource efficiency and water shortage in irrigated crop cultivation (Roy et al., 2020). Machine learning is essential to AgriSens because it analyses the massive amounts of data generated by IoT and remote sensors. According to Roy et al. (2020), ML algorithms can anticipate how much water crops will use, identify abnormalities, and enhance irrigation strategies. Over time, the system learns from prior data, increasing both its accuracy and usefulness in managing water resources. Remote management and optimization of irrigation systems are also made possible by the combination of IoT and ML in AgriSens (Roy et al., 2020). Thus, technologies like AgriSens are crucial resources for maintaining food security and appropriate resource management as water becomes scarcer and agriculture confronts more problems.

MICS for Water Management Using IoT

The recommended control system employs a 4-state switch with manual, automated Internet of Things (IoT) and, ultimately, off-mode operation. A cheap water-level sensor determines how much water is in a field. Based on the needs of the farmer, a flexible and autonomous watering algorithm is provided. The case study by Hadipour et al. (2020) introduces a state-of-the-art technique for water pumps and pump stations that is effectively developed, implemented, and utilized in the agricultural sector. The three primary control systems that comprise MICS are the "electro-pump controller, reservoir water level, and alarm control system" (Hadipour et al., 2020). SMS, which is accessible at any time and from any location, is utilized to administer the whole system under the control of IoT technology (Hadipour et al., 2020). To avoid electrical shocks and mechanical stresses when operating the electro-pump, a soft beginning mechanism was created and considered.

Insights from ML help improve decision-making processes, guaranteeing effective water distribution as well as reducing waste. Additionally, the combination of MICS, IoT, and ML enables remote water infrastructure control, quick reaction to emergencies, and effective resource management (Hadipour et al., 2020). In conclusion, the combination of MICS, IoT, and ML in water management provides a data-driven, proactive, and effective strategy to address water shortage and quality challenges. It improves the capacity to track, examine, and manage water resources, ultimately resulting in more dependable and long-lasting water management systems.

4. SYNERGY UNLEASHED: FUSION OF ML AND IOT IN WATER MANAGEMENT

The most valuable resource may now be controlled in a whole new way thanks to a revolutionary convergence of cutting-edge technology. One of today's most urgent global concerns will be addressed by this ground-breaking combination of ML with the IoT is effective along with sustainable water management. This paradigm change aims to improve resource conservation by utilizing the capabilities of ML algorithms as well as real-time data from IoT devices to optimize water distribution, find leaks, forecast water quality problems, and identify leaks (Torres et al., 2020). This section's methodology strives to unleash hitherto untapped potential for accuracy and efficiency in water management, assuring the sustainability of this precious resource for coming generations.

4.1 Leveraging ML for Data Interpretation and Forecasting in Water Management

The combination of ML and the IoT has changed water management approaches, enabling sustainability and efficiency that were never before possible. By using ML for data interpretation and forecasting, water management systems can precisely decipher complex data patterns to predict water demand and supply fluctuations (Sit et al., 2020). This proactive strategy results in improved resource allocation, less waste, and better decision-making. Additionally, using the distributed ML assistance offered by modern Big Data ecosystems, such as Spark MLlib, the open-source Apache Hadoop setting, and Apache Mahout, may be the best course of action for users in the EWM domain. Theano, Caffe, Torch, and Google TensorFlow are a few examples of open DL libraries and packages that already have hardware acceleration compatibility (Sun & Scanlon, 2019).

The deployment of distributed ML assistance given by modern Big Data ecosystems is particularly advantageous in the area of Enterprise Water Management (EWM). These technologies make it feasible to handle enormous volumes of data in parallel, which speeds up and effectively facilitates the deployment of ML models (Sit et al., 2020). In addition, there are open-source "Deep Learning (DL)" tools and packages that are compatible with hardware acceleration and may be used to apply advanced ML approaches for even more precise predictions and insights. However, ML revolutionizes water management by enabling cutting-edge data interpretation as well as forecasting skills when combined with IoT (Sit et al., 2020). By detecting leaks, ruptures, or contamination early on, one may minimize water loss and the risk of developing waterborne diseases. The significance of anomaly detection stems from the fact that infrequent IoT data irregularities can offer crucial knowledge that can be applied in several sectors, including power, financial services, traffic control, and medicine. It provides an automated approach to spotting potentially dangerous outliers while protecting the data (Nova, 2023).

4.2 Enhancing Real-Time Monitoring and Analysis With IoT and ML

ML algorithms can modify water flow in real time by examining previous usage trends, meteorological information, and infrastructure performance (Arsene et al., 2022). With a focus on the importance of practical examples and application, the paragraph underlines the need for an efficient case study that illustrates the optimization achieved using ML and IoT in water distribution. ML algorithms assess historical consumption trends and meteorological data to adjust water flow in real-time, guarantee equitable distribution, reduce inefficiencies, and react promptly to issues like leaks and pollution. This demonstrates how ML and IoT may transform water management.

One of the key advantages of integrating IoT into water management is the ability to collect real-time data from sensors placed across the water infrastructure. Such sensors might keep an eye on things like water quality, flow rates, pressure, and temperatures (Mohamed et al., 2021). Machine learning technologies must be used extensively to make sense of this vast and complicated data. By examining historical consumption patterns, climate information, and infrastructure efficiency, ML models may precisely predict future water demand (Ryu, 2022). For instance, the network may immediately boost water delivery to match customer demands if an unpredictable heatwave causes an unanticipated rise in demand (Arsene et al., 2022). This would prevent users from suffering problems.

Furthermore, ML systems are capable of spotting data abnormalities like leakage or poisoning issues and notifying users of them. Early identification of such problems can trigger quick response mechanisms, reducing water loss and preserving the health of the public. The case investigation conducted by Arsene et al. (2022) provides an example of the potential benefits of this strategy. They were able to create a water distribution system that was more effective by applying ML algorithms to modify the amount of water flowing in real time using previous consumer patterns and meteorological data (Ryu, 2022). Continuous monitoring and evaluation might be greatly improved by integrating IoT and ML into water management (Mohamed et al., 2021). With the aid of this ground-breaking technique, water utility companies may enhance the distribution of water, adapt to shifting conditions, and take preventative measures against problems like breaches and pollution (Arsene et al., 2022). It is feasible to develop more reliable, effective, and sustainable water management techniques that benefit both people and the environment by utilizing data and sophisticated analytics.

4.3 Case Study: Optimising Water Distribution Through ML and IoT

According to the case study analysis, systems and sensors may utilize ML to gather real-time data and send out alerts about issues, relieving the workload on the personnel and infrastructure that now oversee operations (Bhardwaj et al., 2022). When handled correctly, the infrastructure update may provide new knowledge and achieve efficiency levels that have never been seen before. Additionally, the IoT has made it possible to put sensors throughout the water infrastructure, which is currently continually gathering real-time data on the quality, quantity, and distribution of water. Following that, ML algorithms are used to analyze and evaluate these data points. Along with monitoring water quality, the ML-driven system can alert users in real time when deviations or issues are discovered (Bhardwaj et al., 2022).

This integrated system may also modify and enhance water distribution based on ML predictions. By analyzing historical data and consumption patterns, it can forecast future water demand, allowing for more effective resource allocation and decreasing waste. This forecasting ability is crucial for the efficient management of water resources, especially in areas with unpredictable demand or limited supply (Bhardwaj et al., 2022).

5. ADVANTAGES OF ML AND IOT INTEGRATION IN WATER MANAGEMENT

A revolutionary era in water management has begun as a result of the combination of ML and the IoT. The monitoring, conservation, and optimization of the limited supply of water resources are being revolutionized by the convergence of powerful data analytics and linked devices (Sun & Scanlon, 2019). IoT sensor data may be analyzed using large datasets by ML algorithms, which can then spot trends and abnormalities instantly. This makes early leak detection, effective water distribution, and proactive infrastructure repair possible. Additionally, ML-driven prediction models improve the monitoring of water quality, providing a secure and consistent supply (Sun & Scanlon, 2019). This integration is a crucial component of managing water resources sustainably since it not only saves water but also lowers expenses.

5.1 Efficiency Gains and Reduced Wastage

By utilizing ML algorithms and IoT devices, water management efficacy is increased. Real-time ML analysis of IoT data allows for exact demand predictions and water allocation optimization. Rapid leak detection and repair save waste, improve distribution efficiency, and encourage preservation by preventing unanticipated issues (Togneri et al., 2019).

One advantage of using ML algorithms with IoT devices is their ability to analyze massive volumes of real-time data. ML systems can accurately anticipate water demand by continuously monitoring variables, including weather, usage trends, and water quality (Omambia et al., 2022). The distribution of water resources is effective and productive because of this forecasting capability. This reduces operational costs associated with repairing significant damage caused by concealed leaks, additionally minimizing water loss (Gupta et al., 2020). ML-driven water management systems also play a crucial part in encouraging water conservation. They aid in preventing water shortages and supply interruptions by using predictive analytics to avoid unanticipated issues proactively. This helps conserve this priceless resource while also benefiting customers by assuring a steady water supply (Togneri et al., 2019).

Figure 7. ML in water management

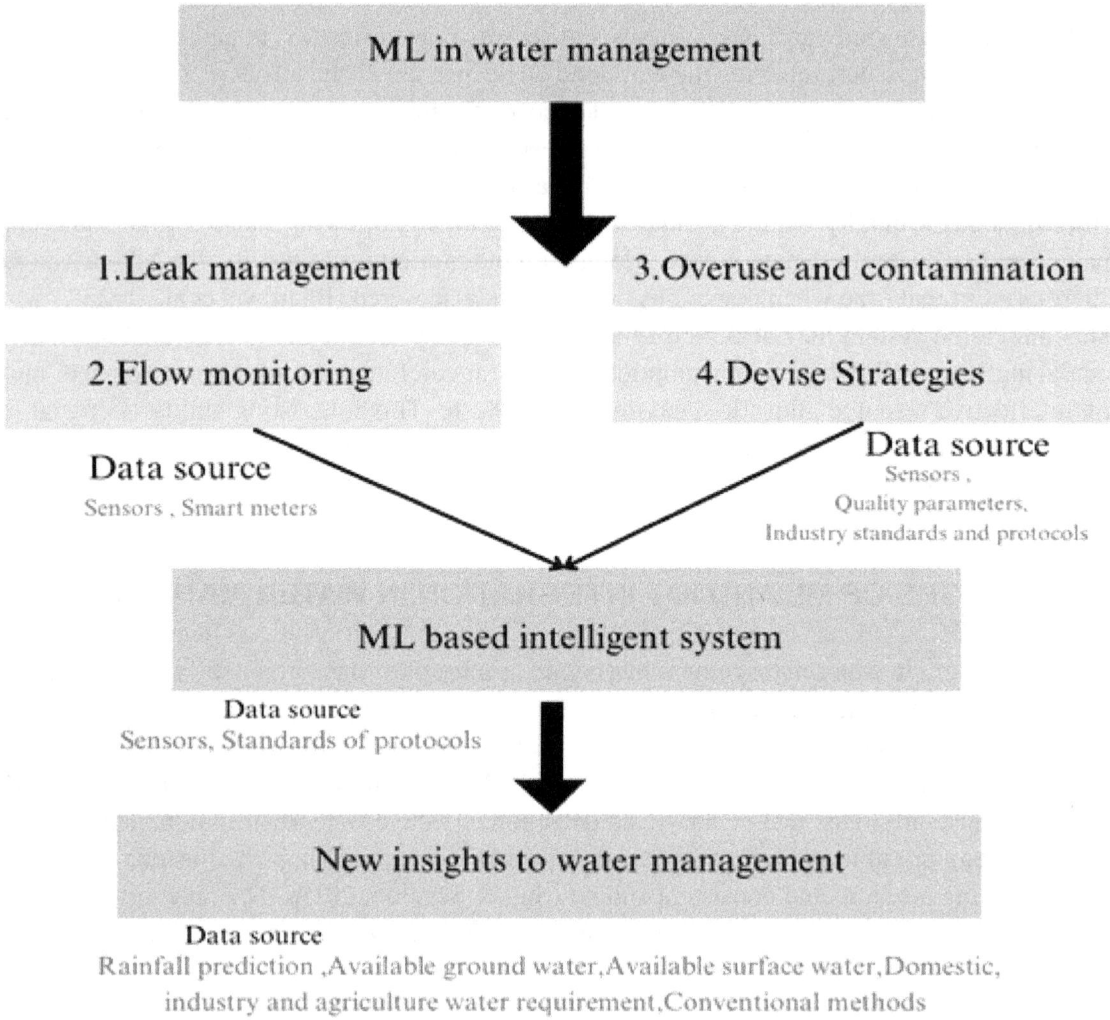

However, the combination of ML algorithms with IoT devices gives a comprehensive strategy for managing water resources. It improves productivity, decreases waste, and encourages sustainable water practices, eventually helping the environment as well as the economy (Gupta et al., 2020). The potential for additional increases in water management efficiency is still bright as technology develops.

5.2 Environmental Preservation and Sustainability

Collaboration between ML and IoT is essential for protecting the environment and ensuring sustainable water supplies. While ML forecasts ecological consequences to support adaptive tactics, IoT's real-time monitoring quickly detects pollution and quality changes. This proactive strategy maintains the health of aquatic life and protects aquatic ecosystems (Krishnan et al., 2022).

Examples include identifying crop species, managing agricultural systems sustainably at the farm and landscape scales, and developing policies. There are several alternatives available with both GIS and remote sensing (RS) techniques (Mondejar et al., 2021). The Analytical Hierarchy Process is combined with RS-GIS, fuzzy reasoning, and multicriteria assessment to create a better database and road plan (Mondejar et al., 2021). Furthermore, Massive datasets and intricate environmental interactions may be analyzed using ML to provide useful insights that guide adaptive tactics. One can make well-informed judgments, manage resources more effectively, and reduce possible environmental impact thanks to this predictive power. IoT, on the other hand, is essential for data collecting and real-time monitoring (Krishnan et al., 2022). IoT sensors positioned in different ecosystems can quickly identify changes in pollution, temperature, and water quality. When conditions deviate from normal, quick intervention is made possible by this instant input (Krishnan et al., 2022). IoT helps to maintain aquatic life and safeguard delicate aquatic habitats by doing this.

5.3 Proactive Water Resource Management for Urban and Rural Areas

The IoT and ML have been combined to produce vital tools for proactively managing water resources in various situations, including urban and rural landscapes. By using data-driven insights to estimate demand precisely in urban settings, this synergy helps to prevent shortages during periods of high use (He et al., 2021). To effectively meet urban demand, it also assists in water distribution optimization.

IoT and ML technologies play a game-changing role in rural communities, which frequently struggle with water scarcity issues (He et al., 2021). By facilitating efficient irrigation techniques and providing thorough data, they enable stakeholders to make informed choices on how to distribute water for agricultural activities (Krishnan et al., 2022). Therefore, by coordinating practices with unique urban and rural demands and developing resource resiliency in changing water supply situations, the combination of ML and IoT significantly helps sustainable water management.

6. OVERCOMING ADOPTION CHALLENGES: SECURITY, SCALABILITY, AND FRAMEWORKS

Leveraging ML and IoT can revolutionize water monitoring and conservation at a time when water resources are becoming more limited (Togneri et al., 2019). This idea does, however, face certain challenges. In-depth analysis of security issues, such as data privacy as well as network weaknesses, is done while emphasizing risk-reduction measures. The chapter also discusses the problems with scalability that develop when systems expand and change. It also explores selecting the appropriate frameworks for reliable and effective ML and IoT applications to guarantee successful as well as sustainable integration into water management practices.

6.1 Security Considerations in ML and IoT Applications

IoT and ML together offer risks that need strict security measures. Protection from unauthorized access and data breaches is crucial as data travels between connected devices and analytic platforms. For the protection of the privacy and security of critical water management data, this part examines encryption methods, authentication mechanisms, and intrusion detection systems (Lowe et al., 2022).

To preserve the security and privacy of critical water management data, a multilayered security framework must be established. Encryption, which encrypts data so that only authorized persons may decrypt it, is a vital component (Togneri et al., 2019). To protect data during transmission and storage, sophisticated cryptographic algorithms and other encryption methods are needed. Mechanisms for authentication are also necessary. These controls ensure that only authorized users or devices interact with sensitive data and control systems by authenticating individuals or devices trying to access the system (He et al., 2021). To prevent unauthorized access, reliable authentication methods are essential. Also serving as a watchful keeper, intrusion detection systems continuously scan the network for anomalies and potential threats.

6.2 Scalability Challenges and Solutions

Scalability problems are caused by the resource-intensive ML algorithms and the abundance of IoT-generated data. This section examines methods for overcoming scalability issues, including the use of edge computing, the addition of cloud services, and the use of data compression. It also looks at workable methods for data processing and storage, making it simple to expand ML and IoT applications (Piemontese et al., 2021).

Additionally, the IoT has made it possible to put sensors throughout the water infrastructure, which is currently continually gathering real-time data on the quality. This approach enhances the efficiency of real-time data processing and decision-making (Piemontese et al., 2021). Cloud services also provide the scalability necessary to manage huge datasets, resulting in dependable and consistent performance. Scalability issues must be resolved via data compression techniques (Gupta et al., 2020). Additionally, as water management systems expand and change, optimizing data processing and storage techniques supports the smooth development of ML and IoT applications. Edge computing, cloud services, data compression, and IoT integration are required in combination with other technologies to overcome scalability issues in the context of ML and IoT integration (Piemontese et al., 2021). Furthermore, creating decision support tools for water management using machine learning can be implemented as a solution. These systems may simulate many scenarios and offer recommendations, assisting decision-makers in selecting the most effective and long-lasting course of action. Since the watershed management "decision support system (DSS)" was created as a cerebral system to help watershed managers allocate water resources as efficiently as possible, the scientific rigor and applicability of watershed management could both be directly impacted by the simulation results of the structure (Zhai et al., 2020). Before wastewater is delivered to freshwater bodies, this kind of device helps to treat it. The sensors make it simple to study factors such as turbidity, pH, and temperature. This is crucial for conducting agricultural tasks safely.

6.3 Robust Frameworks for Successful ML and IoT Implementation

To successfully implement ML and IoT solutions, solid frameworks must be used to accelerate development, deployment, and maintenance. This requires looking at structures that allow for simple changes, optimized data flow, and seamless interaction. Examples from the real world offer helpful insights into recommended procedures. By improving the effectiveness of the development, deployment, and maintenance processes, these frameworks enable ML and IoT's full revolutionary potential in managing water resources (Piemontese et al., 2021). This empowerment makes making smart, data-driven decisions easier, resulting in better resource management and sustainable practices. With an emphasis on scalabil-

Figure 8. Framework for water management using ML and IoT

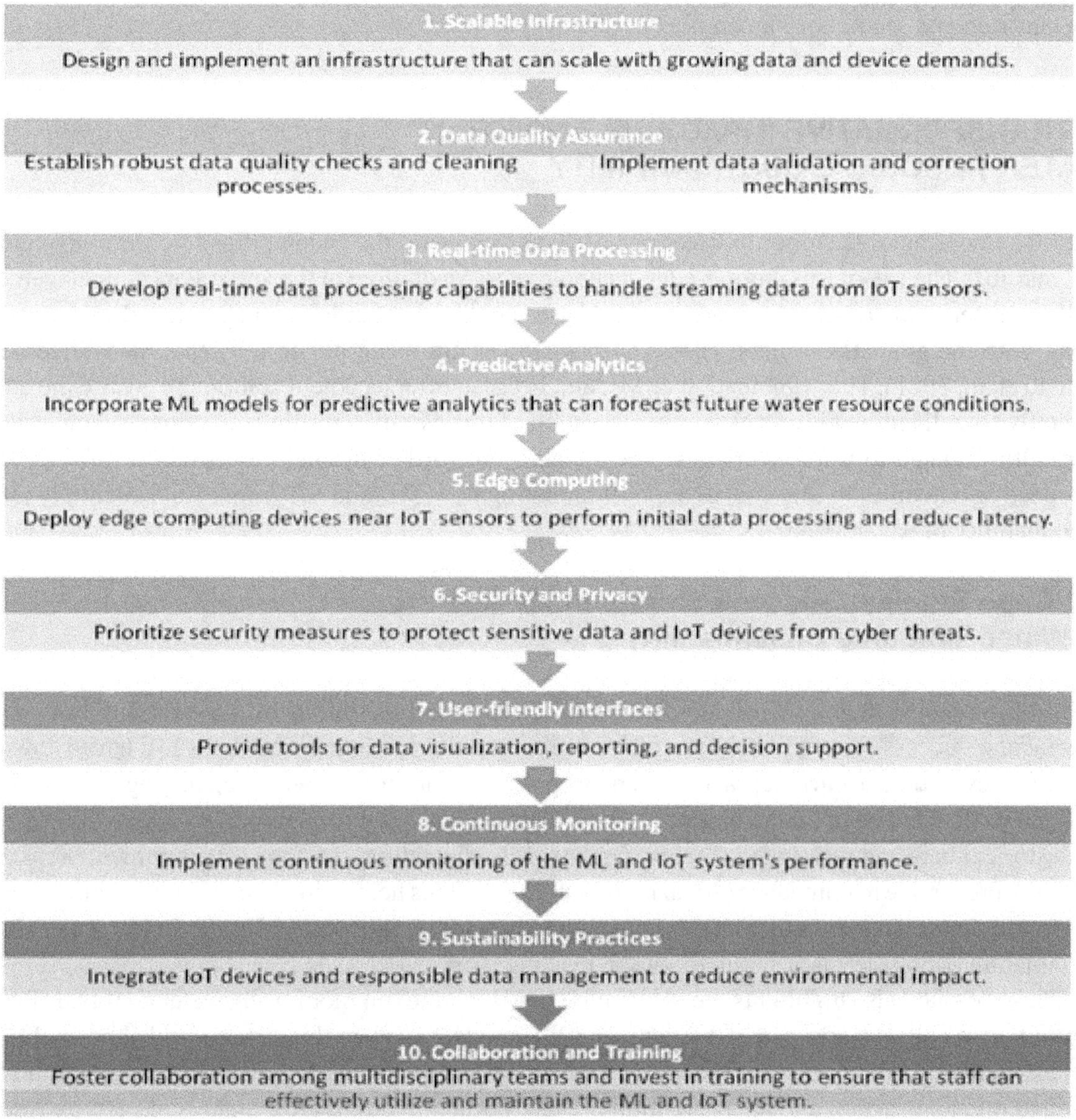

ity, data quality, real-time processing, predictive analytics, security, user-friendliness, sustainability, and collaboration, the framework that follows, which was inspired by research by Piemontese et al. (2021), offers a structured approach to leveraging ML and IoT for effective water resource management:

Moreover, ML assesses sensor data on water quality to find pollutants, pathogens, and heavy metals that can be used. Thus, it is necessary to keep an eye on variables like temperature, turbidity, and TDS to ensure that the water has been appropriately treated before being released. IoT-based smart water quality monitoring systems enhance inspection procedures and thereby lessen the requirement for manual intervention. For the protection of the environment and public health, it is essential to quickly identify problems with water quality. IoT smart meters may be used to track water quality in various locations (Sagan et al., 2020). By doing this, utilities can guarantee that their water, even in isolated places, is nutrient-rich and secure for human use.

7. TRANSFORMATIVE IMPACT: ML AND IOT IN WATER RESOURCE SUSTAINABILITY

To solve important difficulties in water resource management, the chapter digs into the convergence of ML and IoT. The use of this cutting-edge technology to revolutionize the surveillance, management, as well as preservation of water resources is addressed in this chapter. Efficiency gains, decreasing waste, along with the guarantee to gain access to clean water are made possible by this innovative strategy (Ighalo et al., 2021). The combination of ML and IoT also offers a potent solution to ensure the environmentally friendly utilization as well as preservation of this essential resource as climate change worsens difficulties related to water shortages. The chapter will further discuss case studies, difficulties, and prospective outcomes in this crucial area, illuminating how ML and IoT might fundamentally alter the sustainability of water resources.

7.1 Case Studies: Success Stories of ML and IoT in Water Resource Sustainability

According to the article by Aivazidou et al. (2021), the combination of ML and the IoT has produced spectacular success stories in the area of water resource sustainability. ML and IoT have shown their ability to revolutionize current practices in metropolitan environments where managing water is particularly difficult. Predictive maintenance for water infrastructure is one example of success. When trained on historical data and real-time IoT sensor inputs, ML algorithms can forecast equipment breakdowns, enabling preventive maintenance (Aivazidou et al., 2021). This helps conserve water by minimizing water waste and decreasing downtime. Table 2 explains the success stories of ML and IoT in water resource sustainability, considering a summary of the existing works:

The case study also highlights the application of ML and IoT in monitoring water quality. IoT sensors continuously collect water parameter data, and ML models analyze this data to find abnormalities and contamination occurrences. Rapid detection demonstrates the technology's influence on public health by enabling quick action and the prevention of waterborne infections (Krishnan et al., 2022). Additionally, water distribution is optimized using ML-driven demand forecasts, ensuring that resources are distributed effectively. This promotes sustainability by cutting down on both the needless use of energy and water resources.

However, the study's demonstration of the joint application of ML and IoT in urban water management offers a compelling success story (Aivazidou et al., 2021). To achieve sustainable water management in urban settings, it increases infrastructure resilience, boosts water quality, and optimizes resource allocation.

Table 2. Summary of the existing works for success stories of ML and IoT in water resource sustainability

Method	Advantages	Disadvantages	Outcomes
ANN- Artificial Neural Networks RNN- Recurrent Neural Networks Bi-LSTM- Bidirectional long short-term memory. LSTM- Long short-term memory GRU- Gated Recurrent Unit	An effective and efficient model for stream flow	Low accuracy Further to help experts, managers, and officials.	Correlation Coefficient (CC):0.85% Mean Absolute Error (MAE):13.4% Root Mean Square Error (RMSE):21.16% Nash Sutcliffe Efficiency Coefficient (NS):0.65
Hybrid Model Convolutional Neural Network (CNN) Long Short-term Memory (LSTM)	CNN for predicting the water level LSTM for monitoring the water quality Considered three water quality parameters such as, Total Nitrogen (TN), Total organic carbon (TOC), Total Phosphorus (TP)	Used Limited data set Not concentrated on parameters like chlorophyll, algae, dissolved oxygen, and fecal bacteria	NS:0.75 MSE:0.055 TOC:0.832 TN:0.987 TP:0.899
U-Net, TensorFlow Libraries CNN	The proposed method determines the center pivot irrigation systems effectively	The proposed model is deployed on a short area and consumes more time	Accuracy:99% Precision:99% Recall: 88%
SVM(Support Vector Machine) SVR (Support Vector Regression) Radial Basis Function Kernel Random Forest Regression	Proposed IOT smart system for automating the agriculture industry	It does not support dynamic systems Limited data set Low accuracy	Accuracy:81.6%
Deep learning neural network models Belief Rule Based Model (BRDM)	low power consumption, low cost, and high detection accuracy	It works for only a small area Not considered parameters such as Dissolved Solid, Dissolved Oxygen Chemical oxygen demand	Temperature:46.19 celsius Ph value:4.28

7.2 Potential Future Developments and Innovations

Real-time optimization of water distribution networks will be offered via AI-powered devices, which will cut waste and boost effectiveness (Krishnan et al., 2022). According to Ighalo et al. (2021), the Internet of Things is also expected to be more significant in water quality monitoring. IoT must be allowed for remote infrastructure monitoring and management. Pumps, valves, and other system parts may all be adjusted by operators in immediate time to improve system performance and speed up responses to emergencies. Numerous sensors are utilized by IoT-enabled systems for water management to gather real-time data on resource utilization (Sun & Scanlon, 2019). These gadgets send the data that has been collected to the user's internet program. This knowledge enables consumption pattern analysis and promotes more sensible water use.

The use of AI in water management is highlighted by Krishnan et al. (2022), who expect it to continue to be crucial in predictive modeling and decision-making. Even more advanced ML algorithms will make it possible to anticipate water-related problems like droughts and floods more precisely. IoT device implementation throughout aquatic environments will increase due to sensor and network infrastructure developments. These sensors will be able to autonomously transmit and analyze data and detect changes in water quality (Ighalo et al., 2021). AI integration will promote adaptive management tactics and enable speedier reactions to pollution problems.

Additionally, predictive analytics, automated control systems, and cutting-edge sensor networks will all be included in these platforms, which will ultimately result in more resilient and sustainable methods of managing water resources (Ighalo et al., 2021). Technology has the potential to address issues with pollution and water shortages as it develops, opening the path for a more sustainable and water-secure future.

7.3 Implications for Water Management Policies and Strategies

The IoT has made it possible to put sensors throughout the water infrastructure, which is currently continually gathering real-time data on the quality. According to Krishnan et al. (2022), the inclusion of AI and ML into policies and plans for water management will have implications for such policies. These developments enable more proactive and efficient management of water resources. By employing AI and ML, policymakers and water managers may raise the overall sustainability of water systems, better allocate resources, and make smarter decisions. The survey by Sun and Scanlon (2019) emphasized the predictive capacities of AI, which enable the forecasting of water supply and quality. Planning for and responding to water-related problems, such as droughts and pollution disasters, requires forethought. Additionally, as stated by Krishnan et al. (2022), the real-time monitoring capabilities of AI and IoT systems enable quick detection and reaction to water quality and quantity changes. This instantaneous feedback loop allows authorities to act quickly to safeguard ecosystems and water supplies. Furthermore, by analyzing enormous databases and spotting trends that influence long-term planning, AI and ML can help create sustainable policies. The preservation and fair distribution of water resources are prioritized in choices made by policymakers using this data-driven approach.

In conclusion, incorporating AI and ML into water management policies and plans offers a revolutionary route towards more efficient and sustainable management of water resources. While minimizing environmental consequences and preserving the availability of this essential resource for future generations, these technologies offer the capabilities required to manage present and future water concerns.

8. CONCLUSION

In conclusion, ML and IoT applications for managing water resources have a dynamic environment, which this research proposal has explored in depth. A thorough introduction to the subject was followed by an examination of how ML may aid in making data-driven choices. Despite the benefits, difficulties, including security, scalability, and appropriate frameworks, were recognized as significant barriers to overcome. However, it is impossible to understate the game-changing effects of combining ML with IoT to achieve sustainable water resource management. This summary of important findings highlights the continual development of such innovations and sets.

It becomes more evident by moving forward that coordinated actions are required to clear the path for a sustainable water future. It can revolutionize how to manage water by embracing the confluence of ML and IoT and making educated decisions that preserve this priceless resource for future generations.

The constant advancement of ML and IoT in water management represents a path toward more effective and sustainable resource use. These technologies' uses will probably increase as they develop, allowing for increasingly more exact forecasts and real-time monitoring. Moreover, improvements in ML algorithms and data analytics will increase their capacity to analyze massive volumes of data, providing deeper insights into water management. With increasing agility and efficacy, this development promises to solve urgent water concerns, including shortage and pollution.

For a sustainable water future, there is a loud call to action. The integration of ML and IoT has the potential to significantly alter water management. Collective efforts are essential to get past obstacles and realize these technologies' full potential. To ensure the responsible use and reservations of the limited water resources, it is vital to use innovative and data-driven solutions. To achieve sustainability, there must be a shared commitment to using ML and IoT, strengthening decision-making abilities, and ensuring the legacy of clean water for future generations. These measures can improve decision-making, improve resource allocation, and eliminate waste. A sustainable water future is eventually ensured through benefits including fair water distribution, decreased energy use, and protection of pure water for future generations.

REFERENCES

Ahansal, Y., Bouziani, M., Yaagoubi, R., Sebari, I., Sebari, K., & Kenny, L. (2022). Towards Smart Irrigation: A Literature Review on the Use of Geospatial Technologies and Machine Learning in the Management of Water Resources in Arboriculture. *Agronomy (Basel)*, *12*(2), 297. doi:10.3390/agronomy12020297

Aivazidou, E., Banias, G., Lampridi, M., Vasileiadis, G., Anagnostis, A., Papageorgiou, E., & Bochtis, D. (2021). Smart Technologies for Sustainable Water Management: An Urban Analysis. *Sustainability (Basel)*, *13*(24), 13940. doi:10.3390u132413940

Arsene, D., Predescu, A., Pahonțu, B., Chiru, C. G., Apostol, E.-S., & Truică, C.-O. (2022). Advanced Strategies for Monitoring Water Consumption Patterns in Households Based on IoT and Machine Learning. *Water (Basel)*, *14*(14), 2187. doi:10.3390/w14142187

Bedi, S., Samal, A., Ray, C., & Snow, D. (2020). Comparative evaluation of machine learning models for groundwater quality assessment. *Environmental Monitoring and Assessment*, *192*(12), 776. Advance online publication. doi:10.100710661-020-08695-3 PMID:33219864

Bhardwaj, A., Dagar, V., Khan, M. O., Aggarwal, A., Alvarado, R., Kumar, M., Irfan, M., & Proshad, R. (2022). Smart IoT and Machine Learning-based Framework for Water Quality Assessment and Device Component Monitoring. *Environmental Science and Pollution Research International*, *29*(30), 46018–46036. Advance online publication. doi:10.100711356-022-19014-3 PMID:35165843

Chiu, M.-C., Yan, W.-M., Bhat, S. A., & Huang, N.-F. (2022). Development of smart aquaculture farm management system using IoT and AI-based surrogate models. *Journal of Agriculture and Food Research*, *9*, 100357. doi:10.1016/j.jafr.2022.100357

Elbeltagi, A., Srivastava, A., Deng, J., Li, Z., Raza, A., Khadke, L., Yu, Z., & El-Rawy, M. (2023). Forecasting vapor pressure deficit for agricultural water management using machine learning in semi-arid environments. *Agricultural Water Management*, *283*, 108302. doi:10.1016/j.agwat.2023.108302

ESA. (2023). *European Space Agency - Earth Observation.* [Online]. Retrieved on 07 September 2023 from https://www.esa.int/Applications/Observing_the_Earth

Fox, I. (2019). Institutions for water management in a changing world. In *Water In A Developing World* (pp. 9–24). Routledge. https://digitalrepository.unm.edu/cgi/viewcontent.cgi?article=3454&context=nrj

Fu, G., Jin, Y., Sun, S., Yuan, Z., & Butler, D. (2022). The role of deep learning in urban water management: A critical review. *Water Research, 223*, 118973–118973. doi:10.1016/j.watres.2022.118973 PMID:35988335

Geetha, S., & Gouthami, S. (2016). Internet of Things enabled real-time water quality monitoring system. *Smart Water, 2*(1), 1. Advance online publication. doi:10.118640713-017-0005-y

Gupta, A. D., Pandey, P., Feijóo, A., Yaseen, Z. M., & Bokde, N. D. (2020). Smart Water Technology for Efficient Water Resource Management: A Review. *Energies, 13*(23), 6268. doi:10.3390/en13236268

Gupta, J., Pathak, S., & Kumar, G. (2022). Deep Learning (CNN) and Transfer Learning: A Review. *Journal of Physics: Conference Series, 2273*(1), 012029. doi:10.1088/1742-6596/2273/1/012029

Hadipour, M., Derakhshandeh, J. F., & Shiran, M. A. (2019). An experimental setup of a multi-intelligent control system (MICS) of water management using the Internet of Things (IoT). *ISA Transactions*. Advance online publication. doi:10.1016/j.isatra.2019.06.026 PMID:31285060

He, C., Liu, Z., Wu, J., Pan, X., Fang, Z., Li, J., & Bryan, B. A. (2021). Future global urban water scarcity and potential solutions. *Nature Communications, 12*(1), 4667. doi:10.103841467-021-25026-3 PMID:34344898

HESS. (2023). *Hydrology and Earth System Sciences Journal.* https://www.hydrol-earth-syst-sci.net/

Ighalo, J. O., Adeniyi, A. G., & Marques, G. (2020). Internet of Things for Water Quality Monitoring and Assessment: A Comprehensive Review. *Artificial Intelligence for Sustainable Development: Theory, Practice and Future Applications*, 245–259. doi:10.1007/978-3-030-51920-9_13

Kamaruidzaman, N. S., & Rahmat, S. N. (2020). Water Monitoring System Embedded with Internet of Things (IoT) Device: A Review. *IOP Conference Series. Earth and Environmental Science, 498*(1), 012068. doi:10.1088/1755-1315/498/1/012068

Kanade, P., & Prasad, J. P. (2021). Arduino-based Machine Learning and IoT Smart Irrigation System. *International Journal of Soft Computing and Engineering, 10*(4), 1–5. doi:10.35940/ijsce.D3481.0310421

Krishnan, S. R., Nallakaruppan, M. K., Chengoden, R., Koppu, S., Iyapparaja, M., Sadhasivam, J., & Sethuraman, S. (2022). Smart Water Resource Management Using Artificial Intelligence—A Review. *Sustainability (Basel), 14*(20), 13384. doi:10.3390u142013384

Kshirsagar, P. R., Manoharan, H., Selvarajan, S., Althubiti, S. A., Alenezi, F., Srivastava, G., & Lin, J. C.-W. (2022). A Radical Safety Measure for Identifying Environmental Changes Using Machine Learning Algorithms. *Electronics (Basel), 11*(13), 1950. doi:10.3390/electronics11131950

Kumar, T. M. V., Firoz, C. M., Bimal, P., Harikumar, P. S., & Sankaran, P. (2020). Smart water management for smart Kozhikode metropolitan area. *Advances in 21st Century Human Settlements*, 241–306. doi:10.1007/978-981-13-6822-6_7

Lowe, M., Qin, R., & Mao, X. (2022). A Review on Machine Learning, Artificial Intelligence, and Smart Technology in Water Treatment and Monitoring. *Water (Basel), 14*(9), 1384. doi:10.3390/w14091384

Mohamed, E. S., Belal, A. A., Abd-Elmabod, S. K., El-Shirbeny, M. A., Gad, A., & Zahran, M. B. (2021). Smart farming for improving agricultural management. *The Egyptian Journal of Remote Sensing and Space Sciences, 24*(3), 971–981. Advance online publication. doi:10.1016/j.ejrs.2021.08.007

Mondejar, M. E., Avtar, R., Diaz, H. L. B., Dubey, R. K., Esteban, J., Gómez-Morales, A., Hallam, B., Mbungu, N. T., Okolo, C. C., Prasad, K. A., She, Q., & Garcia-Segura, S. (2021). Digitalization to Achieve Sustainable Development Goals: Steps Towards a Smart Green Planet. *The Science of the Total Environment, 794*, 148539. doi:10.1016/j.scitotenv.2021.148539 PMID:34323742

Narendran, S., Pradeep, P., & Ramesh, M. V. (2017). An Internet of Things (IoT) based sustainable water management. *2017 IEEE Global Humanitarian Technology Conference (GHTC).* 10.1109/GHTC.2017.8239320

Nova, K. (2023). AI-Enabled Water Management Systems: An Analysis of System Components and Interdependencies for Water Conservation. *Eigenpub Review of Science and Technology, 7*(1), 105-124. https://studies.eigenpub.com/index.php/erst/article/download/12/11

Omambia, A., Maake, B., & Wambua, A. (2022, May). Water quality monitoring using IoT & machine learning. In 2022 IST-Africa Conference (IST-Africa) (pp. 1-8). IEEE. doi:10.23919/IST-Africa56635.2022.9845590

Piemontese, L., Kamugisha, R., Tukahirwa, J., Tengberg, A., Pedde, S., & Jaramillo, F. (2021). Barriers to scaling sustainable land and water management in Uganda: A cross-scale archetype approach. *Ecology and Society, 26*(3), art6. Advance online publication. doi:10.5751/ES-12531-260306

Roy, S. K., Misra, S., Raghuwanshi, N. S., & Das, S. K. (2020). AgriSens: IoT-based dynamic irrigation scheduling system for water management of irrigated crops. *IEEE Internet of Things Journal, 8*(6), 5023–5030. doi:10.1109/JIOT.2020.3036126

Ryu, J. H. (2022). UAS-based real-time water quality monitoring, sampling, and visualization platform (UASWQP). *HardwareX, 11*, e00277. doi:10.1016/j.ohx.2022.e00277 PMID:35509896

Sagan, V., Peterson, K. T., Maimaitijiang, M., Sidike, P., Sloan, J., Greeling, B. A., Maalouf, S., & Adams, C. (2020). Monitoring inland water quality using remote sensing: Potential and limitations of spectral indices, bio-optical simulations, machine learning, and cloud computing. *Earth-Science Reviews, 205*, 103187. doi:10.1016/j.earscirev.2020.103187

Salam, A., & Salam, A. (2020). Internet of things in water management and treatment. *Internet of Things for Sustainable Community Development: Wireless Communications. Sensory Systems*, 273–298. doi:10.1007/978-3-030-35291-2_9

Sit, M., Demiray, B. Z., Xiang, Z., Ewing, G. J., Sermet, Y., & Demir, I. (2020). A comprehensive review of deep learning applications in hydrology and water resources. *Water Science and Technology, 82*(12), 2635–2670. doi:10.2166/wst.2020.369 PMID:33341760

Sun, A. Y., & Scanlon, B. R. (2019). How can Big Data and machine learning benefit environment and water management: A survey of methods, applications, and future directions. *Environmental Research Letters*, *14*(7), 073001. doi:10.1088/1748-9326/ab1b7d

Togneri, R., Kamienski, C., Dantas, R., Prati, R., Toscano, A., Soininen, J.-P., & Cinotti, T. S. (2019). Advancing IoT-Based Smart Irrigation. *IEEE Internet of Things Magazine*, *2*(4), 20–25. doi:10.1109/IOTM.0001.1900046

Torres, A. B. B., da Rocha, A. R., Coelho da Silva, T. L., de Souza, J. N., & Gondim, R. S. (2020). Multilevel data fusion for the Internet of things in smart agriculture. *Computers and Electronics in Agriculture*, *171*, 105309. doi:10.1016/j.compag.2020.105309

World Bank. (2023). *World Bank Group - International Development, Poverty, & Sustainability*. World Bank. https://www.worldbank.org/

Zhai, Z., Martínez, J. F., Beltran, V., & Martínez, N. L. (2020). Decision support systems for agriculture 4.0: Survey and challenges. (2020). *Computers and Electronics in Agriculture*, *170*, 105256. doi:10.1016/j.compag.2020.105256

Zhu, M., Wang, J., Yang, X., Zhang, Y., Zhang, L., Ren, H., Wu, B., & Ye, L. (2022). *A review of the application of machine learning in water quality evaluation*. Eco-Environment & Health. doi:10.1016/j.eehl.2022.06.001

Chapter 2
A Comprehensive Exploration of Machine Learning and IoT Applications for Transforming Water Management

Mandeep Kaur
https://orcid.org/0000-0001-8054-1605
Chitkara University Institute of Engineering and Technology, Chitkara University, India

Righa Tandon
https://orcid.org/0000-0002-5953-5355
Chitkara University Institute of Engineering and Technology, Chitkara University, India

Rajni Aron
NMIMS University, India

Htet Ne Oo
https://orcid.org/0000-0003-2910-8608
Chitkara University Institute of Engineering and Technology, Chitkara University, India

Heena Wadhwa
https://orcid.org/0000-0002-2029-5921
Chitkara University Institute of Engineering and Technology, Chitkara University, India

Ramandeep Sandhu
https://orcid.org/0000-0003-2595-4030
Lovely Professional University, India

ABSTRACT

Water scarcity and environmental concerns have become pressing issues in the modern world, necessitating innovative approaches to water management. Global issues including water scarcity and environmental concerns now require creative and sustainable approaches to managing water resources. This chapter will examine how the internet of things (IoT) and cutting-edge technologies like machine learning (ML) are revolutionizing the way that water management is done. In this chapter, the effective uses of machine learning in water resource analysis will be examined. Forecasting water demand requires the use of ML algorithms, which help water managers predict consumption trends with accuracy. Predictive analytics can also be used to evaluate the distribution and availability of water, providing information on how to allocate and optimize water resources. The chapter concluded with revolutionary potential of machine learning and the internet of things in modernizing water management practices globally.

DOI: 10.4018/979-8-3693-1194-3.ch002

1. INTRODUCTION

Water management has seen a dramatic change recently as a result of the convergence of Machine Learning (ML), the Internet of Things (IoT), and environmental stewardship. The complex interactions between human activity, climate change, and the availability of freshwater resources have sparked a search for novel approaches that may effectively handle the problems associated with water management. Through a thorough investigation of the synergistic potential of ML and IoT applications, this chapter covers the world of revolutionizing water management. The fusion of ML and IoT offers promising solutions to optimize water usage, improve water resource monitoring, and encourage more effective and eco-friendly practices across numerous sectors as the globe faces increasing water-related concerns. IoT devices, like smart water meters and leak detection systems, make it easier to gather enormous volumes of data about water, enabling proactive leak detection and water resource management. IoT technologies provide remote monitoring of water infrastructure, ensuring quick reactions to faults and disturbances, reducing waste, and promoting sustainable practices. The importance of real-time data on decision-making processes will be emphasized through case studies demonstrating efficient IoT integration in water management. Utilizing IoT technologies, smart irrigation controllers and soil moisture monitoring allow for effective water distribution to crops, increasing yields while preserving precious water resources (Mishra & Tyagi, 2022).

1.1. Background and Significance of Water Management Challenges

Every aspect of human existence and ecological harmony is closely entwined with water, the source of all life and a crucial natural resource. But in the twenty-first century, a number of issues confronting the world's water supplies necessitate the development of novel, technologically advanced solutions. In order to comprehend the crucial need for cutting-edge approaches like ML and IoT, this section first offers a background on the history and significance of these water management concerns. Figure 1 shows various challenges faced by water management.

- **Escalating Water Scarcity:** Along with a growing urbanization and industrialization, the world's population is still expanding at an unheard-of rate. As a result, there is a greater need for water for industrial activities, energy production, agricultural, and drinking and sanitation. With over 2 billion people already residing in places experiencing water stress, this growing demand has made water scarcity worse in many areas. Water management strategies that are effective and sustainable are more important as water scarcity grows more severe (Manny, 2023; Sugam et al., 2023).
- **Climate Change and Variability:** Climate change has introduced a new layer of complexity to the water management equation. Altered precipitation patterns, melting glaciers, and shifting weather extremes have disrupted the natural balance of water availability. Prolonged droughts in certain areas, coupled with sudden intense rainfall in others, pose challenges for traditional water management strategies. Adapting to these changing climatic conditions requires agile and data-driven approaches that can anticipate and respond to such variations (Apa et al., 2023; Elbeltagi et al., 2020; Zhang et al., 2021).
- **Aging Infrastructure and Inefficient Practices:** In many parts of the world, water infrastructure is aging and in need of significant upgrades. Traditional water distribution systems often suffer from leaks, inefficiencies, and lack of real-time monitoring capabilities. These inefficiencies not

Figure 1. Water management challenges

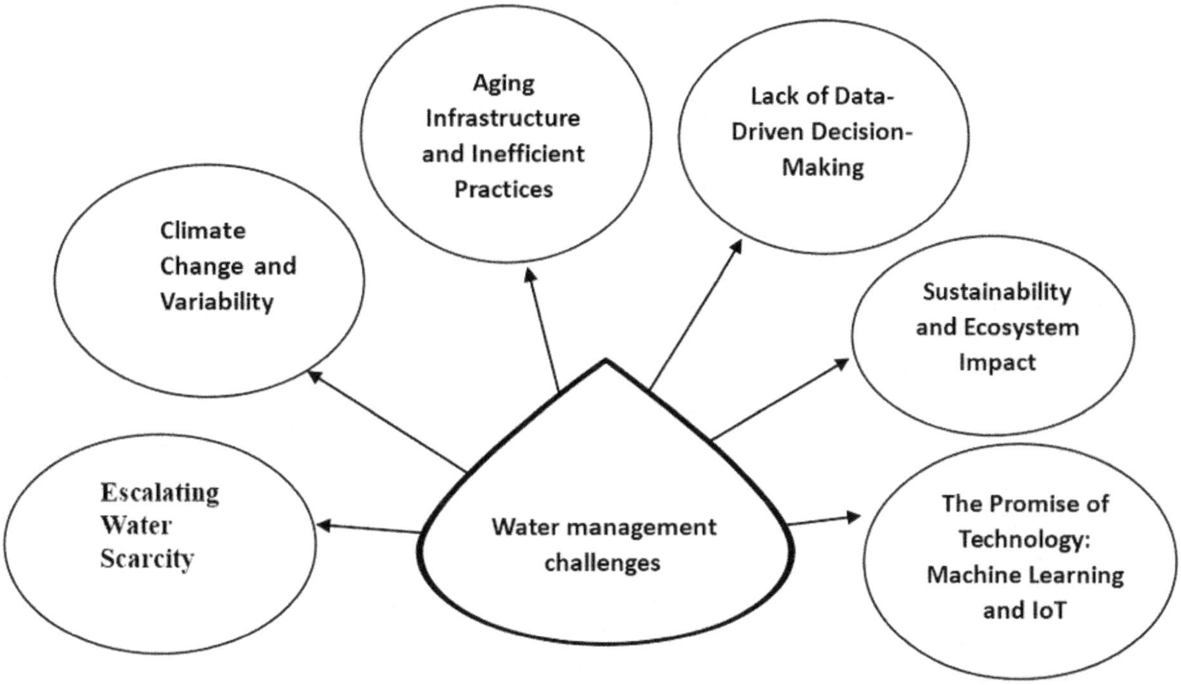

only result in substantial water losses but also contribute to increased energy consumption and operational costs. Modernizing water infrastructure to reduce losses and enhance operational efficiency is a priority, demanding innovative solutions that can provide insights and optimizations in real-time (Nova, 2023).

- **Lack of Data-Driven Decision-Making:** Historically, water management decisions have often been made using limited data and static models. The absence of real-time monitoring and predictive capabilities has hindered the ability to respond proactively to emerging challenges. Inadequate data-driven decision-making has led to inefficient resource allocation, delayed responses to critical events like droughts or floods, and missed opportunities for sustainable water use (Ghoochani et al., 2023; Musa et al., 2023).
- **Sustainability and Ecosystem Impact:** The impact of water management extends beyond human needs, influencing ecosystems, biodiversity, and ecological health. Poor water management practices can lead to habitat degradation, pollution of water bodies, and depletion of aquatic life. Achieving sustainable water management requires a holistic approach that considers both human needs and the ecological health of water systems (Sahoo & Goswami, 2024).
- **The Promise of Technology: ML and IoT:** Addressing these multifaceted water management challenges necessitates a paradigm shift in how to monitor, understand, and respond to water dynamics. The emergence of advanced technologies, particularly ML and the IoT, holds the promise to revolutionize water management. By enabling real-time data collection, analysis, prediction, and optimization, these technologies offer the potential to enhance water resource utilization, reduce waste, and foster sustainable practices (Chinnappan et al., 2023).

The subsequent section will explore deeper into how ML and IoT applications are being harnessed to transform water management practices, contributing to a more resilient and sustainable future for water resources worldwide.

1.2. The Role of Technology in Addressing Water-Related Issues

To guarantee that communities have a steady supply of clean, safe drinking water, efficient water distribution and availability are essential. Water distribution system management and optimization heavily rely on predictive analytic, which is supported by ML approaches. Here's a more thorough explanation of this subsection:

- **Hydrological Modelling:** It is a complex process that is essential to regulating and forecasting the availability of water. It entails the integration of several data sources, such as satellite imaging, topographical maps, meteorological data, and on-ground sensors. In order to integrate and analyse this data and find the patterns and trends that guide the hydrological model, ML methods are essential. ML-enabled real-time monitoring adds real-time data, such rainfall readings, river flow rates, and groundwater levels, to the model on a continuous basis. This makes it possible to react quickly and dynamically to changing circumstances, even extreme occurrences like floods and droughts (Keller et al., 2023; Ali et al., 2023).
- **Optimal Distribution:** Minimizing waste and satisfying customer expectations both depend on the effective distribution of water within a network. In order to achieve this efficiency, ML-enabled predictive analytics is essential. One important component is demand forecasting, which uses ML algorithms to project future water demand based on past consumption trends, population growth, economic development, and environmental concerns. Utility companies may manage their operations, including pump schedules and storage tank levels, with the help of accurate demand projections, guaranteeing a steady supply of water. ML algorithms are used by smart water distribution systems, which are a component of optimal distribution, for real-time control and monitoring (Elshaboury & Marzouk, 2022; Grbčić et al., 2021).

2. ML FOR WATER RESOURCE ANALYSIS

The revolutionary potential of ML for resource assessment and management is the main topic of this section. It offers insight into how ML methods have become essential resources for addressing the intricate issues surrounding water resource management. In-depth discussions of several subjects pertaining to ML's use to water resource analysis are found in this section. First, the importance of ML as an effective tool for utilizing data and drawing knowledgeable conclusions in the subject of water management is highlighted.

2.1. ML Applications in Water Demand Forecasting

The application of ML for demand prediction represents a major advancement in the field of water management. In this part we explore how ML techniques can be applied to predict future water demands with outstanding accuracy. The historical basis for water demand forecasting was statistical models, which often struggled to capture the complexities of the real-world factors influencing water consumption. ML uses an enormous amount of data to make predictions about things like population growth, weather patterns, historical purchase trends, economic indicators, and even social occurrences. Water utilities and authorities can predict water demand on several timescales, ranging from immediate demands to long-range planning, because these models are very good at finding intricate patterns and relationships in data. Water suppliers and customers can interact with each other using these networks, encouraging water conservation during periods of peak demand (Mishra & Tyagi, 2022).

2.2. Predictive Analytics for Water Availability and Distribution

Predictive analytic driven by ML is critical to ensuring a consistent water supply and efficient distribution. This section examines how ML can be used to predict water availability and distribute it optimally in complex water systems. The first portion deals with hydrological modelling, wherein data from numerous sources, including satellite imagery, rainfall records, river flow measurements, and groundwater levels, is fed into multiple intelligence models. These models forecast reservoir capacities, groundwater levels, and river flows with accuracy. In order to effectively manage water supplies, ML algorithms may analyze this data in real-time and dynamically adjust forecasts (Nova, 2023).

2.3. ML-Based Water Quality Assessment and Contamination Detection

Ensuring that water quality remains beneficial is crucial for public health, and ML plays an increasingly important role in this regard. This article examines how contamination detection and water quality assessment based on ML have evolved into essential instruments for protecting water sources. First, sensor data analysis is covered, in which ML models examine data from sensors measuring water quality. These sensors track a number of variables continuously, including temperature, turbidity, chemical composition, and pH levels. When water quality deviates from acceptable levels, ML algorithms identify patterns and abnormalities and send out alarms. In order to avert health emergencies, this real-time monitoring enables prompt reactions to possible threats to the quality of the water (Ghoochani et al., 2023).

2.4. Case Studies Illustrating Successful ML Implementations in Water Resource Analysis

The purpose of this section is to give real-world examples and practical insights into how ML has been utilized successfully to handle difficulties related to water resource management. This section offers actual, verified examples of how ML has been effectively applied to many facets of water resource analysis. These case studies are essentially in-depth analyses of particular projects or efforts where ML techniques and technologies were applied to address actual water-related issues in the real world. Figure 2 represents various case studies that illustrate successful ML implementation in water resource analysis.

- **Water Demand Forecasting:** The case studies illustrate the exceptional precision and advantages of ML in projecting future water requirements through water demand forecasting. In most of these real-world instances, water usage patterns that fluctuate are managed by regions or utilities. ML models take into account a wide range of variables, such as past water usage data, population growth, economic indicators, and meteorological conditions. With thorough examination, these

Figure 2. Case studies in water resources using ML

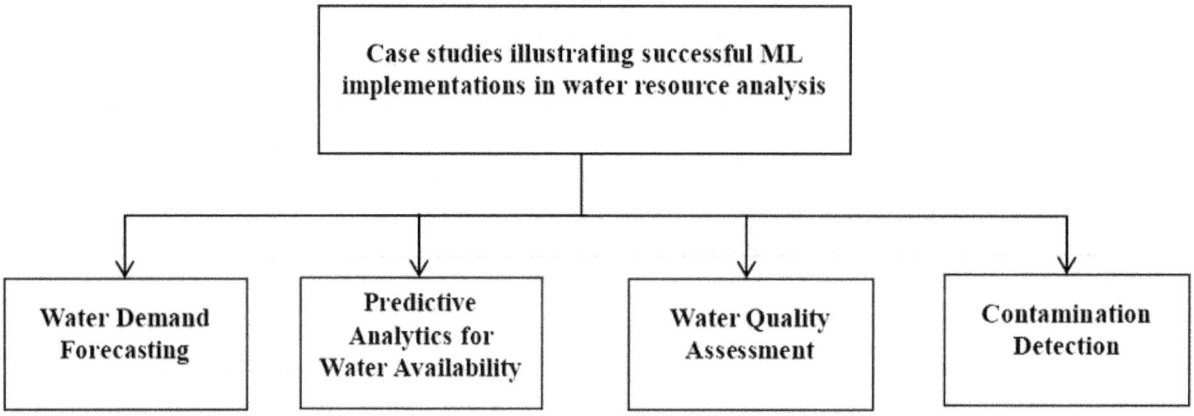

models produce accurate projections of water use across various time intervals, ranging from short-term emergency need to long-term planning. Water utilities may plan ahead and distribute resources more effectively thanks to ML-driven projections, as demonstrated by these case studies (Khan et al., 2023; Nova, 2023).

- **Predictive Analytics for Water Availability:** Predictive analytics for water availability case studies show how ML may transform the way this vital resource is managed. These cases are usually from areas where there is a risk of water scarcity or environmental instability. ML models generate reliable hydrological models by combining a variety of data sources, such as weather forecasts, river flow data, and satellite imagery. With great accuracy, these models forecast reservoir capacities, groundwater levels, and river flows. Algorithms for ML continuously adjust these predictions to situations that change by evaluating data in real time. The way this dynamic method improves water resource management is demonstrated through the case studies (Chinnappan et al., 2023; Nova, 2023).
- **Water Quality Assessment:** These case studies focus mostly on the use of ML in assessing water quality. These instances frequently center on areas where it is critical to preserve high water quality for the general public's health. Data from water quality sensors, which track variables like pH, turbidity, temperature, and chemical composition continually, is analyzed by ML models. These programmes detect trends and abnormalities that might indicate alterations in the quality of the water by analyzing data in real time. The ML-enabled water quality evaluation provides early detection of deviations from acceptable criteria, as demonstrated by these case studies (Uddin et al., 2023a).
- **Contamination Detection:** The case studies highlight the critical role ML plays in protecting water supplies in the context of pollution detection. These instances usually concern situations in which prompt detection of pollutants, including heavy metals, bacteria, or chemicals, is critical. In data on water quality, ML algorithms are trained to identify patterns that point to contamination incidents. By providing early warnings through ML-driven contamination detection, authorities can take prompt action to stop polluted water from reaching consumers, as demonstrated by these case studies (Gong et al., 2023).

3. LITERATURE REVIEW

The management of water resources plays a crucial role in promoting sustainability and safeguarding the environment. In light of the escalating need for freshwater resources and the mounting challenges posed by water scarcity and contamination, there is a pressing need for innovative solutions. The convergence of ML and the IoT has emerged as a paradigm-shifting development with significant potential in the domain of water management. This literature review examines prominent research and advancements pertaining to the utilization of ML and IoT technologies in the field of water management. The review primarily concentrates on the contributions of ML and IoT in the areas of data gathering, predictive analysis, and instantaneous decision-making.

ML techniques, namely deep learning models, have demonstrated potential in the prediction of water quality. The efficacy of deep learning in predicting water quality parameters is underscored in the research conducted by Uddin et al. (2023b). These models have the capability to analyze past data, inputs from sensors, and weather conditions in order to forecast instances of contamination, hence enabling the implementation of proactive management approaches.

The utilization of deep learning techniques, specifically unsupervised learning methods, enables the provision of precise predictions. The authors in Solanki et al. (2015), utilized data obtained from the Chaskaman River in Maharashtra, India, demonstrates that deep learning methodologies, notably denoising autoencoders and deep belief networks, exhibit superior performance compared to supervised learning approaches. This study assesses the performance of unsupervised learning algorithms by utilizing error metrics such as mean absolute error and mean square error. The findings demonstrate the efficacy of these algorithms in predicting water quality parameters and their capacity to effectively handle variations in data.

The authors of Barzegar et al. (2020) examined the crucial undertaking of monitoring water quality, specifically in the Small Prespa Lake located in Greece. The purpose of this study is to forecast the concentrations of dissolved oxygen (DO) and chlorophyll-a (Chl-a). In order to accomplish this objective, the research proposes independent deep learning (DL) models, specifically the long short-term memory (LSTM) and convolutional neural network (CNN) models. Additionally, a novel hybrid model, CNN-LSTM, is proposed, which integrates both DL techniques. The results of this study indicate that deep learning models, particularly the hybrid approach, have the capacity to improve the accuracy of water quality prediction within the framework of lake management.

The authors in Rizal et al. (2023) investigates the urgent matter of river water contamination and emphasizes the necessity of employing sophisticated technologies for precise monitoring and prediction of water quality indicators. This study centres around the Langat River in Malaysia and use the Adaptive Neuro-fuzzy Inference System (ANFIS) as a deep learning predictive model to anticipate six specific metrics related to the quality of river water. The assessment of the model's performance is conducted by employing metrics such as root mean square error (RMSE) and the determination coefficient (R2). The results indicate that ANFIS, namely Model 5, exhibits outstanding predictive ability, as evidenced by a noteworthy R2 value of 0.9712. Furthermore, the model's efficacy is shown by the low root mean square error (RMSE) values observed in the training, testing, and checking datasets, which are 0.0028, 0.0144, and 0.0924, respectively. In conclusion, the research effectively demonstrates the use of ANFIS as a beneficial instrument for forecasting various water quality metrics within the Langat River setting.

Chandra Sekhar et al. (2023) highlights the urgent matter of water pollution, which has a substantial role in the prevalence of numerous waterborne illnesses and is a significant factor in global mortality rates.

The article posits a solution to the issue by suggesting the implementation of a cost-effective real-time water quality monitoring system that leverages the capabilities of the IoT. The major goal of the system is to observe and track essential physical and chemical characteristics, such as temperature, humidity, pH, and turbidity. The framework that has been designed encompasses a diverse range of sensors that possess the capability to measure various water characteristics. These sensors are all under the supervision of a central controller, which utilizes the ATMEGA328 model as its core. The system facilitates remote access to the gathered sensor data over a Wi-Fi network, hence augmenting its operational effectiveness. The process of automation is made possible through the utilization of a microcontroller, which establishes communication with a personal computer via Wi-Fi. Furthermore, all constituent elements of the system are coupled by means of an Arduino ATMEGA328 micro controller. This novel and economically efficient technology exhibit significant potential for mitigating water quality issues and guaranteeing the availability of potable water.

4. IOT SENSORS AND NETWORKS FOR REAL-TIME WATER MONITORING

In-depth discussion of the IoT revolutionary impact on real-time water resource monitoring is provided in this section. In order to ensure the effectiveness, dependability, and sustainability of water management practices, it examines how IoT sensors and networks have become crucial. Fundamentally, this section illustrates how the incorporation of IoT technology has given rise to a new phase of data-driven decision-making in the water resource management domain.

4.1. IoT Devices for Collecting Water-Related Data

From distant river basins to metropolitan water distribution networks, it showcases the wide range of IoT sensors and gadgets that have been thoughtfully included into water systems. Important characteristics including temperature, turbidity, pressure, water flow rates, and water quality can all be measured using these sensors. These case studies offer powerful illustrations of how IoT sensors continuously gather and send data to centralized monitoring systems. Water authorities can now obtain never-before-seen insights on the condition and behavior of water resources because to the high-resolution, dynamic, and real-time data generated. These real-world examples highlight how IoT-driven data collection improves the accuracy and timeliness of water monitoring, empowering stakeholders to decide in ways that maximize resource allocation, infrastructure upkeep, and water usage (Bassine et al., 2023; Kaur & Aron, 2022b).

4.2. Smart Water Meters and Leak Detection Systems

This section focuses on leak detection systems and smart water meters, which are significant technological advancements that have an impact on the management of water infrastructure. As seen in the case studies, smart water meters are placed in commercial, industrial, and residential settings. They are different from typical meters in that they track water usage continually and send real-time data wirelessly to utility companies (Głomb et al., 2023).

4.3. Remote Monitoring of Water Infrastructure

This section highlights a critical IoT application: remote monitoring of water infrastructure. The technique entails the thoughtful positioning of IoT sensors and gadgets in key water system components, such as distribution networks, pumps, valves, and storage tanks. These sensors provide centralized control centres with real-time data on parameters including pressure, temperature, flow rates, and equipment status. The resilience and efficiency of water infrastructure are improved through remote monitoring, as these case studies demonstrate. In order to identify abnormalities, anticipate maintenance requirements, and react quickly to problems, maintenance staff can remotely monitor the condition and functionality of infrastructure components in real-time. In order to minimize downtime, expensive repairs, and service interruptions, predictive analytics and ML algorithms, as demonstrated in these examples, evaluate sensor data to predict when equipment may need servicing (Głomb et al., 2023; Bolick et al., 2023).

4.4. Case Examples Showcasing Effective IoT Integration in Water Management

A collection of real-world case studies that provide compelling evidence of the successful integration of IoT in water resource management are presented in this section. The use of IoT sensors to identify variations in water quality, smart water meters to identify leaks, and remote monitoring of water distribution networks to increase dependability and efficiency are just a few of the applications covered by these cases. Some cases i.e. Citywide Water Distribution Optimization (Singapore) (Marques dos Santos et al., 2023a), Smart Agriculture Water Management (California, USA) (Gong et al., 2023), Water Quality Monitoring (Thames Water, UK) (Butler et al., 2023), Flood Prediction and Mitigation (Netherlands) (Lambrechts et al., 2023), each provide concrete examples of how IoT technologies have changed the way that water management is done.

5. DATA INTEGRATION AND DECISION SUPPORT SYSTEMS

This section explores how data integration and decision support systems are essential for maximizing the use of water resources. It highlights how real-time insights from the seamless integration of data from IoT devices and ML outputs empower water managers to make better decisions and adjust to changing situations. In addition, this section emphasizes how feasible it is to implement decision support systems that apply ML and IoT data to improve decision-making, which in turn leads to more resilient and sustainable water management plans.

5.1. Combining ML Outputs and IoT Data for Comprehensive Analysis

Merging insights from these two cutting-edge technologies is crucial, as the part on merging ML outputs and IoT data for complete analysis emphasizes. As previously said, ML offers data-driven forecasts and predictive analytics, and IoT sensors provide real-time data on water system status. Water managers can examine the dynamics of water resources more comprehensively by combining ML outputs with IoT data. By using real-time IoT data, ML models may provide predictions that improve forecast accuracy and timeliness. These case studies show how accurate decision-making is made possible by a more thorough understanding of water systems, which is made possible by this integration (Chinnappan et al., 2023).

5.2. Developing Decision Support Systems for Water Managers

This section explores how data integration and decision support systems are essential for maximizing the use of water resources. It highlights how real-time insights from the seamless integration of data from IoT devices and ML outputs empower water managers to make better decisions and adjust to changing situations. The creation and implementation of decision support systems (DSS) customized to meet the requirements of water managers become the main emphasis of this section (Optoelectronics, 2023).

5.3. Real-Time Data-Driven Insights for Adaptive Water Management Strategies

The significance of real-time data in building adaptive water management plans is emphasized in this section. Real-time data gathering and analysis is essential for water management to make quick decisions and successfully adjust to constantly changing conditions when it comes to the combination of IoT and ML.

5.3.1. Prompt Reaction to Water Scarcities

Water managers can locate water shortages or surpluses promptly by continuously monitoring variables including water levels, quality, and consumption patterns in real-time. This data is analyzed by ML algorithms, which produce forecasts and enable fast modifications to water distribution, allocation, and conservation plans. This guarantees the effective use of available water resources and the early detection and resolution of water shortages before they become serious problems (Nova, 2023).

5.3.2. Scheduling Irrigation Adaptively

Weather forecasts and real-time data from IoT sensors in the fields can help with adaptive irrigation scheduling in precision agriculture. ML algorithms analyze this data to calculate the amount and timing of irrigation needed for each crop. The irrigation schedule can be dynamically modified to prevent over- or under-irrigation, hence enhancing crop health and preserving water resources, in the event of unforeseen weather changes (Optoelectronics, 2023).

5.3.3. Prompt Reaction to Contamination Incidents

Sensors measuring water quality keep an eye on things like turbidity, pH, and chemical composition all the time. Real-time notifications are sent upon detection of departures from the norm. When possible, contamination events arise, water authorities can act quickly to ensure public health hazards are kept to a minimum and drinking water safety is maintained (Nova, 2023).

5.3.4. Streamlining Operations of Dams and Reservoirs

Real-time information on water levels, outflow, and inflow is essential in areas with reservoirs and dams. When processing this data, ML algorithms take previous trends and weather forecasts into account. Water managers can balance water supply and flood management by modifying dam operations in real-time to reduce flooding hazards during periods of heavy rainfall or to strategically release water during droughts (Shumilova et al., 2023).

5.3.5. Improving Leak Identification and Infrastructure Upkeep

IoT sensors on water pipelines and distribution networks keep an eye out for anomalies and leaks all the time. Real-time identification of these problems by ML models can notify maintenance teams. Not only does prompt leak detection save water, it also prolongs the life of vital water infrastructure (Nova, 2023).

5.4. Demonstrative Cases Highlighting the Collaboration Between ML and IoT in Decision-Making

This section compiles real-world examples that demonstrate how ML and the IoT work together to enhance decision-making. The way in which the incorporation of these technology improves water management practices is demonstrated by example like these. Figure 3 represents the various examples of collaboration between ML and IoT in decision making.

5.4.1. Predictive Maintenance in Manufacturing (Industry 4.0)

IoT sensors are installed on machinery in a manufacturing plant to track its performance in real-time, gathering information on temperature, vibration, and other operational characteristics. ML algorithms examine this constant flow of sensor data to forecast the likelihood of equipment failure. In order to minimize downtime and increase production efficiency, maintenance personnel receive notifications and repairs are arranged before faults occur (Rosati et al., 2023).

5.4.2. Smart Grid Management (Energy Sector)

Smart grid networks collect data on electricity flow, voltage, and grid characteristics by placing IoT devices on power lines, transformers, and substations. By processing this data, ML algorithms forecast moments of peak demand and possible errors. Utility firms make good use of these insights to optimize the distribution of electricity, avert blackouts, and integrate renewable energy sources (Hasan et al., 2023).

5.4.3. Medical Monitoring (IoT) Healthcare Devices

Heart rate, sleep habits, and activity levels are just a few of the health-related data that wearable IoT devices, like fitness trackers and smartwatches, regularly gather. By analyzing this data, ML algorithms give consumers and healthcare providers access to real-time health insights. Abnormal data patterns in emergencies set off alarms that facilitate prompt medical attention and proactive illness treatment (Alshammari, 2023).

5.4.4. Traffic Management (Cities Smart)

An large network of IoT sensors that are thoughtfully positioned on roads, traffic signals, and automobiles continuously gathers a lot of data about traffic flow, congestion, and accidents in smart cities. Then, using powerful ML techniques, this real-time data is utilized to uncover hidden patterns and trends. These algorithms provide the best routes for commuters to go across cities in addition to forecasting future traffic patterns. Furthermore, this data-driven strategy even reaches the infrastructure itself, as traffic

Figure 3. Collaboration between ML and IoT in decision making

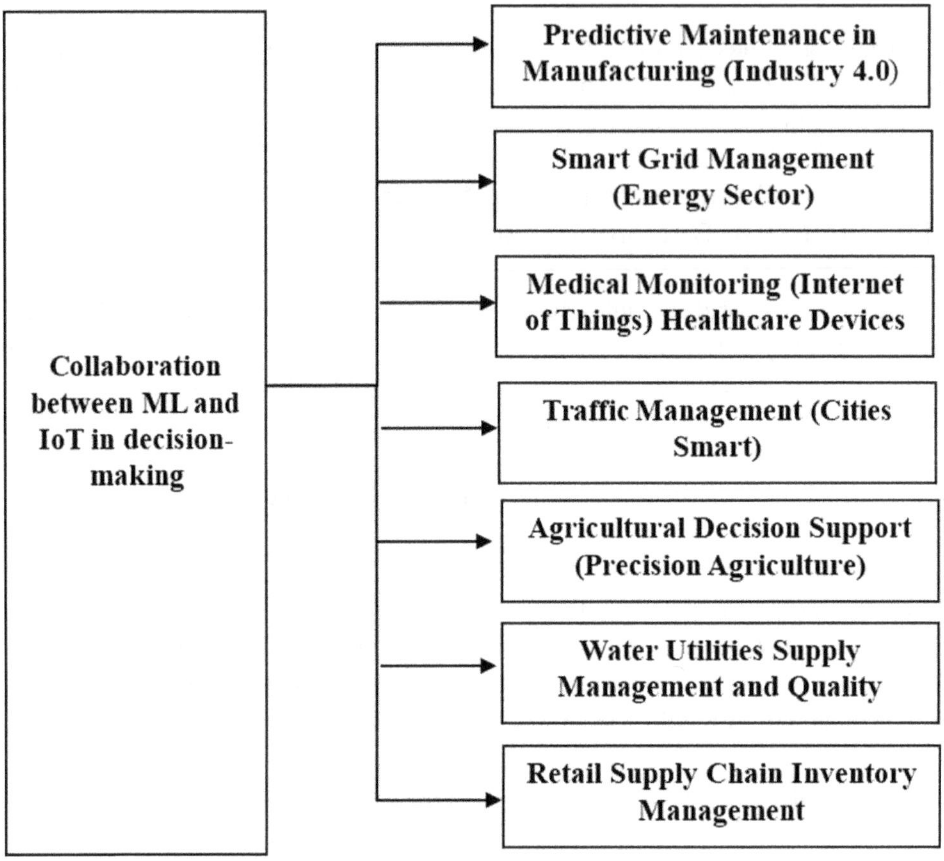

lights are able to dynamically modify their operation in response to the constantly changing conditions they identify. For the benefit of all city dwellers and commuters, this integrated system is an essential instrument in the continuous endeavor to improve overall traffic management, lessen congestion, and make urban transportation more sustainable and efficient (Kaur & Aron, 2022a; Tandon et al., 2022b, 2022a, 2023; Verma et al., 2022).

5.4.5. Agricultural Decision Support (Precision Agriculture)

It involves the use of IoT sensors to monitor crop health, weather, and soil moisture on a farm, while drones are used to take overhead photos. ML models incorporate this data to predict agricultural yields, optimize irrigation plans, and identify disease outbreaks early. To maximize output while preserving resources, farmers make data-driven decisions (Nova, 2023).

5.4.6. Water Utilities Supply Management and Quality

Chemical levels, turbidity, pH, and other water quality parameters are continuously monitored in reservoirs and distribution networks by IoT sensors. By processing this data, ML algorithms forecast trends in water quality and contamination threats. Water utilities make real-time modifications to distribution and treatment procedures based on these findings, guaranteeing a reliable and safe supply of water (Bassine et al., 2023).

5.4.7. Retail Supply Chain Inventory Management

IoT sensors monitor stock levels in warehouses and retail locations, gathering information on product availability, demand, and shelf life. By optimizing inventory replenishment and minimizing overstocking and understocking problems, ML algorithms evaluate this data. Retailers make wise choices to increase efficiency and consumer happiness.

5.5. Smart Irrigation Systems and Precision Agriculture

With an emphasis on how the convergence of IoT and ML algorithms is transforming crop management techniques and agricultural water utilization, this section explores the revolutionary role of smart irrigation systems and precision agriculture. The statement highlights the ways in which these innovations are improving crop yields and agricultural sustainability in addition to making the most use of moisture resources. Figure 4 shows the Smart Irrigation Systems that is explained in this section.

5.5.1. Implementing ML Algorithms and IoT in Agricultural Water Usage

Modern agriculture uses sophisticated methods that are examined in the subsection on using IoT and ML algorithms to optimize agricultural water utilization. It emphasizes how irrigation methods and water allocation are greatly enhanced by ML algorithms that are powered by data gathered from IoT sensors and devices. Applying the right amount, at the right time, and on the right crop requires careful planning, which these tools help farmers achieve (Uddin et al., 2023a).

5.5.2. Smart Irrigation Controllers and Soil Moisture Monitoring

Modern agricultural practices need the use of smart irrigation controllers and soil moisture monitoring devices. The impact of these technologies on crop health and irrigation efficiency is explored in detail in this section. In order to make sure that crops receive the ideal amount of moisture, smart irrigation controllers, are made to autonomously modify water distribution based on real-time data. Monitoring soil moisture, which is usually made possible by IoT sensors, is also very important in this situation. At different depths, these sensors measure the soil moisture content continually and send the information to the central control system. From then, ML algorithms analyze this data to pinpoint irrigation requirements. The case studies in previous sections depict situations in which farmers have used these technologies and observed notable enhancements in crop quality and productivity, all while preserving water resources (Sarmas et al., 2022).

Figure 4. Smart irrigation system

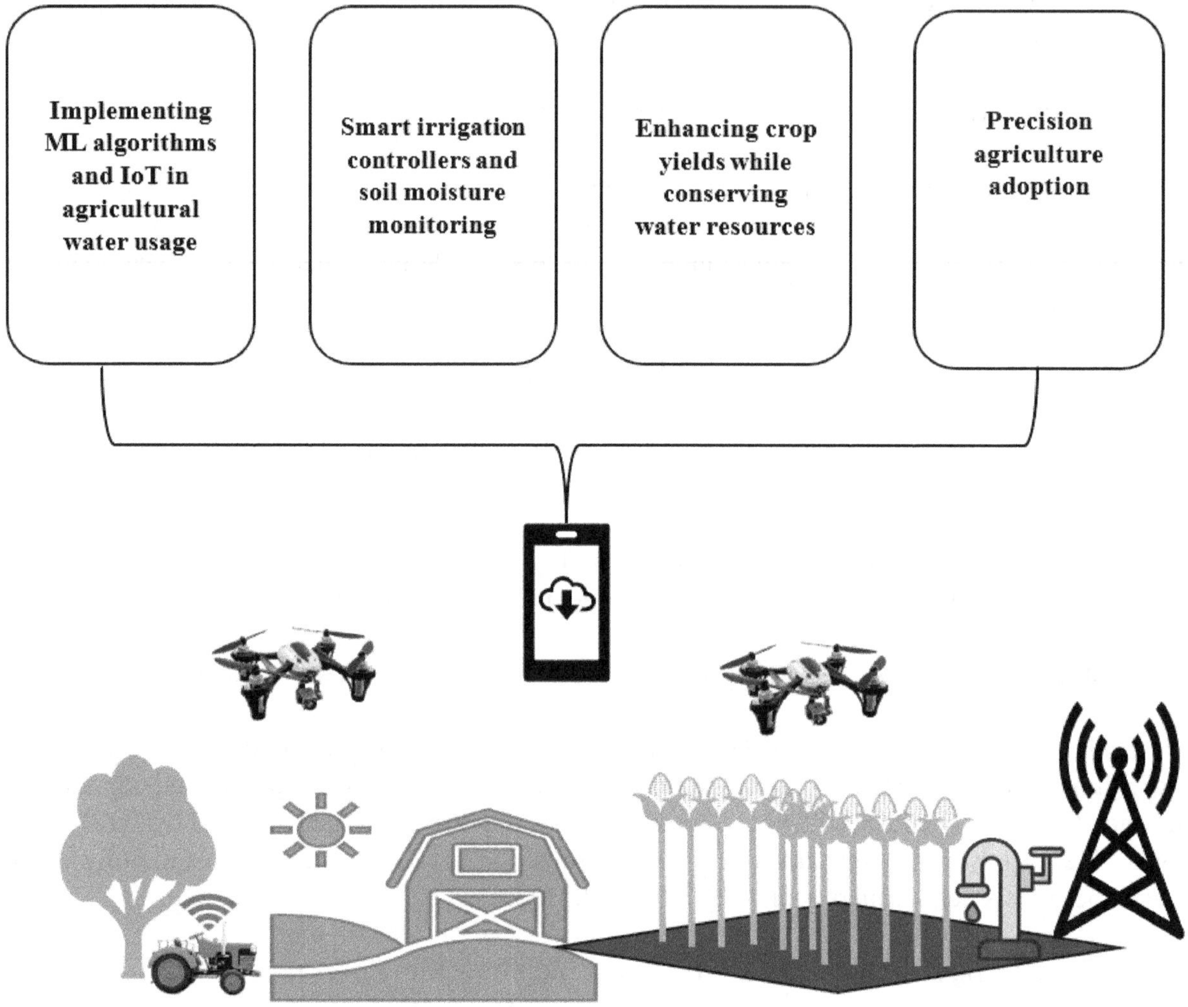

5.5.3. Enhancing Crop Yields While Conserving Water Resources

This section focuses on the double advantages of precision agriculture and smart irrigation systems: increased crop yields and sustainable water resource management. It investigates how more effective water usage leads to higher agricultural productivity when ML-driven decision-making and IoT-enabled monitoring work together. These technologies lessen the risk of over- or under-irrigation, minimizing crop stress and water waste, by precisely delivering the appropriate amount of water when and where it's needed. These methods help farmers make more money in addition to helping to conserve water over the long run, which is crucial in areas where there is a water shortage (Nova, 2023).

5.5.4. Precision Agriculture Adoption

Implementing ML and IoT technology in agriculture has proven to have positive effects. The revolutionary potential of these technologies is often demonstrated by exhibiting actual farms and agricultural businesses that have adopted precision agriculture techniques and smart irrigation systems. Farm sustainability as a whole, agricultural yields, and water use efficiency are all significantly improved in these success stories. In addition to boosting farmers' profits, they highlight how the use of these technologies has promoted sustainable and ethical management of water resources. Precision agriculture practices promise to ensure food security and protect essential water resources for future generations (Nova, 2023).

5.6. Reinforcement Learning-Optimization Hybrid for Water Resource Allocation

In order to efficiently manage and distribute water resources, a novel strategy known as the "Reinforcement Learning-Optimization Hybrid for Water Resource Allocation" combines two potent methodologies: optimization and reinforcement learning (RL). The goal of this hybrid strategy is to address the dynamic and complicated problems related to the distribution of water resources in several contexts, including urban water supply, agricultural, and environmental conservation (Khalilpourazari & Hashemi Doulabi, 2022). This algorithm is explained as follows:

- **Artificial Intelligence (AI):** Reinforcement learning is a ML paradigm in which an agent has the ability to maximize a cumulative reward by learning to make a series of decisions (actions) in a given environment. RL is used to develop a framework for decision-making when it comes to the distribution of water resources. An environment simulating the distribution of water resources is interacted with by the RL agent. It acts (giving water to various regions, for example) and gets feedback (rewards or punishments) according to the results of its deeds. The RL agent acquires ideal policies over time, which direct decision-making in the water allocation procedure. As the agent develops additional expertise and experiments with various tactics, these policies are modified iteratively (Khalilpourazari & Hashemi Doulabi, 2022).
- **Hybrid Strategy:** To capitalize on the advantages of both methodologies, the hybrid approach blends reinforcement learning with optimization. While optimization refines and maximizes the policies produced by RL, reinforcement learning (RL) offers flexibility and the capacity to learn from data and experience. Through experimentation with various allocation procedures, the RL agent gains knowledge through its interactions with the water allocation environment. Due to the intricacy of the issue, RL may not always be able to identify the globally best solution on its own. The taught policies are further optimized by means of optimization techniques (Figueiredo et al., 2021; Khalilpourazari & Hashemi Doulabi, 2022).

Algorithm: Hybrid Reinforcement Learning-Optimization for Allocating Water Resources

```
Input: - Environment for allocating water resources (states, actions, incen-
tives)
```

```
- Deep Q-Network hyperparameters and the RL algorithm
- Algorithms for optimization, such as genetic algorithms
- Time horizon (T) for simulation
- The starting point (s_0)
Required Output: - An optimal approach for allocating water resources
1. Set up the RL agent:
   - Set state-action pair Q-values to arbitrary initial values.
   - Assign s = s_0 as the initial state.
   - Establish an exploration strategy for the RL agent, such as ε-greedy.
2. Set up the settings for optimization:
   Determine the population size to be optimized.
   Establish the termination conditions (convergence, maximum generations,
etc.).
3. Launch the RL-Optimization loop: - Continue until convergence is reached or
a predetermined halting condition is satisfied.
4.1. RL Exploration: - Apply RL policy to choose an action (allocation of wa-
ter) in accordance with the existing state (s).
       - Take note of the final state (s') and the instant reward (r).
       - Utilizing the RL algorithm, update Q-values.
4.2. Optimization: - Create a population of solutions (potential water distri-
bution strategies).
       - Use the policy of the RL agent for T time steps to assess the perfor-
mance of each solution.
       - Using RL-learned policies, choose the best-performing solutions.
4.3. Refinement: - To fine-tune chosen solutions, apply the optimization meth-
od (e.g., genetic algorithms).
       - Using crossover and mutation, create a new population of solutions.
       - Apply the RL agent's policy for T time steps to assess the new solu-
tions.
4.4. Policy Update: - Modify the RL policy in accordance with the best-per-
forming optimization results.
       - Modify the exploration plan as necessary.
5. Output: - The ultimate water resource allocation policy that was optimized
using the hybrid RL-Optimization approach.
End.
```

This algorithm describes how to learn and improve the policy for allocating water resources using an iterative approach that combines RL exploration and optimization. After the RL agent has surveyed the surroundings and determined what to do, the optimization algorithm adjusts these policies to maximize efficiency. In dynamic water management scenarios, this hybrid approach looks for the best and most flexible way to allocate resources.

6. SMART CITIES AND SUSTAINABLE WATER MANAGEMENT

This section explores the intersection of technology, water resource management, and urban development, with a focus on the idea of "smart cities." It looks at how cities are using technology more and more to drive creative solutions to address the complicated issues brought on by population expansion, urbanization, and sustainable water resource management. These solutions cover a wide range of tactics, including improved data collecting via IoT sensors, data analysis through ML algorithms, and redesigned water distribution networks, more effective treatment procedures, and the encouragement of sustainable water practices. In summary, this section highlights the ways in which cities throughout the world are utilizing technology and innovative strategies to guarantee the robust and sustainable administration of water resources in the framework of Smart Cities.

6.1. IoT-Driven Solutions for Urban Water Distribution and Consumption

The optimization of urban water distribution systems through the use of the IoT is examined in this subsection. Water infrastructure in smart cities includes strategically positioned IoT sensors and devices in pipelines, reservoirs, and treatment plants. Real-time data on water flow, quality, and distribution is gathered by these sensors. The efficient distribution of water, the avoidance of leaks, the reduction of water loss, and the increased dependability of the urban water supply are all achieved by analyzing this data (Manny, 2023).

6.2. ML Applications in Optimizing Water Treatment Processes

This section focuses on how ML might improve the effectiveness of water treatment systems in smart cities. To anticipate changes in water quality and improve treatment procedures, ML algorithms are utilized. Water treatment facilities can make real-time adjustments to their processes for best outcomes by using ML models to analyze data from several sources, such as historical records and water quality sensors. These models can spot patterns, anomalies, and probable contaminants (Chinnaappan et al., 2023).

6.3. Smart Water Grids and Demand-Responsive Systems

Water distribution systems in smart cities frequently have sophisticated control mechanisms installed. The ability of these systems to adjust in real-time to shifting demand patterns guarantees that water is delivered exactly where and when it is needed. Variable pricing is another feature of demand-responsive systems that can be used to promote wise water use during times of high demand (Nova, 2023).

6.4. Examining the Environmental and Economic Benefits of Smart City Water Initiatives

This advantages for the environment of less water waste and better water quality, which support the health of ecosystems and lower energy use. It also takes into account the financial benefits of intelligent water management, such as lower operating costs, longer-lasting infrastructure, and better living conditions for city dwellers (Sahoo & Goswami, 2024).

7. ETHICAL AND SOCIETAL IMPLICATIONS OF ML AND IOT TECHNOLOGIES IN WATER MANAGEMENT

The use of ML and IoT technologies in the field of water management has clearly introduced significant transformational potential. Nevertheless, it is imperative to acknowledge and address the ethical and societal implications that emerge from the incorporation of these powerful instruments.

7.1. Factors and Consequences of Responsible and Sustainable Water Management Practices

7.1.1. Data Privacy and Security

The privacy and safety of data gathered by IoT sensors and analyzed by ML algorithms is one of the most important ethical issues. The risk of data breaches and unauthorized access grows as IoT networks constantly collect data from different water sources. Making sure that strong data security, access control, and following data protection rules are in place is very important. Also, there needs to be a clear set of rules for how private information about water quality is handled (Fu et al., 2022).

7.1.2. Inequality and Access

Everyone should be able to get the benefits of advanced water quality tracking, no matter how much money they have or where they live. Access to clean water and the technologies used for tracking are at the center of ethical concerns. There should be efforts to close the digital gap and give people who aren't getting enough help the chance to use IoT and ML solutions to improve how they manage water (Rizal et al., 2023).

7.1.3. Accountability and Openness

ML systems are having a bigger impact on water management decisions, so it's important to make sure that decisions are clear and accountable. People with a stake in the matter should know how these programmes work and what data they use. To fix algorithmic bias, which can unintentionally keep environmental crimes or differences in water quality going, there should be ways to hold people accountable (Khan et al., 2023).

7.1.4. Effects on the Environment

ML and the IoT can help find problems with water quality early on. However, making and using IoT sensors and devices can add to electronic trash and carbon emissions. For ethical reasons, lowering these technologies' effects on the environment through eco-friendly production, smart energy use, and proper removal is important (Taneja et al., 2020).

7.1.5. Involvement and Agreement From the Community

It is the right thing to do to include local communities in the installation of IoT sensors and the data gathering processes. Residents must give their prior informed consent and be involved in the community to make sure they understand what these technologies are for and how they help handle water quality. Also, ways for people to give feedback should be set up so that community concerns can be addressed and monitoring methods can be changed as needed (Ghoochani et al., 2023).

7.1.6. Working Together Across Disciplines

Data scientists, water engineers, environmentalists, and politicians need to work together to effectively deal with the ethical and social issues that arise. A multidisciplinary strategy can help build an ethical framework that supports the goals of sustainable water management and takes into account the needs and values of all stakeholders (Khan et al., 2023).

7.1.7. Following the Rules

IoT and ML technologies should follow the rules and laws that are already in place for water quality. As technology improves, ethical concerns include the need to follow the rules set by regulatory bodies and work with those bodies to make changes to the rules.

7.1.8. Education and Making People Aware

Lastly, it is important to teach people about the moral and social effects of using AI and IoT in water management and make them more aware of these effects. This gives people the power to make smart choices, hold stakeholders responsible, and shape the right way to use these tools for the good of society and the environment (Sahoo & Goswami, 2024).

7.2. Comparative Analysis of ML and IoT Solutions in Water Management Against Other Available Technologies

ML and IoT solutions are compared to traditional methods in order to find the most efficient and effective ways to handle water. This section compares the pros, cons, and special features of ML and IoT technologies to traditional ways of managing water.

In terms of monitoring water quality, this table shows how ML and IoT solutions stack up against traditional approaches and standard sensor networks. The technology or method chosen should depend on the needs, the available funds, and the need to monitor water quality in real time and find problems early on.

8. CHALLENGES AND FUTURE DIRECTIONS

The journey of incorporating ML and the IoT into water management practices is critically examined in this section, which also highlights the opportunities and obstacles that still need to be addressed.

Table 1. Comparative analysis of water quality monitoring technologies

Characteristics	ML and IoT-Based Solutions	Traditional Approaches	Conventional Sensor Networks
Methodology Followed	Real-time monitoring of data with IoT sensors, ML algorithms(Chinnappan et al., 2023)	Manual water sample collection and lab testing(Kang et al., 2017)	Data collection is done through sensors(Javaid et al., 2022)
Accuracy and Precision	Accuracy is good for predictive models	High accuracy and precision	Moderate accuracy
Cost and Resource Efficiency	Lower operational costs	High operational costs and resource-intensive	Moderate operational costs
Scalability and Accessibility	Highly scalable and accessible	Limited geographical coverage	Limited scalability
Response Time and Detection	Early detection and quick response	Delayed results	Delayed response
Impact on environment	Efficient data gathering has less of an effect on the environment	Emissions from transportation	Less damage to the earth
Flexibility and Adaptability	Able to adapt to changing water quality conditions	Static methods	Some adaptability
Data Volume and Analysis	Deals with a lot of data and uses ML to analyze it	Small amount of data, review manually	Not enough tools for analyzing data
Maintenance and Reliability	Maintenance on a regular basis and predictability through prediction models	High reliability with little care	Maintenance is average, and dependability is average

8.1. Challenges

This chapter highlights a number of significant challenges faced by integrating ML and the IoT within the context of water management.

8.1.1. Water Data Collection and Analysis: Ethical Considerations

As the potential of IoT sensors and ML algorithms are harnessed to gather and analyze vast amounts of water data, ethical issues become more pressing. It's critical to address issues with data ownership, data sources' informed consent, and data usage openness. The necessity of ethical frameworks that are strong enough to direct the management of data related to water is discussed in this subsection, which also emphasizes the importance of responsible data collection practices. The significance of guaranteeing that the advantages of new technologies are equal and accessible is also emphasized, particularly in marginalized or disadvantaged populations (Marques dos Santos et al., 2023b).

8.1.2. IoT and ML Integration-Related Security and Privacy Concerns

Security and privacy issues take on a new level with the combination of ML and IoT technology. ML algorithms handling sensitive data require protection against unauthorized access, and IoT devices, which are frequently linked in large networks, can be targets for cyberattacks. The nuances of data encryption techniques, privacy protection methods, and cybersecurity precautions are covered in detail in this section. In order to protect against potential breaches, data manipulation, and misuse, it emphasizes how important it is to address these issues with diligence and preserve the security and integrity of water-related data (Ghoochani et al., 2023).

8.1.3. Augmenting IoT and ML Solutions for More Comprehensive Water Management Uses

Although ML and IoT applications for water management have proven useful, scaling these solutions for wider and more extensive use comes with its own set of difficulties. This article delves into the necessity of making large investments in infrastructure, standardizing communication protocols, and fostering interoperability across various IoT devices and ML platforms. In addition to aiding metropolitan areas, it explores methods for spreading the use of these revolutionary technology to undeserved rural groups and locations (Khan et al., 2023).

8.1.4. Encouraging Developments and Prospective Research Paths

Looking ahead, this segment offers a taste of the fascinating opportunities and future directions that the fields of ML and IoT water management are exploring. Promising developments are mentioned, such as the creation of increasingly complex ML algorithms for more precise water quality prediction. The potential of edge computing to process data closer to its source, lowering latency and improving real-time decision-making, is also taken into consideration. Emerging technologies like Blockchain are integrated for safe and transparent data management. To promote innovation and further develop the sector, the subsection calls for data scientists, engineers, and specialists in water management to maintain their interdisciplinary work (Khan et al., 2023; Zhao et al., 2023).

8.2. Future Directions

8.2.1. Ethical Considerations in Water Data Collection and Analysis

The increasing utilization of IoT sensors and ML algorithms in water data collection necessitates the prioritization of ethical considerations. Robust ethical frameworks pertaining to the proper management of data should be focused in future research. In future works researchers should focus on obtaining consents from data sources, transparent use of data, and addressing data ownership properly (Marques dos Santos et al., 2023b).

8.2.2. Security and Privacy Concerns

The convergence of IoT with ML presents novel security and privacy concerns. Future research should aim to investigate advanced data encryption techniques, privacy protection approaches, and Cyber security measures in order to ensure the security of sensitive water-related data. Ensuring the prevention of unauthorized access, data tampering, and misuse is of paramount importance in upholding the security and integrity of the data acquired through these technologies (Shumilova et al., 2023).

8.2.3. Holistic Approach to Water Management Applications

In order to facilitate the broader use of IoT and ML solutions in the domain of water management, it becomes imperative to address the obstacles associated with scalability. Future research should prioritize the allocation of significant resources towards the enhancement of infrastructure, the establishment of standardized communication protocols, and the facilitation of interoperability among diverse IoT devices and ML platforms. This will facilitate the broader implementation of these technologies, yielding advantages for both urban and rural regions (Lambrechts et al., 2023).

8.2.4. Encouraging Advances and Potential Research Directions

This encompasses the advancement of more advanced ML algorithms aimed at accurately predicting water quality. In addition, edge computing holds significant potential as a technology that may effectively mitigate latency and improve the efficiency of real-time decision-making processes. The exploration of integrating emerging technologies, such as Blockchain, for the purpose of safe and transparent data management, is warranted. Collaborative interdisciplinary collaboration between data scientists, engineers, and water management specialists will be important to promote innovation in this field (Sugam et al., 2023).

9. CONCLUSION

This chapter concludes with a call to action for different stakeholders and emphasizes the revolutionary potential of combining ML and the IoT in water management. It covers how ML and IoT are being applied in real-time water monitoring, decision support systems, smart irrigation, and water resource analysis, as well as how they fit into the larger picture of smart cities. It highlights the importance of these technological developments in tackling urgent problems related to water management. It also focuses on the ways in which the combination of ML and the IoT not only improves the precision of forecasts pertaining to water resources, but also supports water conservation, better upkeep of infrastructure, and better decision-making. It motivates academics, policymakers, and industry players to acknowledge the importance of ML and the IoT in the context of water management. In terms of ML and IoT applications for water management, it encourages more study, instruction, and training for academics. It serves as a reminder to legislators of the significance of developing laws and other policies that encourage the appropriate use of new technologies for the good of society and the environment.

REFERENCES

Ali, M. H., Popescu, I., Jonoski, A., & Solomatine, D. P. (2023). Remote Sensed and/or Global Datasets for Distributed Hydrological Modelling: A Review. *Remote Sensing (Basel)*, *15*(6), 1–43. doi:10.3390/rs15061642

Alshammari, H. H. (2023). The IoT healthcare monitoring system based on MQTT protocol. *Alexandria Engineering Journal*, *69*, 275–287. doi:10.1016/j.aej.2023.01.065

Apa, A. D., Boenish, R., & Kleisner, K. (2023). *Effects of climate change and variability on large pelagic fish in the Northwest Atlantic Ocean : Implications for improving climate resilient management for pelagic longline fi sheries.* doi:10.3389/fmars.2023.1206911

Barzegar, R., Aalami, M. T., & Adamowski, J. (2020). Short-term water quality variable prediction using a hybrid CNN–LSTM deep learning model. *Stochastic Environmental Research and Risk Assessment, 34*(2), 415–433. doi:10.100700477-020-01776-2

BassineF. Z.EpuleT. E.KechchourA.ChehbouniA. (2023). *Recent applications of ML, remote sensing, and iot approaches in yield prediction: a critical review*. https://arxiv.org/abs/2306.04566

Bolick, M. M., Post, C. J., Naser, M. Z., Forghanparast, F., & Mikhailova, E. A. (2023). Evaluating Urban Stream Flooding with ML, LiDAR, and 3D Modeling. *Water (Basel), 15*(14), 1–25. doi:10.3390/w15142581

Butler, M. J., Yellen, B. C., Oyewumi, O., Ouimet, W., & Richardson, J. B. (2023). Accumulation and transport of nutrient and pollutant elements in riparian soils, sediments, and river waters across the Thames River Watershed, Connecticut, USA. *The Science of the Total Environment, 899*(March), 165630. doi:10.1016/j.scitotenv.2023.165630 PMID:37467973

Chandra Sekhar, K., Venkatesh, B., Reddy, K. S., Giridhar, G., Nithin, K., & Eshwar, K. (2023). IoT-based Realtime Water Quality Management System using Arduino Microcontroller. *Turkish Journal of Computer and Mathematics Education, 14*(02), 783–792.

Chinnappan, C. V., John William, A. D., Nidamanuri, S. K. C., Jayalakshmi, S., Bogani, R., Thanapal, P., Syed, S., Venkateswarlu, B., & Syed Masood, J. A. I. (2023). IoT-Enabled Chlorine Level Assessment and Prediction in Water Monitoring System Using ML. *Electronics (Basel), 12*(6), 1458. Advance online publication. doi:10.3390/electronics12061458

Elbeltagi, A., Aslam, M. R., Malik, A., Mehdinejadiani, B., Srivastava, A., Bhatia, A. S., & Deng, J. (2020). The impact of climate changes on the water footprint of wheat and maize production in the Nile Delta, Egypt. *The Science of the Total Environment, 743*, 140770. doi:10.1016/j.scitotenv.2020.140770 PMID:32679501

Elshaboury, N., & Marzouk, M. (2022). Prioritizing water distribution pipelines rehabilitation using ML algorithms. *Soft Computing, 26*(11), 5179–5193. doi:10.100700500-022-06970-8

Figueiredo, I., Esteves, P., & Cabrita, P. (2021). Water wise - A digital water solution for smart cities and water management entities. *Procedia Computer Science, 181*(2019), 897–904. doi:10.1016/j.procs.2021.01.245

Fu, G., Jin, Y., Sun, S., Yuan, Z., & Butler, D. (2022). The role of deep learning in urban water management: A critical review. *Water Research, 223*, 118973. doi:10.1016/j.watres.2022.118973 PMID:35988335

GhoochaniS.KhorramM.NazemiN.ClassificationS.QualityD. W.ScholarG. (2023). *Uncovering Top-Tier ML Classifier for Drinking Water Quality Detection*. doi:10.20944/preprints202308.1636.v1

Głomb, P., Cholewa, M., Koral, W., Madej, A., & Romaszewski, M. (2023). Detection of emergent leaks using ML approaches. *Water Science and Technology: Water Supply, 23*(6), 2371–2386. doi:10.2166/ws.2023.118

Gong, J., Guo, X., Yan, X., & Hu, C. (2023). Review of Urban Drinking Water Contamination Source Identification Methods. *Energies, 16*(2), 705. Advance online publication. doi:10.3390/en16020705

Grbčić, L., Kranjčević, L., & Družeta, S. (2021). ML and simulation-optimization coupling for water distribution network contamination source detection. *Sensors (Basel), 21*(4), 1–25. doi:10.339021041157 PMID:33562175

Hasan, M. K., Habib, A. A., Islam, S., Balfaqih, M., Alfawaz, K. M., & Singh, D. (2023). Smart Grid Communication Networks for Electric Vehicles Empowering Distributed Energy Generation: Constraints, Challenges, and Recommendations. *Energies, 16*(3), 1140. Advance online publication. doi:10.3390/en16031140

Hayder, G., Kurniawan, I., & Mustafa, H. M. (2021). Implementation of ML methods for monitoring and predicting water quality parameters. *Biointerface Research in Applied Chemistry, 11*(2), 9285–9295. doi:10.33263/BRIAC112.92859295

Javaid, M., Haleem, A., Singh, R. P., Suman, R., & Gonzalez, E. S. (2022). Understanding the adoption of Industry 4.0 technologies in improving environmental sustainability. *Sustainable Operations and Computers, 3*, 203–217. doi:10.1016/j.susoc.2022.01.008

Kang, G., Gao, J. Z., & Xie, G. (2017). Data-driven water quality analysis and prediction: A survey. *Proceedings - 3rd IEEE International Conference on Big Data Computing Service and Applications, BigDataService 2017*, 224–232. 10.1109/BigDataService.2017.40

Kaur, M., & Aron, R. (2022a). A Novel Load Balancing Technique for Smart Application in a Fog Computing Environment. *International Journal of Grid and High Performance Computing, 14*(1), 1–19. doi:10.4018/IJGHPC.301583

Kaur, M., & Aron, R. (2022b). An Energy-Efficient Load Balancing Approach for Fog Environment Using Scientific Workflow Applications. *Lecture Notes in Electrical Engineering, 903*(September), 165–174. doi:10.1007/978-981-19-2281-7_16

Keller, A. A., Garner, K., Rao, N., Knipping, E., & Thomas, J. (2023). Hydrological models for climate-based assessments at the watershed scale: A critical review of existing hydrologic and water quality models. *Science of the Total Environment, 867*, 161209. doi:10.1016/j.scitotenv.2022.161209

Khalilpourazari, S., & Hashemi Doulabi, H. (2022). Designing a hybrid reinforcement learning based algorithm with application in prediction of the COVID-19 pandemic in Quebec. *Annals of Operations Research, 312*(2), 1261–1305. doi:10.100710479-020-03871-7 PMID:33424076

Khan, J., Lee, E., Balobaid, A. S., & Kim, K. (2023). A Comprehensive Review of Conventional, Machine Leaning, and Deep Learning Models for Groundwater Level (GWL) Forecasting. *Applied Sciences (Basel, Switzerland), 13*(4), 2743. Advance online publication. doi:10.3390/app13042743

Lambrechts, H. A., Paparrizos, S., Brongersma, R., Kroeze, C., Ludwig, F., & Stoof, C. R. (2023). Governing wildfire in a global change context: Lessons from water management in the Netherlands. *Fire Ecology, 19*(1), 6. Advance online publication. doi:10.118642408-023-00166-7

Manny, L. (2023). Socio-technical challenges towards data-driven and integrated urban water management: A socio-technical network approach. *Sustainable Cities and Society, 90*, 104360. doi:10.1016/j.scs.2022.104360

Marques dos Santos, M., Caixia, L., & Snyder, S. A. (2023a). Evaluation of wastewater-based epidemiology of COVID-19 approaches in Singapore's 'closed-system' scenario: A long-term country-wide assessment. *Water Research, 244*, 120406. doi:10.1016/j.watres.2023.120406

Marques dos Santos, M., Caixia, L., & Snyder, S. A. (2023b). Evaluation of wastewater-based epidemiology of COVID-19 approaches in Singapore's 'closed-system' scenario: A long-term country-wide assessment. *Water Research, 244*, 120406. doi:10.1016/j.watres.2023.120406 PMID:37542765

Mishra, S., & Tyagi, A. K. (2022). The Role of ML Techniques in IoT-Based Cloud Applications. *IoT*, (February), 105–135. doi:10.1007/978-3-030-87059-1_4

Musa, A. A., Malami, S. I., Alanazi, F., Ounaies, W., Alshammari, M., & Haruna, S. I. (2023). Sustainable Traffic Management for Smart Cities Using Internet-of-Things-Oriented Intelligent Transportation Systems (ITS): Challenges and Recommendations. *Sustainability (Basel), 15*(13), 1–15. doi:10.3390u15139859

Nadkarni, S., Kriechbaumer, F., Rothenberger, M., & Christodoulidou, N. (2020). The path to the Hotel of Things: IoT and Big Data converging in hospitality. *Journal of Hospitality and Tourism Technology, 11*(1), 93–107. doi:10.1108/JHTT-12-2018-0120

Nova, K. (2023). AI-Enabled Water Management Systems: An Analysis of System Components and Interdependencies for Water Conservation. *Eigenpub Review of Science and Technology, 7*(1), 105–124. https://studies.eigenpub.com/index.php/erst/article/view/12

Optoelectronics, S. (2023).. . *Semiconductor Optoelectronics*, *42*(1), 200–211.

Rizal, N. N. M., Hayder, G., & Yussof, S. (2023). *River Water Quality Prediction and Analysis*. In G. H. A. Salih & R. A. Saeed (Eds.), *Deep Learning Predictive Models Approach BT - Sustainability Challenges and Delivering Practical Engineering Solutions* (pp. 25–29). Springer International Publishing.

Rosati, R., Romeo, L., Cecchini, G., Tonetto, F., Viti, P., Mancini, A., & Frontoni, E. (2023). From knowledge-based to big data analytic model: A novel IoT and ML based decision support system for predictive maintenance in Industry 4.0. *Journal of Intelligent Manufacturing*, *34*(1), 107–121. doi:10.100710845-022-01960-x

Sahoo, S. K., & Goswami, S. S. (2024). Theoretical framework for assessing the economic and environmental impact of water pollution: A detailed study on sustainable development of India. *Journal of Future Sustainability*, *4*(1), 23–34. doi:10.5267/j.jfs.2024.1.003

Salam, A. (2020). IoT in water management and treatment. *IoT*, 273–298. doi:10.1007/978-3-030-35291-2_9

Sarmas, E., Spiliotis, E., Marinakis, V., Tzanes, G., Kaldellis, J. K., & Doukas, H. (2022). ML-based energy management of water pumping systems for the application of peak shaving in small-scale islands. *Sustainable Cities and Society*, *82*, 103873. doi:10.1016/j.scs.2022.103873

Shumilova, O., Tockner, K., Sukhodolov, A., Khilchevskyi, V., De Meester, L., Stepanenko, S., Trokhymenko, G., Hernández-Agüero, J. A., & Gleick, P. (2023). Impact of the Russia–Ukraine armed conflict on water resources and water infrastructure. *Nature Sustainability*, *6*(5), 578–586. doi:10.103841893-023-01068-x

Solanki, A., Agrawal, H., & Khare, K. (2015). Predictive Analysis of Water Quality Parameters using Deep Learning. *International Journal of Computer Applications*, *125*(9), 29–34. doi:10.5120/ijca2015905874

Sugam, V., Parthiban, P., Ravikumar, K., Das, I. C., & Ashutosh, D. (2023). Steady-state Assessment of Hydraulic Potential at Water Scarce regions of Agniyar River Basin, India using GMS-MODFLOW. *Disaster Advances*, *16*(5), 38–43. doi:10.25303/1605da038043

Tandon, R., Verma, A., & Gupta, P. K. (2022a). Blockchain enabled vehicular networks: a review. *2022 5th International Conference on Multimedia, Signal Processing and Communication Technologies, IMPACT 2022*. 10.1109/IMPACT55510.2022.10029136

Tandon, R., Verma, A., & Gupta, P. K. (2022b). RVTN: Recommender system for vehicle routing in transportation network. *PDGC 2022 - 2022 7th International Conference on Parallel, Distributed and Grid Computing*, 352–356. 10.1109/PDGC56933.2022.10053267

Tandon, R., Verma, A., & Gupta, P. K. (2023). Nature-inspired whale optimization technique for efficient information exchange in vehicular networks. *Proceedings - 2023 12th IEEE International Conference on Communication Systems and Network Technologies, CSNT 2023*, 833–838. 10.1109/CSNT57126.2023.10134671

Taneja, M., Byabazaire, J., Jalodia, N., Davy, A., Olariu, C., & Malone, P. (2020). ML based fog computing assisted data-driven approach for early lameness detection in dairy cattle. *Computers and Electronics in Agriculture, 171*, 105286. doi:10.1016/j.compag.2020.105286

Uddin, M. G., Nash, S., Rahman, A., & Olbert, A. I. (2023a). A novel approach for estimating and predicting uncertainty in water quality index model using ML approaches. *Water Research, 229*, 119422. doi:10.1016/j.watres.2022.119422

Uddin, M. G., Nash, S., Rahman, A., & Olbert, A. I. (2023b). Performance analysis of the water quality index model for predicting water state using ML techniques. *Process Safety and Environmental Protection, 169*, 808–828. doi:10.1016/j.psep.2022.11.073

Verma, A., Tandon, R., & Gupta, P. K. (2022). TrafC-AnTabu: AnTabu routing algorithm for congestion control and traffic lights management using fuzzy model. *Internet Technology Letters*, *5*(2), 1–6. doi:10.1002/itl2.309

Zhang, X., Rane, K. P., Kakaravada, I., & Shabaz, M. (2021). Research on vibration monitoring and fault diagnosis of rotating machinery based on IoT technology. *Nonlinear Engineering*, *10*(1), 245–254. doi:10.1515/nleng-2021-0019

Zhao, J., Liu, D., & Huang, R. (2023). A Review of Climate-Smart Agriculture: Recent Advancements, Challenges, and Future Directions. *Sustainability (Basel)*, *15*(4), 1–15. doi:10.3390u15043404

Chapter 3
Artificial Intelligence for Water Resource Planning and Management

Richa Saxena
https://orcid.org/0000-0002-9229-9235
Invertis University, India

Vaishnavi Srivastava
Chhatrapati Shahu Ji Maharaj University, India

Dipti Bharti
Darbhanga College of Engineering, India

Rahul Singh
Darbhanga College of Engineering, India

Amit Kumar
Indian Institute of Technology, Ropar, India

Abhilekha Sharma
Noida International University, India

ABSTRACT

In an era marked by growing water scarcity and increasing demand for efficient resource allocation, the integration of artificial intelligence (AI) has emerged as a crucial approach for revolutionizing water resource planning and management. The chapter emphasizes how important water management is to maintaining ecosystems, sustaining human livelihoods, and promoting economic growth. It looks at how AI, which includes machine learning, data analytics, and optimization approaches, acts as a keystone in improving the precision of projections of water availability, allowing stakeholders to make wise decisions in real-time. These programs provide water managers with useful information that they can use to prevent emergencies related to water. The international community may collaborate to achieve sustainable water security by utilizing AI capacity to decode complicated patterns, predict possible outcomes, and optimize resource distribution. It is a necessary step towards a more resilient and water-secure future as difficulties related to water continue to worsen.

DOI: 10.4018/979-8-3693-1194-3.ch003

INTRODUCTION

Water resource management is a significant and intricate problem that impacts economies, human populations, and ecosystems worldwide. Rivers are the main source of water for drinking and farming throughout the entire world. In addition to posing a threat to aquatic life, environmental degradation and pollution of river ecosystems have detrimental consequences on human health (Saxena, 2022 p. 429). The need for freshwater is increasing due to population growth, industrialization, and climate change, thus planning and management of water resources must be effective and efficient (Gleick, 2018 p. 8863). In this context, artificial intelligence (AI) has emerged as a powerful technology that could fundamentally alter the way we monitor, evaluate, and manage our water resources (Sharma et al, 2021 p. 125).

The Role of Artificial Intelligence in Water Resource Management

Among the technologies that fall under the general category of artificial intelligence are machine learning and modelling algorithms. These instruments are capable of managing massive data sets and yielding perceptive outcomes. Among the many crucial functions AI offers in the field of managing water resources are the following:

1. Integration and Analysis of Data: AI can collect, compile, and evaluate data in real-time from a wide range of sources, including sensors, weather stations, and remote sensing. This allows for the ongoing observation of water quantity, quality, and usage patterns. Predictive Modelling: Based on historical data and environmental conditions, machine learning algorithms may predict future water availability, demand, and probable shortages. These models support resource allocation and proactive planning.
2. Optimization and Decision Support: AI-driven optimization techniques facilitate effective water distribution, reservoir management, and infrastructure design. These instruments help water managers make wise choices to increase water use effectiveness.
3. Early Warning Systems: By evaluating real-time data and sending out notifications when anomalies are found, AI can offer early warning systems for floods, droughts, and water quality problems.
4. Resource Conservation: AI aids in resource conservation by reducing water wastage in agriculture and urban areas with smart irrigation systems and leak detection algorithms. Figure 1 illustrates how artificial intelligence (AI) plays a pivotal role in the all-encompassing administration of water resources.

Water Resource Planning's Importance

Planning for Water Resources Effectively Is Essential for Several Reasons

1. Sustainability: Water resource planning ensures that freshwater is used sustainably, which is crucial for the preservation of biodiversity and ecosystems (Poff, 2019 p.25).
2. Economic Stability: Effective planning lowers the danger of water scarcity, which can disrupt businesses, agriculture, and energy production (UN Water, 2018).

Figure 1. AI's place in managing water resources

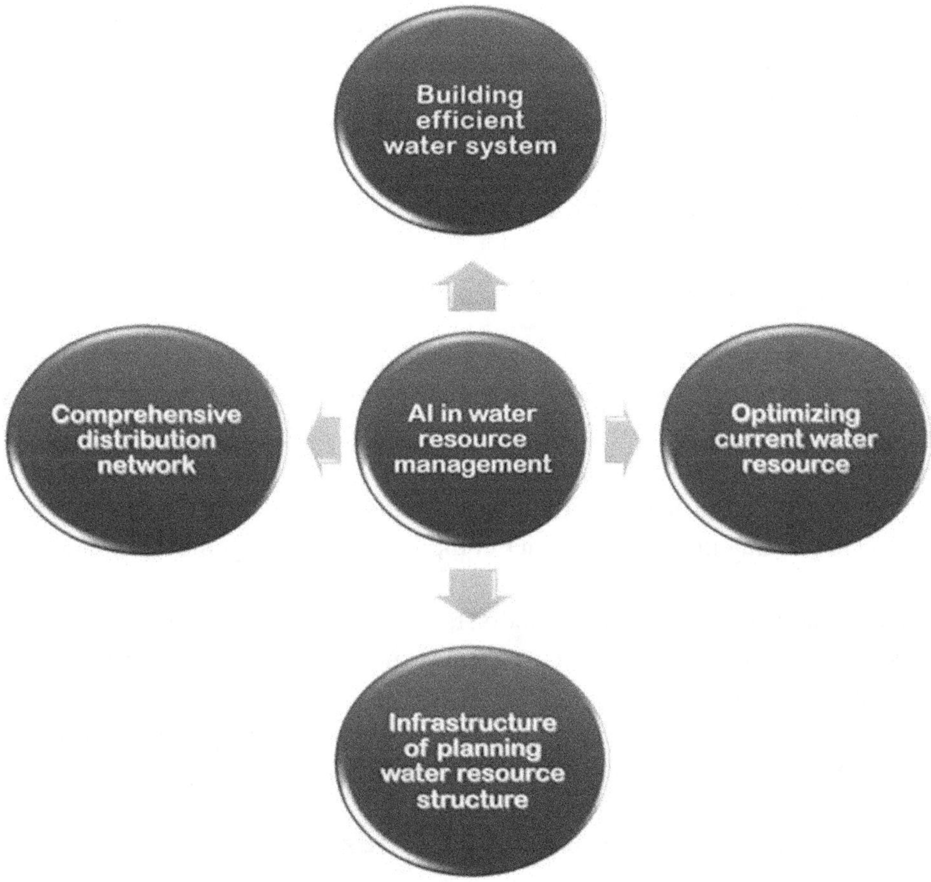

3. This protects economic stability. Planning is necessary to adjust to changing conditions as a result of climate change's increasing variability in rainfall and water availability (Vörösmarty, 2018 p.317).
4. Social Equity: To address the needs of vulnerable people and ensure social justice, equitable water allocation is essential (Srinivasan, 2012).
5. Environmental Protection: According to Puterman, 2018, water resource planning involves efforts to safeguard aquatic ecosystems, ensure species survival, and deliver ecosystem services.

Essentials of Managing Water Resources

The planning, development, use, and preservation of water resources to satisfy the many demands of society while maintaining environmental sustainability constitute the vital field of water resource management. Addressing the mounting issues of pollution, climate change, and water shortages requires an understanding of the principles of water resource management.

Knowing About Water Resources

All freshwater sources, such as lakes, rivers, groundwater, and reservoirs, are considered water resources. Freshwater is essential for many uses, including drinking, agriculture, industry, and ecosystems. A thorough understanding of the hydrological cycle, water availability, quality, and the relationships between environmental factors and human activity are necessary for effective water management.

Obstacles in the Management of Water Resources

The management of water resources is fraught with difficulties:

1. Water Scarcity: As a result of rising water demands and population growth, many regions are experiencing water scarcity, which intensifies competition for scarce resources (UN Water, 2021).
2. Pollution: Water quality and public health are at risk due to industrial, agricultural, and urban activities contaminating water bodies (EPA, 2020).
3. Climate Change: Modified precipitation patterns and an increase in the frequency of extreme weather events affect the availability of water, which complicates the management of water resources (IPCC, 2021).
4. Ecosystem Conservation: To preserve biodiversity and ecosystem services, it is a continuous challenge to strike a balance between human requirements and the preservation of aquatic ecosystems (Dudgeon, 2006 p.163).
5. Infrastructure Aging: Water supply systems are in danger from aging infrastructure, which calls for maintenance and modernization expenditures (AWWA, 2021).

Conventional Methods vs. AI-Powered Methods

Empirical models and historical data have been the foundation of conventional water resource management. On the other hand, AI-driven methods make use of cutting-edge technologies to improve decision-making:

Traditional Approaches: These techniques frequently rely on deterministic models built from historical data and straightforward statistical analysis. Despite their worth, they could find it difficult to adjust to new circumstances.

AI-Driven Approaches: AI can process large datasets and recognize intricate patterns, allowing for more precise forecasts and real-time monitoring (Kroll, 2019). This includes machine learning and data analytics. AI facilitates the integration of various data sources, early warning systems, and resource allocation optimization (Sharma et al, 2021 p.125).

Figure 2 illustrates the integration of AI with systems for water resource management, highlighting the transformative impact of AI-driven approaches on conventional methods. The diagram showcases how AI technologies play a crucial role in enhancing decision-making processes and overall efficiency in managing water resources.

Table 1 provides a comprehensive overview of the various applications of artificial intelligence and how they support key concerns in the planning and administration of water resources. Predictive modelling, real-time monitoring, decision support systems, and other tools are used in these applications.

Figure 2. AI integration with systems for water resource management

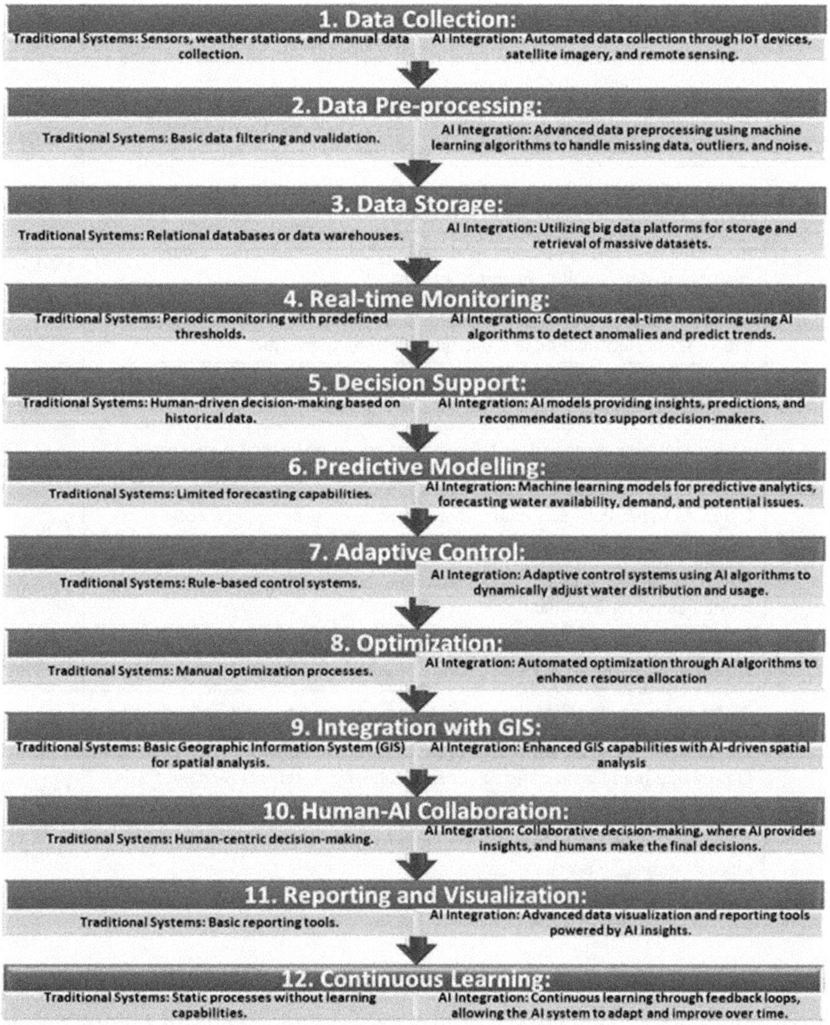

Obtaining and Preparing Data

Pre-processing and data collection are essential phases in the management of water resources that guarantee the availability of precise and trustworthy data for making decisions. This procedure entails gathering information from multiple sources, evaluating its caliber, and getting it ready for analysis.

Water Resource Management Data Sources

1. Monitoring Stations: Real-time data on characteristics like water level, flow rate, temperature, and water quality are collected by monitoring stations that are outfitted with sensors and instruments. Governmental organizations and academic institutions frequently oversee these stations (EPA, 2021).

Table 1. Artificial intelligence applications in water resource planning and management

Application	Description	Reference
Groundwater Level Prediction	AI models that predict groundwater levels based on historical data, aiding in sustainable groundwater management.	(Zheng, 2021).
Reservoir Operation	AI-driven optimization of reservoir release schedules to balance water supply, flood control, and hydropower generation.	(Moubayed, 2022).
Drought Monitoring	AI-based drought early warning systems that analyse meteorological and remote sensing data for drought prediction.	(Han,2019 p.317)
Water Quality Monitoring	Real-time monitoring of water quality using AI-enabled sensors and data analytics to detect pollution and health risks.	(Salehahmadi, 2018 p.534).
Stream flow Forecasting	AI-based streamflow forecasting models that combine weather data, hydrological information, and machine learning techniques.	(Demirel, 2021).
Aquifer Remediation	AI-guided strategies for groundwater and aquifer remediation, optimizing the injection of chemicals to mitigate contamination.	(Dutta et al, 2019 p.554).
Watershed Management	AI-based decision support systems for integrated watershed management, considering land use, climate, and ecological factors.	(Kim et al, 2020).
Water Quality Prediction	Using AI to predict water quality parameters like turbidity, pH, and contaminants based on historical data.	(Smith, 2020, p. 398).
Flood Prediction	AI-driven models for early flood detection and warning systems, analysing rainfall and river level data.	(Afrifa, 2022 p.259).
Water Demand Forecasting	Predicting future water demand patterns to optimize water distribution and allocation.	(Zhang, 2017).
Irrigation Management	AI-based irrigation scheduling, optimizing water usage in agriculture using data from soil sensors and weather forecasts.	(Wang, 2020).
Water Infrastructure Optimization	Optimizing maintenance and repair schedules for water infrastructure systems, reducing leaks and failures.	(Ball, 2017).
Water Resource Allocation	AI-assisted decision-making for equitable allocation of water resources among competing users and sectors.	(Madani, 2019 p. 3183)

2. Remote sensing: Information on surface water bodies, land use, and precipitation patterns can be obtained by satellite and aerial imaging. Large-scale monitoring of changes in water resources is made possible via remote sensing.
3. Social and Economic Data: Understanding the influence of humans on water resources requires an understanding of socioeconomic data, which includes population increase, water demand, and industrial activity (Hejazi, 2013 p.205).
4. Citizen Science: Information on local environmental conditions and water quality can be obtained through crowd sourced data from citizen scientists.

Data Quality and Integrity

Ensuring data quality and integrity is critical for making informed decisions in water resource management:

1. **Accuracy**: Data should accurately represent the parameter being measured, with minimal errors or biases.

2. **Precision**: Precise data have low variability and can consistently measure the same parameter (ISO, 2017).
3. **Completeness**: Data should cover relevant periods and geographic locations comprehensively (EPA, 2021).
4. **Consistency**: Data collected from different sources or over time should be consistent and compatible for integration (Hafezparast, 2021 p.428).
5. **Metadata**: Metadata, such as data sources, collection methods, and quality assurance procedures, should be well-documented to ensure transparency and reproducibility (USGS, 2021).

ARTIFICIAL INTELLIGENCE AND MACHINE LEARNING IN WATER RESOURCE MANAGEMENT

Artificial intelligence (AI) and machine learning have emerged as indispensable instruments for managing water resources, providing effective methods to tackle diverse problems. Here, we examine the various applications of AI and machine learning techniques in particular water resource management fields.

Supervised Learning for Predicting Water Quality

Based on input attributes and historical data, supervised learning algorithms are used to forecast water quality parameters. These algorithms include, for instance:

1. **Support Vector Machines (SVM)**: SVM can classify water quality as suitable or unsuitable based on various parameters, helping in decision-making for water use.
2. **Artificial Neural Networks (ANN)**: ANN models are capable of capturing complex relationships in water quality data, making them suitable for forecasting (Son, 2008 p. 569).

Time Series Forecasting for Streamflow and Rainfall

Time series forecasting is essential for predicting streamflow, rainfall, and other hydrological variables. AI algorithms used for this purpose include:

1. **ARIMA (Autoregressive Integrated Moving Average)**: ARIMA models are frequently used for rainfall and streamflow time series analysis.
2. **Long Short-Term Memory (LSTM)**: LSTM networks excel in capturing temporal dependencies and are used for accurate streamflow forecasting.
3. **Prophet**: Facebook's Prophet is useful for forecasting streamflow, especially when dealing with datasets that have missing values and outliers (Taylor, 2018 p.37).

Unsupervised Learning for Anomaly Detection

Unsupervised learning techniques identify anomalies or unusual patterns in water resource data, which can be indicative of water quality issues or infrastructure problems:

1. **Cluster Analysis**: Clustering algorithms, such as K-means, group similar water quality samples, making it easier to identify anomalies.
2. **Isolation Forest**: Isolation Forest is effective for detecting anomalies in time series data, such as sudden water quality deviations

Reinforcement Learning for Optimal Control

Reinforcement learning (RL) is applied to optimize water resource management decisions and control strategies:

1. **Q-Learning**: Q-learning algorithms can optimize water allocation in reservoir systems by learning from past actions and rewards.
2. **Deep Reinforcement Learning**: Deep RL techniques are used for real-time control of water distribution systems, adjusting valve settings to meet demand and minimize energy consumption.

Using GIS and Remote Sensing for Water Resource Management

Geographic Information Systems (GIS) and remote sensing are two potent technologies that are essential to the management of water resources. They make it possible to gather, evaluate, and visualize geographical data, which offers insightful information for water resource monitoring, assessment, and decision-making.

Using Satellite Images to Track Water

When it comes to maintaining and keeping an eye on water resources, satellite imaging is an invaluable tool. Because it offers a synoptic perspective of wide areas, it's very helpful for:

Water Quality Evaluation According to (Adam, 2019), satellite sensors can identify changes in water quality, including pollution and algal blooms, by monitoring changes in temperature, turbidity, and colour.

Flood Monitoring: By tracking flood extents and assessing flood risk, satellite-based flood mapping and early warning systems aid in disaster management (Di Baldassarre, 2019 p. 6327).

Drought Monitoring: By evaluating soil moisture content and vegetation health, remote sensing can shed light on drought conditions and how they affect the availability of water (Dinku, p. 6577).

Monitoring of Lakes and Reservoirs: The health of aquatic ecosystems, variations in the surface area of the water, and water levels are all determined using satellite photography.

Utilizing GIS in Watershed Analysis

Watershed analysis, which examines the whole drainage area that contributes to a particular water body, heavily relies on GIS. Applications of GIS in this field include:

Hydrological Modelling: GIS is used to simulate surface runoff, define watersheds, and produce digital elevation models (DEMs) that enable the forecast of river flow and flood risk.

Land Use and Land Cover Analysis: Geographic Information System (GIS) makes it easier to track changes in land use and cover within watersheds, which aids in locating pollution sources and evaluating their effects on water quality (Pijanowski, 2002 p.553).Water Quality Mapping: By combining

Figure 3. Data lifecycle in water management

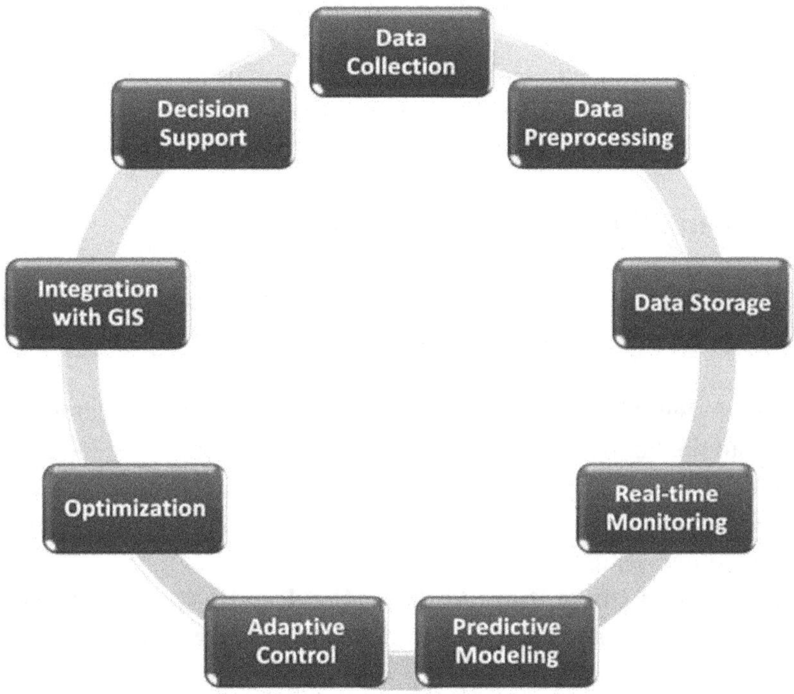

geographic data with water quality data, GIS is used to produce spatially explicit maps of water quality, allowing for targeted water quality management techniques (Zhang,2017). The holistic data lifecycle in water management is illustrated in Figure 3, which covers data collection and integration, processing, decision-making, and implementation. A continuous feedback loop is included to enhance the process of data management over time.

Forecasting and Forecasting for Drought and Flood Control

To effectively manage droughts and floods, predictive modelling is essential since it offers insightful information for both mitigation and response actions. Here, we examine how artificial intelligence (AI) and predictive models can be used in these crucial water resource management areas.

AI-Based Drought Forecasting Models

Anticipating droughts is crucial for proactive management of water resources. Drought conditions are predicted with artificial intelligence algorithms and predictive analytics:

1. **Machine Learning Models**: Algorithms like Random Forest, Support Vector Machines, and Gradient Boosting are trained on historical climate and hydrological data to predict drought severity and duration (Yaseen, 2018 p.130).

2. **Deep Learning**: Convolutional Neural Networks (CNNs) and Recurrent Neural Networks (RNNs) analyse remote sensing data and meteorological variables for more accurate and timely drought predictions (Sharma, 2021 p.125).
3. **Hydrological Models**: Physically-based hydrological models integrated with AI techniques offer insights into groundwater levels, streamflow, and soil moisture, helping anticipate drought impacts (Abbaszadeh, 2020).

Flood Forecasting and Early Warning Systems

Timely flood forecasting and early warning systems are vital for minimizing flood-related damage and protecting lives. AI-driven models are used for flood prediction:

1. **Hydrodynamic Models**: Coupled with machine learning, these models simulate river and rainfall-runoff dynamics to predict river discharge and flood inundation.
2. **Rainfall-Runoff Models**: Machine learning algorithms, including Artificial Neural Networks (ANNs) and Long Short-Term Memory (LSTM) networks, enhance the accuracy of rainfall-runoff modelling for flood forecasting.
3. **Remote Sensing and GIS Integration**: AI-driven flood models incorporate real-time satellite data and GIS information to monitor water levels and trigger early warnings.

Reservoir Management and Decision Support

Effective reservoir management is critical for both flood control and water supply. AI supports decision-making in reservoir operations:

1. **Reinforcement Learning**: Reinforcement learning algorithms optimize reservoir release strategies by considering water inflow, demand, and downstream conditions (Jin, 2021).
2. **Probabilistic Models**: Bayesian networks and probabilistic graphical models help assess reservoir performance under uncertain hydrological conditions (Hao, 2018 p. 1767).
3. **Real-Time Control Systems**: AI-driven control systems adjust reservoir release rates and gate operations in response to changing weather patterns and inflow forecasts.
4. **Multi-Objective Optimization**: AI-based multi-objective optimization techniques balance conflicting objectives, such as flood control, water supply, and hydropower generation (Dutta, 2019 p.554).

Systems for Decision Support and Optimization

Decision support tools and optimization strategies are essential to efficient water resource management. They enable the allocation of resources, the resolution of complex trade-offs, and the implementation of sustainable strategies. Here, we explore how AI-enhanced optimization techniques, multi-objective decision-making, and real-time decision support systems are applied in this context.

AI-Enhanced Optimization Techniques

Artificial intelligence (AI) is revolutionizing optimization techniques in water resource management, allowing for more efficient and adaptive decision-making:

1. **Genetic Algorithms (GA)**: GAs are used to optimize water allocation and distribution systems. AI-enhanced GAs adaptively evolve solutions, considering multiple objectives, constraints, and uncertainties (Zambrano-Bigiarini, 2019 p. 221).
2. **Particle Swarm Optimization (PSO)**: PSO algorithms optimize reservoir operation schedules by mimicking the social behaviour of birds or insects. AI-enhanced PSO enhances the convergence and accuracy of solutions (Nasseri, 2015 p.1069).
3. **Machine Learning in Optimization**: Machine learning models, such as neural networks, are integrated with optimization techniques to enhance their performance. They learn from historical data and adapt to changing conditions (Lan et al., 2020).

Multi-Objective Decision-Making in Water Allocation

Water allocation decisions often involve conflicting objectives, such as water supply, environmental conservation, and hydropower generation. Multi-objective decision-making techniques help strike a balance:

1. **Pareto Optimization**: Pareto-based approaches identify a range of solutions representing trade-offs between conflicting objectives. Decision-makers can then select the most suitable option.
2. **Multi-Objective Evolutionary Algorithms**: Evolutionary algorithms, like NSGA-II (Non-dominated Sorting Genetic Algorithm II), optimize multiple objectives simultaneously. They are applied in reservoir operation and irrigation management (Dutta, 2019 p.554).
3. **Fuzzy Logic Decision Support**: Fuzzy logic systems accommodate uncertainty and vagueness in decision-making. They are used to assess trade-offs in water allocation considering imprecise data.

Real-Time Decision Support Systems

Real-time decision support systems (DSS) provide timely information and recommendations for immediate action:

1. **Hydro Informatics DSS**: These systems integrate real-time sensor data, hydrological models, and AI algorithms to monitor and predict river conditions, facilitating flood forecasting and control.
2. **Reservoir Management DSS**: Real-time DSS for reservoirs use AI to optimize release strategies based on inflow forecasts, weather conditions, and downstream demands.
3. **Sensor Networks**: IoT-based sensor networks provide continuous data on water quality, quantity, and infrastructure health. AI-driven DSS use this data for adaptive management and infrastructure maintenance (Bing, 2017 p. 620).

CASE STUDIES AND APPLICATIONS

AI in Urban Water Supply

AI is revolutionizing urban water supply management, ensuring efficient and sustainable water distribution:

1. **Singapore's Smart Water Grid**: Singapore uses artificial intelligence (AI) systems to forecast patterns in water consumption, optimize water distribution, and identify leaks instantly, saving a lot of water. (Pandey, 2019 p. 136).
2. **Smart Metering in California**: California's water utilities use AI-driven smart meters to monitor water usage, identify anomalies, and promote water conservation among residents (Gude, 2019).

Agricultural Water Management With AI

AI aids in optimizing water use in agriculture, enhancing crop yields while conserving water resources:

1. **Precision Irrigation in India**: Indian farmers employ AI-based precision irrigation systems to supply the ideal amount of water to crops, cutting down on water waste and raising productivity. These systems examine weather, soil, and crop data. (Chandrasekaran, 2020 p. 855).
2. **Remote Sensing in Precision Agriculture**: AI-driven remote sensing technologies enable farmers worldwide to monitor soil moisture levels, assess crop health, and make data-driven irrigation decisions (Zhang, 2017).

Ecosystem Conservation and Restoration

AI contributes to the protection and restoration of ecosystems and aquatic habitats:

1. **Coral Reef Monitoring**: AI-powered underwater drones equipped with cameras and machine learning algorithms monitor coral reefs, assessing their health and aiding in conservation efforts (Smith, 2020 p.398).
2. **River Restoration in Europe**: AI models analyse river flow and habitat data to inform ecological restoration projects, helping restore natural river ecosystems (Paillex, 2017 p. 42).

International Water Resource Management

AI fosters international cooperation and sustainable management of shared water resources:

1. **Trans boundary River Management**: AI-based hydrological models and data-sharing platforms facilitate cooperation among nations sharing trans boundary rivers, promoting equitable and efficient water use (Sawunyama, 2023).
2. **Predictive Analytics for Water Scarcity**: AI-driven predictive models assist international organizations like the UN in identifying regions at risk of water scarcity, enabling proactive intervention (UN Water, 2018).

CHALLENGES AND ETHICAL CONSIDERATIONS

Data Privacy and Security

1. **Lake Mead Water Level Prediction**: An AI model for predicting Lake Mead water levels in the south-western U.S. inadvertently exposed the need for stricter data privacy safeguards when handling sensitive water resource data (UNLV, 2021).

Bias and Fairness in AI Models

1. **Bias in Water Allocation**: AI algorithms used in water allocation decisions in some regions have been criticized for perpetuating historical biases and exacerbating water inequality (Herman, 2021).

FUTURE DIRECTIONS AND RESEARCH OPPORTUNITIES

Emerging Technologies in Water Management

1. **Quantum Computing**: Exploring the potential of quantum computing for solving complex water resource optimization problems, enabling faster and more precise decision-making (Biamonte, 2017 p. 195).
2. **Block chain**: Investigating the use of blockchain technology for transparent and secure water transactions, particularly in water trading and allocation systems (Srivastava, 2018 p. 668).
3. **Nanotechnology**: Researching the application of nanomaterials for water purification and desalination, addressing water quality and scarcity challenges (Nzila, 2020).
4. **IoT and Edge Computing**: Leveraging the Internet of Things (IoT) and edge computing for real-time monitoring and control of water infrastructure, enhancing resilience and efficiency (Al-Fuqaha, 2015 p. 2347).

Interdisciplinary Collaboration for Sustainable Solutions

1. **Hydro Informatics**: Promoting collaboration between water resource experts, data scientists, and AI researchers to develop advanced hydro informatics tools that integrate AI, big data, and domain knowledge for holistic water management (Di Lecce, 2018 p. 1).
2. **Ecological Engineering**: Advancing interdisciplinary research that combines engineering, ecology, and AI to design water systems that mimic natural ecosystems for sustainable water treatment and management (Mitsch, 2012 p. 237).
3. **Social Sciences**: Integrating social sciences into water resource management research to better understand human behaviour, preferences, and perceptions related to water use, ensuring that AI solutions align with societal needs.

Policy and Regulation in AI-Enabled Water Resource Management

1. **AI Governance Frameworks**: Establishing norms and legal frameworks for AI applications in water resource management to handle concerns about data security, accountability, and openness (UNESCO, 2021).
2. **Stakeholder Engagement**: Encouraging stakeholder involvement in shaping water management policies and regulations related to AI technologies to ensure inclusivity and fairness (Dietz, 2003 p. 1907).

The development of ethical and legal frameworks, interdisciplinary collaboration, and emerging technology will propel the area of AI-enabled water resource management forward in the upcoming years. The solution to the world's water problems and the achievement of sustainable water resource management lies in research in these fields.

CONCLUSION

The incorporation of artificial intelligence (AI) into the management of water resources presents considerable potential for tackling the complex issues of water scarcity, degradation of quality, and sustainable distribution. We have examined the various uses of AI in the sector throughout this thorough analysis, emphasizing how it is revolutionizing the management of water resources. One important aspect is how AI helps water resource managers make data-driven decisions by giving them insights into more accurate forecasts, effective resource allocation, and enhanced risk assessment. Additionally, through real-time monitoring and prediction models, AI enhances the resilience and adaptability of water infrastructure by enabling quick responses to natural disasters and shifting weather patterns. Additionally, by improving water utilization in industry, urban supplies, and agriculture, AI significantly contributes to the promotion of sustainable water management by lowering environmental effects. Additionally, it promotes biodiversity, helps with conservation efforts, and makes it easier to monitor and preserve aquatic environments.

The Potential Impact of AI in Shaping Water Resource Management

The potential impact of AI in shaping water resource management cannot be overstated. AI can revolutionize how we approach water-related challenges, offering innovative solutions that were once considered unattainable. It equips us with the tools to:

- Predict and mitigate the effects of climate change on water resources.
- Optimize water allocation and distribution, ensuring equitable access.
- Enhance early warning systems for floods and droughts, saving lives and property.
- Improve water quality monitoring and treatment.
- Foster international cooperation in trans boundary water management.

Call to Action for Researchers and Practitioners

To harness the full potential of AI in water resource management, we issue a call to action:

- Research Innovation: Continue exploring the frontiers of AI, including emerging technologies like quantum computing, blockchain, and nanotechnology, to address water challenges with cutting-edge solutions.
- Interdisciplinary Collaboration: Foster collaboration among water experts, data scientists, environmentalists, and policymakers to develop holistic, sustainable, and ethical AI-driven water management strategies.
- Policy Development: Advocate for the development and implementation of policies and regulations that ensure the responsible and equitable use of AI in water resource management.
- Ethical Considerations: Prioritize ethical considerations, data privacy, and fairness when designing and deploying AI systems for water management.
- Knowledge Sharing: Share knowledge, best practices, and case studies to accelerate the adoption of AI in water resource management on a global scale.

By taking these actions, researchers and practitioners can collectively drive innovation, promote sustainability, and secure the future of our most precious resource: water.

The future of water resource management is entwined with the evolution of AI, and together, we have the power to shape a more water-secure and sustainable world.

REFERENCES

Abbaszadeh, P., Alizadeh, F., & Arabi, M. (2020). Integration of artificial intelligence and physically-based hydrological model for drought prediction under different climatic regions. *The Science of the Total Environment*, *744*, 140664.

Adams, S., & Acheampong, A. O. (2019). Reducing carbon emissions: The role of renewable energy and democracy. *Journal of Cleaner Production*, *240*, 118245. doi:10.1016/j.jclepro.2019.118245

Afrifa, S., Zhang, T., Appiahene, P., & Varadarajan, V. (2022). Mathematical and machine learning models for groundwater level changes: A systematic review and bibliographic analysis. *Future Internet*, *14*(9), 259. doi:10.3390/fi14090259

Al-Fuqaha, A., Guizani, M., Mohammadi, M., Aledhari, M., & Ayyash, M. (2015). Internet of Things: A survey on enabling technologies, protocols, and applications. *IEEE Communications Surveys and Tutorials*, *17*(4), 2347–2376. doi:10.1109/COMST.2015.2444095

AWWA. (2021). *Buried No Longer: Confronting America's Water Infrastructure Challenge*. American Water Works Association.

Ball, J. E., Anderson, D. T., & Chan, C. S. (2017). A comprehensive survey of deep learning in remote sensing: Theories, tools, and challenges for the community. *Journal of Applied Remote Sensing*, *11*(4), 042609–042609. doi:10.1117/1.JRS.11.042609

Biamonte, J., Wittek, P., Pancotti, N., Rebentrost, P., Wiebe, N., & Lloyd, S. (2017). Quantum machine learning. *Nature*, *549*(7671), 195–202. doi:10.1038/nature23474 PMID:28905917

Bing, X., Yu, J., & Chen, J. (2017). A real-time sensor data-driven decision support system for water quality management in an industrial park. *Environmental Monitoring and Assessment, 189*(12), 620. PMID:29124450

Chandrasekaran, S., Khaparde, V., & Seshagiri Rao, G. (2020). Adoption of AI in Indian agriculture: Opportunities and challenges. *AI & Society, 35*(4), 855–866.

Choubin, B., Abdolshahnejad, M., Moradi, E., Querol, X., Mosavi, A., Shamshirband, S., & Ghamisi, P. (2020). Spatial hazard assessment of the PM10 using machine learning models in Barcelona, Spain. *The Science of the Total Environment, 701*, 134474. doi:10.1016/j.scitotenv.2019.134474 PMID:31704408

Datta, B. (2019). Artificial Intelligence for Aquifer Remediation and Management: A Review. *The Science of the Total Environment, 670*, 550–566.

Deb, K., Pratap, A., Agarwal, S., & Meyarivan, T. A. M. T. (2002). A fast and elitist multiobjective genetic algorithm: NSGA-II. *IEEE Transactions on Evolutionary Computation, 6*(2), 182–197. doi:10.1109/4235.996017

Demirel, M. C. (2021). Streamflow Forecasting with Deep Learning: A Case Study in California's American River Basin. *Journal of Hydrology (Amsterdam), 598*, 126444.

Di Baldassarre, G., Sivapalan, M., Rusca, M., Cudennec, C., Garcia, M., Kreibich, H., Konar, M., Mondino, E., Mård, J., Pande, S., Sanderson, M. R., Tian, F., Viglione, A., Wei, J., Wei, Y., Yu, D. J., Srinivasan, V., & Blöschl, G. (2019). Sociohydrology: Scientific challenges in addressing the sustainable development goals. *Water Resources Research, 55*(8), 6327–6355. doi:10.1029/2018WR023901 PMID:32742038

Di Lecce, V., Menoni, S., Mancini, L., & Masseroli, M. (2018). Hydro informatics: Data integration and knowledge discovery for smarter water management. *Environmental Modelling & Software, 101*, 1–4.

Dietz, T., Ostrom, E., & Stern, P. C. (2003). The struggle to govern the commons. *Science, 302*(5652), 1907-1912.

Dinku, T., Ceccato, P., Grover-Kopec, E., Lemma, M., Connor, S. J., & Ropelewski, C. F. (2018). Validation of satellite rainfall products over East Africa's complex topography. *International Journal of Remote Sensing, 29*(18), 6577–6600.

Dudgeon, D., Arthington, A. H., Gessner, M. O., Kawabata, Z. I., Knowler, D. J., Lévêque, C., Naiman, R. J., Prieur-Richard, A.-H., Soto, D., Stiassny, M. L. J., & Sullivan, C. A. (2006). Freshwater biodiversity: Importance, threats, status and conservation challenges. *Biological Reviews of the Cambridge Philosophical Society, 81*(2), 163–182. doi:10.1017/S1464793105006950 PMID:16336747

Dutta, D., Deka, L., & Mandal, D. (2019). Multi-objective optimization in real-time reservoir operation using multi-agent reinforcement learning. *Journal of Hydrology (Amsterdam), 574*, 554–565.

EPA. (2020). *National Water Quality Inventory Report to Congress.* United States Environmental Protection Agency.

EPA. (2021). *Water Quality Monitoring and Assessment*. United States Environmental Protection Agency.

Gleick, P. H. (2018). Transitions to freshwater sustainability. *Proceedings of the National Academy of Sciences of the United States of America*, *115*(36), 8863–8871. doi:10.1073/pnas.1808893115 PMID:30127019

Gude, V. G., Rumbos, P., & Mattson, J. E. (2019). Smart meters for enhanced urban water supply management: A review. *Journal of Water Resources Planning and Management*, *145*(9), 04019037.

Gupta, H. V., Mohtar, R. H., & Pande, S. (2019). The Water Energy Food Nexus: An integrated assessment framework for policy analysis. *Environmental Science & Policy*, *93*, 101–110.

Guyon, I., & Elisseeff, A. (2003). An introduction to variable and feature selection. *Journal of Machine Learning Research*, *3*(Mar), 1157–1182.

Hafezparast, M. (2021). Monitoring groundwater level changes of Mianrahan aquifer with GRACE satellite data. *Iranian Journal of Irrigation and Drainage*, *15*(2), 428–443.

Han, S. (2019). A Machine Learning Approach to Drought Prediction in the Context of the U.S. Drought Monitor. *Environmental Monitoring and Assessment*, *191*(5), 317. PMID:31041530

Hao, Z., Yang, D., Zhao, J., Wang, X., Xu, J., & Li, Z. (2018). Bayesian network-based multi-objective reservoir operation with multi-scenario simulation considering hydrological uncertainty. *Water Resources Management*, *32*(5), 1767–1783.

Harmarneh, S. H., Hani, R. B., & Yaseen, Z. M. (2020). Predicting the water quality index using ensemble-based machine learning models. *Journal of Water Process Engineering*, *37*, 101442.

Hejazi, M., Edmonds, J., Clarke, L., Kyle, P., Davies, E., Chaturvedi, V., Wise, M., Patel, P., Eom, J., Calvin, K., Moss, R., & Kim, S. (2014). Long-term global water projections using six socioeconomic scenarios in an integrated assessment modelling framework. *Technological Forecasting and Social Change*, *81*, 205–226. doi:10.1016/j.techfore.2013.05.006

Herman, J. D., Zeff, H. B., Lamontagne, J. R., Reed, P. M., & Characklis, G. W. (2021). Balancing water allocation under deep uncertainty and evolving infrastructure networks: Lessons from the California Water System. *Environmental Research Letters*, *16*(1), 014006.

IPCC. (2021). *The Physical Science Basis. Intergovernmental Panel on Climate Change*. IPCC.

ISO. (2017). *ISO 5725-1:1994. Accuracy (trueness and precision) of measurement methods and results - Part 1: General principles and definitions*. International Organization for Standardization.

Jin, J., Li, Z., & Xu, C. Y. (2021). Reservoir operation with improved hydrological forecasts using reinforcement learning: A case study in China. *Environmental Modelling & Software*, *137*, 104915.

Kallis, G., Kiparsky, M., & Norgaard, R. B. (2015). Collaborative governance and adaptive management: Lessons from California's CALFED Water Program. *Environmental Science & Policy*, *55*, 1–12.

Kroll, C., Warchold, A., & Pradhan, P. (2019). Sustainable Development Goals (SDGs): Are we successful in turning trade-offs into synergies? *Palgrave Communications*, *5*(1), 140. doi:10.105741599-019-0335-5

Lan, Y., Lee, B. J., Wei, Y., & Zhang, C. (2020). Artificial neural networks for optimizing regional-scale water resource allocation: A framework and case study. *Environmental Modelling & Software*, *131*, 104779.

Liu, F. T., Ting, K. M., & Zhou, Z. H. (2008, December). Isolation forest. In *2008 eighth IEEE International Conference on Data Mining* (pp. 413-422). IEEE. 10.1109/ICDM.2008.17

Madani, K. (2019). Water Resources Allocation: A Comprehensive Review. *Water Resources Management*, *33*(9), 3183–3213.

Mitsch, W. J., Zhang, L., Stefanik, K. C., Nahlik, A. M., Anderson, C. J., Bernal, B., Hernandez, M., & Song, K. (2012). Creating wetlands: Primary succession, water quality changes, and self-design over 15 years. *Bioscience*, *62*(3), 237–250. doi:10.1525/bio.2012.62.3.5

Moubayed, A., Shami, A., & Ibrahim, A. (2022). *Intelligent Transportation Systems Orchestration: Lessons Learned & Potential Opportunities.* arXiv preprint arXiv:2205.14040.

Nasseri, M., Mahdavi, M., & Shahcheraghi, H. (2015). Multi-objective particle swarm optimization for real-time operation of reservoir systems considering fuzzy operation rules. *Water Resources Management*, *29*(4), 1069–1087.

Nzila, A., Al-Ayoubi, S., Al-Gharabli, S., & Sayadi, S. (2020). Nanotechnology applications for the removal of biological and chemical contaminants from water. *Environmental Technology & Innovation*, *17*, 100589.

Paillex, A., Enters, T., Angelini, C., & Bruder, A. (2017). Machine learning and ecological modelling for river habitat management. *Ecological Modelling*, *346*, 42–54.

Pandey, G., Pathak, N., & Chatterjee, D. (2019). Implementation of artificial intelligence for urban water supply system: A case study of Singapore. *Procedia Computer Science*, *152*, 136–143.

Pijanowski, B. C., Brown, D. G., Shellito, B. A., & Manik, G. A. (2002). Using neural networks and GIS to forecast land use changes: A land transformation model. *Computers, Environment and Urban Systems*, *26*(6), 553–575. doi:10.1016/S0198-9715(01)00015-1

Poff, N. L., Brown, C. M., Grantham, T. E., Matthews, J. H., Palmer, M. A., Spence, C. M., Wilby, R. L., Haasnoot, M., Mendoza, G. F., Dominique, K. C., & Baeza, A. (2016). Sustainable water management under future uncertainty with eco-engineering decision scaling. *Nature Climate Change*, *6*(1), 25–34. doi:10.1038/nclimate2765

Puterman, E., Weiss, J., Lin, J., Schilf, S., Slusher, A. L., Johansen, K. L., & Epel, E. S. (2018). Aerobic exercise lengthens telomeres and reduces stress in family caregivers: A randomized controlled trial Richter Award Paper 2018. *Psychoneuroendocrinology*, *98*, 245–252. doi:10.1016/j.psyneuen.2018.08.002 PMID:30266522

Salehahmadi, Z. (2018). Artificial Neural Network and Support Vector Machine for Prediction of Water Quality Parameters. *Environmental Monitoring and Assessment*, *190*(9), 534. PMID:30128706

Sawunyama, L., Oyewo, O. A., Seheri, N., Onjefu, S. A., & Onwudiwe, D. C. (2023). Metal oxide functionalized ceramic membranes for the removal of pharmaceuticals in wastewater. *Surfaces and Interfaces*, *38*, 102787. doi:10.1016/j.surfin.2023.102787

Saxena, R., Hardainiyan, S., Singh, N., & Rai, P. K. (2022). Prospects of microbes in mitigations of environmental degradation in the river ecosystem. In *Ecological Significance of River Ecosystems* (pp. 429–454). Elsevier. doi:10.1016/B978-0-323-85045-2.00003-0

Sharma, S. K., Ghosh, S., & Bhattacharya, B. (2021). A review of recent developments in artificial intelligence for groundwater management. *Journal of Hydrology (Amsterdam)*, *590*, 125–487.

Shekhar, S., Evans, M. R., Kang, J. M., & Mohan, P. (2011). Identifying patterns in spatial information: A survey of methods. *Wiley Interdisciplinary Reviews. Data Mining and Knowledge Discovery*, *1*(3), 193–214. doi:10.1002/widm.25

Shekhar, S., Li, W., & Zhang, P. (2011). *Encyclopedia of GIS*. Springer Science & Business Media.

Smith, J. E., Brainard, R. E., Carter, G., Grinham, A., de Carvalho, R. L., Petus, C., ... Shaw, E. (2020). Monitoring coral reefs using artificial intelligence: A feasibility assessment. *Remote Sensing in Ecology and Conservation*, *6*(4), 398–411.

Son, T. C., Kim, H. S., & Lee, W. (2008). Prediction of river water quality by artificial neural network model. *Water Science and Technology*, *58*(3), 569–576.

Srinivasan, S., Hoffman, N. G., Morgan, M. T., Matsen, F. A., Fiedler, T. L., Hall, R. W., Ross, F. J., McCoy, C. O., Bumgarner, R., Marrazzo, J. M., & Fredricks, D. N. (2012). Bacterial communities in women with bacterial vaginosis: High-resolution phylogenetic analyses reveal relationships of microbiota to clinical criteria. *PLoS One*, *7*(6), e37818. doi:10.1371/journal.pone.0037818 PMID:22719852

Srivastava, M., Singh, R. K., & Sharma, A. (2018). Privacy-preservation blockchain with edge computing for secure IoT in agriculture. *Procedia Computer Science*, *132*, 668–675.

Sulea, T., Rohani, N., Baardsnes, J., Corbeil, C. R., Deprez, C., Cepero-Donates, Y., Robert, A., Schrag, J. D., Parat, M., Duchesne, M., Jaramillo, M. L., Purisima, E. O., & Zwaagstra, J. C. (2020, January). Structure-based engineering of pH-dependent antibody binding for selective targeting of solid-tumor microenvironment. *mAbs*, *12*(1), 1682866. doi:10.1080/19420862.2019.1682866 PMID:31777319

Taylor, S. J., & Letham, B. (2018). Forecasting at scale. *The American Statistician*, *72*(1), 37–45. doi: 10.1080/00031305.2017.1380080

UNESCO. (2021). *Ethics of Artificial Intelligence*. UNESCO. Retrieved from https://en.unesco.org/themes/ethics-artificial-intelligence

UNLV. (2021). *AI algorithm predicts Lake Mead levels*. UNLV News Center. Retrieved from https://www.unlv.edu/news/release/ai-algorithm-predicts-lake-mead-levels

USGS. (2021). *Data Quality Information: U.S. Geological Survey Guidelines and Practices*. United States Geological Survey.

Vörösmarty, C. J., Osuna, V. R., Cak, A. D., Bhaduri, A., Bunn, S. E., Corsi, F., ... Uhlenbrook, S. (2018). Ecosystem-based water security and the Sustainable Development Goals (SDGs). *Ecohydrology & Hydrobiology*, *18*(4), 317–333. doi:10.1016/j.ecohyd.2018.07.004

Water, U. N. (2018). *The United Nations World Water Development Report 2018: Nature-Based Solutions for Water.* UNESCO. Retrieved from https://www.unwater.org/publications/the-united-nations-world-water-development-report-2018-nature-based-solutions-for-water/

Water, U. N. (2021). *Water Scarcity*. United Nations World Water Development Report.

Yaseen, Z. M., Abba, A. H., & Deo, R. C. (2018). Drought prediction: A challenge to meet in Northern Sudan. *The Science of the Total Environment*, *621*, 130–144.

Zambrano-Bigiarini, M., & Baez-Villaneuva, O. M. (2019). Characterizing meteorological droughts in data scare regions using remote sensing estimates of precipitation. In *Extreme Hydroclimatic Events and Multivariate Hazards in a Changing Environment* (pp. 221–246). Elsevier. doi:10.1016/B978-0-12-814899-0.00009-2

Zhang, C., & Song, Y. (2017). Spatial analysis of water quality in the Haihe River Basin in China. *Scientific Reports*, *7*, 40123.

Zheng, W. (2021). Deep Learning for Groundwater Level Prediction in an Arid Region: Case Study in the Middle Reaches of Heihe River Basin, Northwest China. *Journal of Hydrology (Amsterdam)*, *596*, 125758.

Chapter 4
Forecasting Weather and Water Management Through Machine Learning

Inzimam Ul Hassan
https://orcid.org/0000-0001-7475-7039
Vivekananda Global University, India

Zeeshan Ahmad Lone
Vivekananda Global University, India

Swati Swati
Lovely Professional University, India

Aya Gamal
Damietta University, Egypt

ABSTRACT

A useful tool for making data-driven decisions is machine learning. Numerous issues, such as those pertaining to weather forecasting and water management, can be resolved using it. Machine learning may be applied to water management to optimise water distribution, anticipate floods and droughts, and predict water demand. Machine learning may be used to track storms, anticipate weather trends, and provide early warnings. The most recent developments in machine learning for weather forecasting and water management will be reviewed in this chapter. It will cover the potential and difficulties of applying machine learning to various domains and offer illustrations of effective uses.

INTRODUCTION

Today's world faces many significant issues, chief among them being weather and water management. In addition to rising demand on water supplies due to urbanisation and population increase, climate change is making weather events more extreme and unpredictable. With the use of machine learning

DOI: 10.4018/979-8-3693-1194-3.ch004

(ML), we can more precisely predict weather patterns and water patterns, which will improve resource management and lessen the dangers associated with climate change. By analysing past meteorological data and applying sophisticated algorithms, machine learning models are able to produce forecasts and insights that are more precise (Dehghanisanij et al., 2022).

According to a Pennsylvania study, machine learning considerably enhanced analogue weather forecasting when it was used to evaluate surface wind speed and sun irradiation predictions (Pennsylvania State University). Using total column water vapour as a predictor, another study project showed how machine learning may be used to accurately provide storm reforecasts (Phys.org).

Machine learning has been used in the field of water management for a number of purposes, such as optimising water usage and real-time water treatment systems. In order to improve information extraction of significant environmental events in water resources, researchers have employed machine learning to address issues in water treatment and management systems (MDPI).

Additionally, by examining data on water usage, pressure, and other factors, machine learning models have been used to optimise system performance. All things considered, machine learning has demonstrated tremendous promise for transforming water management and weather forecasting. Machine learning models can offer more precise forecasts and insights by utilising sophisticated algorithms and historical data, which can enhance decision-making in applications related to water management and weather (Gino Sophia et al., 2020).

Briefly Introduce the Significance of Water Management and Weather Forecasting

For water resources to be used sustainably and conserved, water management is essential. In order to efficiently control the distribution, use, and supply of water, policies and regulations must be put into place. Water management is important for a number of reasons. First of all, it assists in supplying water to meet the growing demand brought on by industrialization and population expansion. We can guarantee a consistent and adequate water supply for a number of industries, including domestic use, industry, and agriculture, by effectively managing our water resources.

The second reason is that water management is essential for reducing the effects of water-related calamities such as droughts and floods. There is different condition which arise from the extreme weather conditions, to overcome the damage from these phenomenon, we implement some measure preventions which includes control flood and storage water system

Moreover, by protecting rivers, lakes, and wetlands from pollution, water management contributes to the protection of ecosystems (Lowe et al., 2022).

However, weather forecasting is necessary in order to anticipate and understand weather conditions and patterns. Reliable weather forecasts offer important information that businesses, governments, and people can use to make decisions and take the necessary action. There are several ways in which weather forecasting is important. First of all, it aids in weather preparation, including outfit selection and outdoor activity scheduling. Weather forecasts can also help businesses by allowing them to modify their supply chains and operations.

Furthermore, weather forecasting is essential for both preparedness and response to disasters. Authorities can prevent fatalities and minimise property damage by forecasting severe weather events like storms, hurricanes, and heatwaves and issuing timely warnings and orders for evacuation. By offering information on the best times to plant, safe routes to travel, and ways to manage energy consumption, weather forecasting also benefits a number of industries, such as transportation, energy, and agriculture (Sun & Scanlon, 2019).

In our society, weather forecasting and water management are very important. While weather forecasting assists us in anticipating and responding to weather-related events, water management guarantees the sustainable use and conservation of water resources. The environment, the economy, and individual well-being are all benefited by both factors. By putting into practise efficient water management plans and making use of reliable weather forecasts, we can improve our ability to withstand and adjust to shifting environmental conditions.

Pilers of Water Management

As indicated in figure 1 it shows how we can do the water management in efficient way. As water is very necessary for the life, we need to manage it very efficiently.

Experience and knowledge: This is very important to understand how water tell us about how we the uses of water are interconnected and how to recycle the water efficiently.

Public and Private Participation: Both public and Private participation is very important for the water management. The decision for water management should be transparent and inclusive.

Technology: Through technology we can increase the efficiency of water management, reduce the pollution, and monitor the water resources, we need to choose the right technology for this.

Social factors: Water managers should make policies that everybody should get the fresh and clean water keeping different factors in mind like poverty, gender inequality etc.

Figure 1. Water management

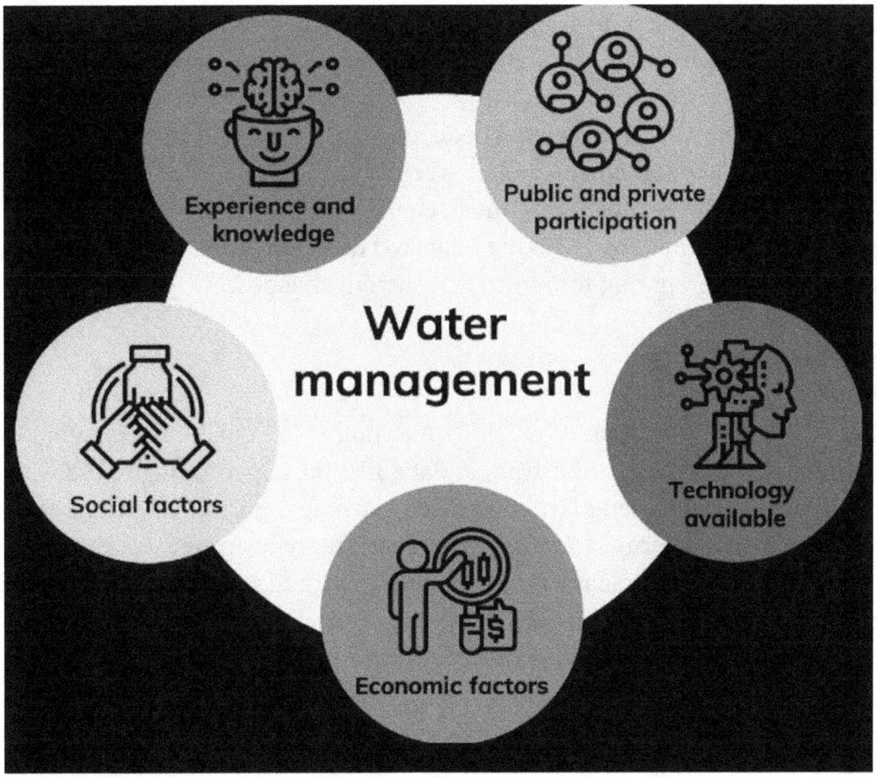

Highlight the Challenges and Importance of Making Informed Decisions in These Domains

Decision-making challenges related to weather and water management:

Making wise choices about water and weather management is a difficult task. There are several issues that require attention, such as:

- Uncertainty: Because weather and water systems are so dynamic and intricate, it is challenging to make firm predictions about their behaviour. Managing these resources can be challenging when decisions are made in this uncertain environment.
- Data accessibility: Accurate data is necessary to make well-informed decisions about water and weather management. But gathering and exchanging data can be costly and difficult.
- Tools for making decisions: To help integrate data from multiple sources and produce insights that can be used to make well-informed decisions, decision-making tools are necessary.
- Communication: It's critical to clearly and succinctly inform stakeholders of the risks and uncertainties related to weather and water forecasts and management decisions.

Making intelligent choices regarding weather and water management is crucial.

Ensuring economic prosperity, preserving the environment, and preserving life and property all depend on well-informed decisions regarding weather and water management. Making educated choices can assist in:

- Lessen the likelihood of natural disasters: People can prepare for and evacuate from natural disasters like hurricanes and floods with the aid of precise and timely weather forecasts.
- Make the most use of water possible: Making wise decisions about water management can help guarantee that there is enough water to meet everyone's needs while also safeguarding the environment.
- Enhance water quality: Making educated decisions can lessen pollution in water and enhance the quality of drinking water and other water sources.
- Reduce the effects of climate change: Climate change is increasing the extreme and unpredictability of weather and water patterns. Making educated decisions can assist us in reducing the effects of climate change and preparing for already-occurring changes.

Role of Machine Learning

Artificial intelligence (AI) in the form of machine learning (ML) enables computers to learn without explicit programming. Large datasets of historical data, like those pertaining to the weather, water use, or customer transactions, can be used to train machine learning algorithms. After being trained, the algorithms can be used to find patterns in the data or forecast future occurrences. Three essential elements are involved in machine learning fundamentals: models, data, and learning algorithms. Since data provides the information from which the models learn, it is the cornerstone of machine learning. This data can be unlabeled, in which case the machine discovers patterns and structures on its own, or labelled, in which case it has predetermined categories or results. Mathematical representations of the relationships and patterns found in data are called models in machine learning. In order to reduce errors and enhance predictions or decisions, learning algorithms are used to train these models. These algorithms modify the model's parameters in response to the input data (Alloghani et al., 2019).

How Does Machine Learning Work?

In order to generate predictions or uncover latent relationships between the variables in the data, machine learning algorithms first look for patterns in the data. To predict the possibility of rain on a particular day, for instance, a machine learning algorithm could be trained using a dataset of historical weather data. In order to generate predictions, the algorithm would look for patterns in the data, such as the correlation between temperature, humidity, and wind speed.

Types of Machine Learning

Supervised learning, unsupervised learning, and reinforcement learning are the three primary categories of machine learning as clearly indicated in Figure 2.

- Supervised learning: In supervised learning, an output is known for every data point in the dataset that the algorithm is trained on. After learning to map inputs to outputs, the algorithm can be applied to forecast the values of fresh data points.
- Unsupervised learning: The algorithm is trained via unsupervised learning using an unlabelled dataset. Without any prior knowledge of the outputs, the algorithm gains the ability to recognise patterns in the data. Data clustering, anomaly detection, and the discovery of unobserved correlations between variables are all possible with unsupervised learning.
- Reinforcement learning: Through interactions with its surroundings, the algorithm picks up new skills in reinforcement learning. The algorithm receives rewards for decisions it makes that result in the desired outcomes and receives penalties for decisions it makes that result in the undesirable outcomes. Robots and other agents are frequently trained to carry out tasks in the real world through the use of reinforcement learning.

Figure 2. Categorization of machine learning

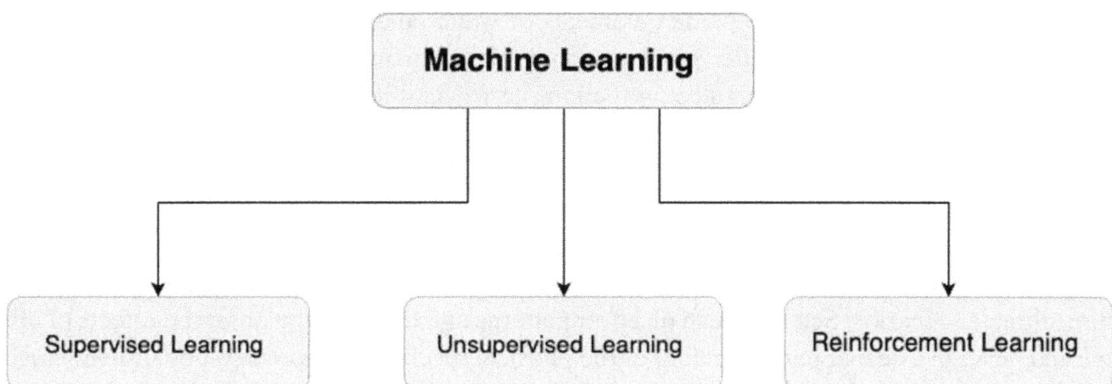

Applications of Machine Learning

ML is used in a wide variety of applications, including:

- Weather forecasting: More accurate weather patterns are predicted through the use of ML algorithms.
- Water management: To maximise water use and enhance water quality, ML algorithms are applied.
- Customer relationship management (CRM): To forecast consumer behaviour and spot sales opportunities, ML algorithms are applied.
- Fraud detection: To identify fraudulent transactions and other forms of fraud, machine learning algorithms are employed.
- Medical diagnosis: Doctors can diagnose diseases with the help of machine learning algorithms.
- Recommendation systems: These systems use machine learning algorithms to suggest movies, products, and other content to users (Alloghani et al., 2019).

Application of ML in Analyzing Large Datasets Related to Weather and Water

Significant use of machine learning (ML) has been observed in the analysis of big water and weather datasets. Machine learning techniques have been applied to weather analysis to improve forecasting precision and automate data processing. For example, ML algorithms have been used in conjunction with topological data analysis and ML techniques to identify atmospheric rivers in climate data. This makes it possible to comprehend weather patterns better and makes extreme precipitation events easier to predict. Furthermore, large-scale circulation patterns linked to meteorological events, like Midwest extreme precipitation, have been analysed using ML models. Convolutional neural networks (CNNs) are employed in these models to improve prediction model speed by extracting pertinent information from large-scale weather datasets.

Machine learning techniques have shown great promise in a variety of applications within the domain of water data analysis. ML helps groundwater analysts evaluate and forecast the quality and pollution of their groundwater. Groundwater level prediction and modelling using machine learning (ML) has produced accurate and effective data models for water resource management. Furthermore, using artificial intelligence techniques like support vector machines (SVM) and neural networks, machine learning (ML) algorithms have been used to predict elements of water quality, such as dissolved oxygen levels and chlorophyll concentrations. Better water resource management and monitoring are made possible by these ML-based techniques, guaranteeing the long-term viability and high calibre of water systems (Aguilera et al.,2019).

ML is essential to comprehending and mitigating the effects of climate change in the context of climate change analysis. The effects of climate change on various crop types have been analysed using machine learning techniques, which has aided in the development of mitigation and adaptation plans. Furthermore, machine learning models have been applied to refine climate models and raise the accuracy of future climate scenarios. Scientists can discover patterns and trends in the observed effects of climate change and develop a deeper understanding of the physical mechanisms causing it by utilising machine learning algorithms. In efforts to mitigate the effects of climate change and adapt to it, this makes predictions more accurate and decision-making more informed (Inoue et al., 2017).

Here are some specific examples of how ML is being used to analyze large datasets related to weather and water:

- Weather forecasting: ML algorithms are applied to increase the precision of weather predictions, particularly those made for the near future. From past weather data, machine learning algorithms can learn to find patterns that can be used to forecast the weather in the future. ML algorithms, for instance, can be used to forecast the possibility of rain, snow, or other meteorological phenomena.
- Water management: To maximise water use and enhance water quality, ML algorithms are applied. Machine learning algorithms, for instance, can be used to forecast water demand, locate leaks in water distribution networks, and create water management strategies that reduce water waste.
- Disaster risk assessment: The likelihood of natural disasters like floods and droughts is determined using machine learning algorithms. Machine learning algorithms possess the ability to recognise patterns linked to natural disasters by learning from past data. Plans for evacuation and early warning systems can then be created using this information.
- Research on climate change: ML algorithms are used to examine how weather and water patterns are affected by climate change. ML algorithms, for instance, can be used to forecast the effects of climate change on extreme weather events, sea levels, and rainfall patterns (Zhu eta al., 2022).

The following are some benefits of using machine learning to analyze large datasets related to weather and water:

- Enhanced accuracy: Big datasets may be used to train machine learning algorithms to identify patterns that are difficult or impossible for people to identify. This might lead to better catastrophe risk assessment, enhanced water management, and more precise weather and water forecasts.
- Timeliness: Accurate weather and water predictions, as well as early warning systems for natural disasters, are made possible by machine learning algorithms' ability to handle massive datasets quickly.
- Low cost: ML can reduce the cost which is related to weather, water forecasting also it can reduce cost in water management

What are the difficulties which we face in applying ML related weather and water datasets

- Quality of Data: because of the quality of training data we have the quality of ML algorithms are limited . As a result, having access to reliable weather and water data is crucial.
- Interpretability of the model: The way machine learning models generate predictions can be hard to understand. As a result, it could be challenging to determine whether the forecasts are accurate and to spot any possible model biases.
- Computational resources: Training and running machine learning algorithms can demand a large amount of computing power. For organisations with limited resources, this could be difficult.

All things considered, using machine learning (ML) to analyse massive datasets pertaining to weather, water, and climate change provides insightful information and useful tools for enhancing the precision of forecasts, the management of water resources, and climate change mitigation techniques. Researchers and practitioners can harness the power of these enormous datasets to solve important problems and make wise decisions for a more sustainable future by utilising machine learning techniques (Miro et al., 2021).

Weather Forecasting With ML

Weather forecasting accuracy has been greatly improved by machine learning (ML). Through the application of deep learning algorithms to analogue weather forecasting, scientists have been able to extrapolate more accurate future predictions from historical weather data. Promising outcomes have been observed in the prediction of numerical weather conditions and general weather patterns by machine learning models used for weather forecasting. These models are capable of picking up on subtle relationships and non-linear patterns in data that conventional forecasting models might overlook. Furthermore, machine learning algorithms are able to quickly and more accurately predict the weather by analysing large datasets and finding relevant patterns and relationships. The application of machine learning (ML) to weather forecasting has the potential to transform the industry and raise the efficacy and efficiency of existing systems (Jakaria et al., 2020).

Machine learning (ML) enhances weather forecasting accuracy by:

- Finding patterns in past weather data that are hard or impossible for people to find. Large datasets of historical weather data can be used to train machine learning algorithms to find patterns that correspond to various weather conditions. An ML algorithm might discover, for instance, that a particular combination of temperature, humidity, and wind speed is linked to a higher risk of thunderstorms. The machine learning algorithm can utilise these patterns to forecast future weather conditions once it has learned them.
- Adding more data sources to models used for weather forecasting. ML algorithms can be used to add new data sources, like radar, satellite imagery, and social media data, to weather forecasting models in addition to historical weather data. This has the potential to increase weather forecast accuracy, particularly for short-term forecasts.
- Issuing more precise and timely alerts for severe weather occurrences. Large weather data sets can be processed quickly by ML algorithms, leading to more precise and timely warnings for extreme weather events like hurricanes, tornadoes, and floods.

Different examples how ML is being used to enhance weather forecasting accuracy:

- Using machine learning (ML), the National Oceanic and Atmospheric Administration (NOAA) is able to forecast hurricanes with greater accuracy. Compared to conventional forecasting techniques, NOAA's ML models can incorporate a larger range of data sources, which enhances forecast accuracy, particularly for longer-term forecasts.
- The California Department of Water Resources forecasts supply and demand for water using machine learning. Plans for water management are created using this data in an effort to guarantee that there is sufficient water available to fulfil everyone's needs in California.
- Phoenix, Arizona is optimising its water distribution system with machine learning. As a result, the system is now more efficient and there is less water waste.

Figure 3. Weather forecasting using ML

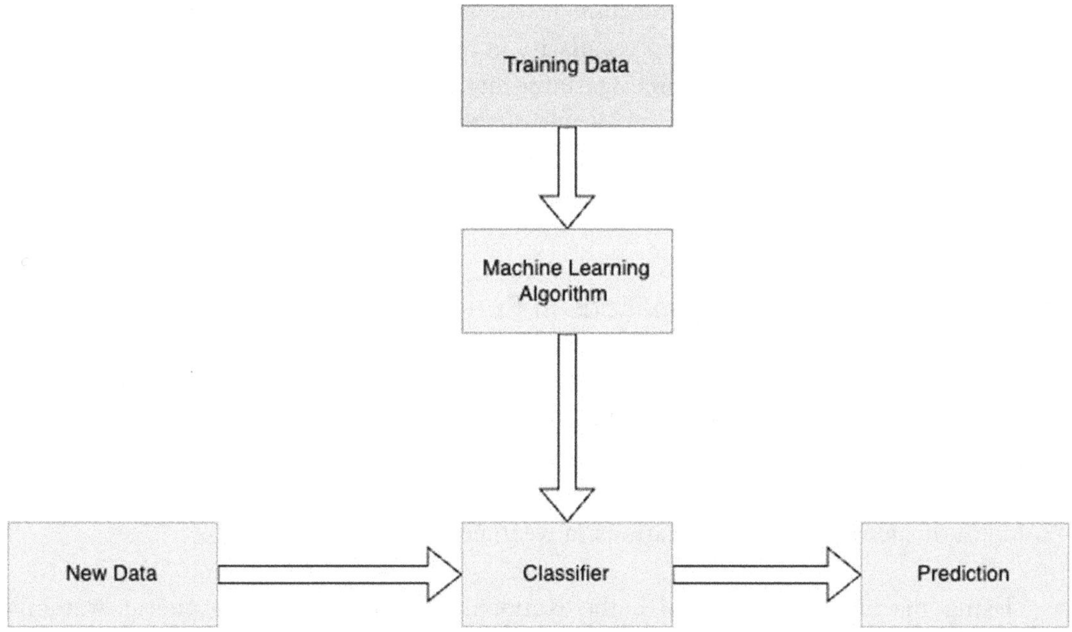

- ◦ The insurance sector is utilising machine learning to create risk models for weather-related events that are more accurate. This makes it possible for insurance companies to appropriately price their goods and expedite the processing of claims.

Discuss Specific ML Techniques Used in Weather Prediction

With the ability to increase forecast accuracy and offer more in-depth understanding of weather patterns, machine learning techniques have grown in significance in the field of weather prediction. Data-driven ML models are one particular ML technique used in weather prediction. These models find patterns and relationships between various variables, including temperature, pressure, humidity, and wind speed, by analysing vast amounts of historical weather data. ML models are able to forecast future weather conditions by examining these patterns.

Ensemble modelling is another machine learning method used in weather forecasting. Using this method, several machine learning models are combined to produce a prediction that is more reliable and accurate. For instance, combining an artificial neural network with a Random Forest model is a popular strategy. By utilising the advantages of each individual model, this ensemble model raises the performance of predictions overall (Hewage et al. 2020).

Additionally, other machine learning models which we use in whether forecasting:

- Support vector machines, or SVMs, are a class of supervised learning algorithms that are applicable to regression and classification problems. SVMs are frequently used in weather prediction to categorise various weather patterns, including tornadoes, hurricanes, and thunderstorms.
- Neural networks: Inspired by the composition and operations of the human brain, neural networks are a class of machine learning algorithms. Numerous tasks, such as pattern recognition, regres-

sion, and classification, can be accomplished with neural networks. Because they can identify intricate patterns in data that other machine learning algorithms find challenging or impossible to identify, neural networks are gaining popularity in the field of weather prediction.

- Deep learning: This kind of machine learning makes use of multi-layered artificial neural networks. By removing features at various levels of abstraction from the data, deep learning models are able to identify intricate patterns in the data. Promising outcomes in weather prediction have been demonstrated by deep learning models, particularly in short-term forecasts (Hewage et al. 2020).

Examples of Successful ML Applications in Forecasting

The accuracy of weather prediction models has improved thanks to the promising results of machine learning applications in weather forecasting. Weather forecasts can be more accurate thanks to machine learning (ML) models that examine and identify patterns in historical data.

Some examples of successful ML applications in weather forecasting:

- Using machine learning (ML), the National Oceanic and Atmospheric Administration (NOAA) is able to forecast hurricanes with greater accuracy. Compared to conventional forecasting techniques, NOAA's machine learning models can incorporate a larger range of data sources, including radar, social media, and satellite imagery. This results in increasing prediction accuracy for the projects which runs for the long time.
- The ML is used improve the accuracy of short-term weather forecasts the by the European Centre for Medium Range weather forecasts (ECMWF). They are using the data which is gathered from the radar, satellites and also from other sources so that they can find the pattern in the data. By this they can predict weather for upcoming days with more precision.
- Dark Sky is a firm that uses machine learning to produce hyperlocal weather forecasts. Dark Sky's machine learning models can forecast the weather to the nearest block. Users may get the most accurate forecasts for their location thanks to this.
- Tomorrow is utilising machine learning (ML) to produce climate science-based forecasts for the upcoming 15 days. Climate change and other factors are factored into tomorrow's machine learning models, which produce more accurate forecasts than conventional forecasting techniques.

Weather forecasting systems are becoming more accurate, efficient, and effective thanks to the powerful tool known as machine learning. We should anticipate seeing even more creative and useful uses of ML in weather forecasting as the technology advances (Broad et al. 2007).

ML is also being utilised to create new weather forecasting services and products in addition to the aforementioned. For instance, some businesses are utilising machine learning to create early warning systems for severe weather phenomena like tornadoes and floods. Other businesses are utilising machine learning (ML) to create weather forecasting products customised for particular sectors, like transportation and agriculture (Dadhich et al., 2021).

Water Management With ML

Machine learning (ML) for water management is a quickly developing field with lots of potential and advantages. Three categories of machine learning applications exist for water management: prediction, clustering, and reinforcement learning. When it comes to identifying dissolved oxygen depletion events that endanger aquatic life in urban streams, for example, machine learning techniques can improve knowledge extraction of significant environmental events in water resources (Rozos, 2019). Furthermore, machine learning (ML) is widely used in wastewater treatment, water quality assessment, and monitoring and prediction of water quality. Machine learning (ML) algorithms have the potential to enhance system efficiency and support water conservation initiatives by evaluating data pertaining to pressure, usage, and other variables. Using the Internet of Things (IoT) for water management is another related. The whole water supply chain benefits from enhanced control and transparency thanks to IoT-based smart water management systems. These systems use real-time monitoring of water levels, quality parameters, and flow through the use of IoT devices and sensors. Improved industrial equipment efficiency, increased process transparency in the water supply chain, and the capacity to provide utilities and consumers with data on water consumption are just a few advantages of IoT-powered smart water management systems. Industries and cities can effectively manage water resources and address the challenges of water scarcity by incorporating IoT technology into water management. Furthermore, water quality is monitored using machine learning techniques(Duerr et.al, 2018). Various components of water quality can be predicted by using the ML techniques like MLP and SVM. These models facilitate improved management of water resources by enabling precise predictions of river water quality. In order to identify threats to water quality and prevent water-related health risks, machine learning (ML) algorithms can be used to develop models for water quality assessment based on datasets containing a variety of features. To estimate the water quality index (WQI) and evaluate overall water quality, sophisticated AI algorithms can also be used. We can have the great results if we intergrade the ML and IoT technologies into the water management. Machine learning algorithms have the potential to improve knowledge extraction, optimise system performance, and aid in water conservation initiatives. The water supply chain is made more transparent and controlled by IoT-based smart water management systems, and precise water quality prediction and assessment are made possible by machine learning techniques. These technological developments could help with issues related to water scarcity and enhance the management of water resources as a whole (Adamowsli, 2008).

ML contributes to optimizing water resource management.

- **Forecasting water demand:** By training machine learning models on past data, one can forecast future water demand. Plans for water management that guarantee there is enough water to meet demand can be created using this information.
- **Leak detection:** Water distribution systems leaks can be found using ML models. This saves money and lessens the waste of water.
- **Improving water distribution:** Water distribution systems can have their water distribution optimised with the use of machine learning models. This makes it easier to make sure that water is efficiently delivered to the areas that need it most.
- **Water quality monitoring:** Real-time water quality monitoring is possible with ML models. This facilitates the swift detection and resolution of issues with water quality.

Figure 4. Water management using ML

- **Predicting extreme weather events:** Droughts and floods, for example, can be predicted using machine learning models. Plans for managing water resources can be created using this information to reduce the risks associated with these occurrences (Bata et al., 2020).

ML Applications in Water Management

Here are some specific examples of ML applications in water management:

I. Phoenix, Arizona is optimising its water distribution system with machine learning. Machine learning models are employed to forecast water consumption and detect possible issues within the system, like leaks and breaks in the pipes. With the use of this data, the system's operation can be modified to guarantee that water is delivered to areas that require it most effectively and that it is resilient to interruptions.

    I. The business HydroIQ is creating a real-time platform for managing water distribution that optimises water quality and distribution through machine learning. The platform forecasts water demand and detects possible issues by utilising data from sensors located throughout the water distribution system. Real-time operating recommendations for the system are generated using this data.

- The WaterSmart research project is creating machine learning (ML)-based tools to assist in making decisions about the distribution of water in rural areas in real time. The project is creating instruments to forecast water consumption, find leaks, and enhance water quality. Small water utilities should be able to afford and use these tools easily.
- Farmers Edge is a company that uses machine learning to help farmers maximise their water usage. Farmers Edge creates customised irrigation schedules for every field by utilising machine learning to analyse data from soil moisture sensors, weather stations, and other sources.
- Xylem is utilising machine learning to create innovative water treatment technologies. Xylem develops novel materials and water treatment processes using machine learning.

All things considered, ML has a lot of potential to greatly enhance water management. ML can assist us in using water more sustainably and efficiently by enhancing our capacity to forecast water supply and demand, optimise water allocation, minimise water waste, safeguard water quality, and upgrade water infrastructure (Chou et al., 2018).

Predictive Accuracy and Adaptability

Our everyday lives depend heavily on weather forecasting because it gives us important information about the weather that is coming. The forecasting of weather has become much more accurate over time. About 80% of the time, a seven-day forecast can accurately predict the weather, according to a National Oceanic and Atmospheric Administration study. Farmers have been able to increase crop yields thanks to this increase in forecast accuracy, which has also helped people prepare for storms and other catastrophic events (Antunes et al., 2018).

The necessity of adjusting to the effects of climate change on water resources is becoming increasingly apparent in terms of adaptability in water management. Ensuring the sustainable use of water and safeguarding ecosystems requires the integration of adaptive water management techniques. Water management adaptation options are being developed with consideration for the possible effects and hazards of climate change. This covers methods to defend water quality against impending dangers like rising stormwater runoff, erosion, and algal blooms (Bougadis et al., 2005).

Moreover, resilience can be improved by fostering synergies between climate change adaptation and water management. This entails figuring out viable substitutes and combining environmental water management with other initiatives. It is believed that adopting integrated and adaptive approaches to water management will fundamentally change how water resources are used and how climate change is mitigated.

Weather forecasting's predictive accuracy has increased dramatically, making it possible to better prepare for weather events. Simultaneously, climate change presents challenges that must be addressed in order to ensure the sustainable use of water resources, and adaptability in water management is essential. Key tactics for attaining resilience and effective water resource management include the integration of adaptive approaches and the development of synergies between water management and climate change adaptation.

Dynamic Adaptation: Enhancing Predictive Accuracy in Changing Conditions

In order to produce more accurate predictions, ML models can adjust in a variety of ways to changing conditions.

Using online learning is one method. Online learning algorithms can adjust to shifts in the underlying data distribution because they can learn from fresh data as it becomes available.

ML models can also use ensembles to adjust to changing conditions. One kind of machine learning model that integrates several different models into one is called an ensemble. Because ensembles can learn from a larger range of patterns in the data, they are frequently more accurate than individual ML models.

Lastly, transfer learning allows ML models to be adjusted to changing circumstances. Through the use of a technique called transfer learning, machine learning models can be trained on data that is similar to but distinct from the data on which they will be making predictions. This can be helpful when gathering a lot of data for a particular task that the machine learning model is being used for is costly or difficult (Rasp et al., 2020).

Challenges and Ethical Considerations

In the fields of water and weather, machine learning (ML) has become a potent instrument that can be used to enhance resource management, decision-making, and forecasting. Nevertheless, there are a number of difficulties with putting it into practise.

Challenges

- Data quality: The quality of ML models depends on the data they are trained on. As a result, having access to reliable weather and water data is crucial. This can be difficult, particularly in rural and developing nations.
- Interpretability of the model: Deciphering the predictions made by machine learning models can be challenging. As a result, it could be challenging to determine whether the forecasts are accurate and to spot any possible model biases.
- Robustness of the model: Machine learning models are susceptible to adversarial attacks, which are inputs intended to deceive the model into generating false predictions. Creating machine learning models that can withstand adversarial attacks is crucial.
- Algorithmic bias: The data used to train machine learning models can contain biases. This may result in models that forecast inequitably or discriminatorily. In machine learning models, algorithmic bias must be recognised and reduced.
- Privacy: Sensitive information about specific people can be deduced using ML models. It is critical to create ML models that safeguard individuals' privacy (Ridwan et al., 2021).

Ethical considerations

- Fairness: ML models ought to be impartial and fair. This implies that no specific group of people should be discriminated against by the models.
- Transparency: ML models should be developed and used transparently. This implies that the models' development and application should be transparent to the general public.
- Accountability: Those who create and apply machine learning models should be held responsible through a well-defined procedure. This involves taking responsibility for the models' fairness, accuracy, and privacy.

- Beneficence: ML models ought to be applied constructively. This indicates that rather than being used to harm people, the models should be used to better their lives.

How to address the challenges and ethical considerations

- Data quality: Gathering data from various sources and meticulously cleaning and curating the data are crucial steps in addressing the problem of poor data quality. Utilising data validation techniques is crucial in guaranteeing the accuracy and representativeness of the data.
- Model interpretability: Using interpretable machine learning algorithms and creating explainable artificial intelligence tools are crucial to overcoming the problem of model interpretability. To find any biases or mistakes in the ML models' predictions, visual analysis is also crucial.
- Model robustness: Developing ML models that are resistant to adversarial attacks is crucial in order to address the problem of model robustness. Ensembles of ML models or adversarial training methods can be used for this.
- Algorithmic bias: Recognising and reducing bias in the data and ML algorithms is crucial to addressing the issue of algorithmic bias. This can be accomplished by auditing ML models for bias and by applying strategies like debiasing algorithms.
- Privacy: Developing ML models that safeguard individuals' privacy is crucial to addressing the problem of privacy. Techniques like federated learning and differential privacy can be used to achieve this.

Fairness, transparency, accountability, and beneficence: It's critical to develop and apply ML models responsibly in order to address the ethical considerations of fairness, transparency, accountability, and beneficence. This entails creating moral standards for applying ML models and incorporating the general public in their creation and application (Almikaeel et al., 2022).

Human-Machine Collaboration

In the domains of water management and forecasting, cooperation between human experts and machine learning (ML) systems is critical. Combining human judgement with machine learning models can improve forecasting accuracy and efficacy. To get the best results and increase confidence in the forecasts, careful cooperation between scientists, forecasters, and forecast users is essential. Through this partnership, the advantages of human intuition and domain expertise can be combined with the processing capacity and data processing skills of machine learning systems (Wee et al., 2021). Additionally, it guarantees that human experts correctly comprehend and interpret the constraints and uncertainties of machine learning models. Similar to this, human experts and machine learning systems working together is essential to water management. To create discrimination techniques for determining the water status of plant roots in greenhouses, phenotyping and machine learning techniques can be integrated. Water management systems' complexity and difficulties can be solved by combining IoT technologies with AI, ML, and deep learning. Intelligent models can be embedded to optimise water resource allocation, improve water quality monitoring, and strengthen water conservation efforts thanks to the cooperation of human experts and machine learning systems (sun & Scanlon, 2019). Accurate predictions and a deeper comprehension of environmental phenomena depend on the cooperation of ML systems and human experts in environmental forecasting. Artificial intelligence techniques that transform our comprehension and forecasting of environmental shifts depend on the cooperation of machine learning models and human specialists. Predicting future outcomes in a variety of domains, such as weather and geopolitics, appears promising when utilising machine learning models and crowdsourcing human forecasts. Understanding and forecasting changes in the environment is greatly aided by the combination of quantitative analysis supplied by machine learning systems and qualitative knowledge provided by human experts. In general, to fully utilise the strengths of both sides, human experts and machine learning systems must work together in the areas of environmental forecasting, water management, and forecasting. While machine learning (ML) systems offer computational power, data analysis capabilities, and pattern recognition, human expertise provides domain knowledge, intuition, and interpretability. Better decision-making, more precise forecasting, and efficient resource allocation are all made possible by this collaboration. By working together, it will be possible to fully utilise the potential of machine learning systems while making sure that ethical and human judgement are taken into account when making decisions (Ben-Bouallegue et al., 2023).

Some examples of successful partnerships in weather and water management.

- Through the Alliance for Hydromet Development, the World Meteorological Organisation (WMO) and the Global Water Partnership (GWP) are collaborating to enhance water management. The main goal of combining together is to increase the development in hydrometeorological services. These services can provide the data which is very necessary to the water management.
- In the US there are different companies which are working together for the improvement of weather forecasting and water management. Also the companies lile National Oceanic and Atmospheric Administration(NOAA) and National Weather Service(NWS) are working together to create a model that can predict the droughts and flood in the US.
- To develop water management model for the California different departments came together to work on the same model. They have developed an innovative solution for the sustainable water resource management.
- To increase the water quality and its distribution the HydroIO developed a model which provides the data at real-time. They have also combined with different firms to improve the infrastructure for the distribution of water in the city. The city of Phoenix is utilising the HydroIQ platform to reduce water waste and increase the efficiency of the water distribution system.
- To create more precise risk models for weather- and water-related events, the insurance industry is collaborating with the community that forecasts weather and water. Insurance companies benefit from this partnership by having more precise product pricing and faster claim payouts.

These are only a handful of the numerous fruitful collaborations that are enhancing water and weather management globally. In order to address the problems of climate change, water scarcity, and other water-related hazards, these collaborations are crucial.

There are numerous other fruitful collaborations between human experts and machine learning systems in the weather and water management domains in addition to the ones mentioned above. For instance, to create new machine learning (ML) algorithms for weather forecasting, the National Centre for Atmospheric Research (NCAR) is collaborating with Google AI. Also, Microsoft AI and the World Resources Institute (WRI) are collaborating to create new machine learning (ML) tools for water management.

The state of the art in weather and water management is being advanced by these collaborations. Together, ML systems and human experts are creating new approaches to optimise water resources, minimise water waste, and forecast weather and water conditions.

Case Studies and Applications

Case Study 1

Utilising ML to Enhance Hurricane Predictions

Compared to conventional forecasting techniques, NOAA's machine learning models can incorporate a greater range of data sources, such as:

- Satellite imagery: Satellite imagery sheds light on hurricane intensity and structure.
- Radar data: Information about hurricane-related wind and rainfall patterns can be found in radar data.
- Data from social media: It is possible to use social media data to monitor hurricane movements and pinpoint areas that are vulnerable to flooding.

Using this data, NOAA's machine learning models are trained to identify patterns linked to the formation and motion of hurricanes. Forecasts of the future hurricanes' track, intensity, and rainfall are then produced using the models.

Compared to conventional forecasting techniques, NOAA's machine learning models were able to predict Hurricane Laura's track and intensity in 2020 with a higher degree of accuracy. This made it possible for local authorities to warn people in a timely and accurate manner, potentially saving lives.

For instance, Hurricane Laura was expected to reach landfall as a Category 4 storm near Lake Charles, Louisiana, according to NOAA's ML models. Since this forecast came true, local authorities were able to evacuate the area's residents before the hurricane made landfall.

Case Study 2

ML-Based Water Distribution System Optimisation

Using machine learning (ML), HydroIQ's real-time water distribution management platform forecasts water demand and locates possible system issues like leaks and pipe breaks. The platform makes use of numerous data sources, such as:

Data from sensors installed in the water distribution system; Weather data; Real-time data on water pressure and flow; Historical data on water demand.

Using this data, HydroIQ's machine learning models are trained to identify trends linked to both water demand and system malfunctions. After that, the models are used to forecast water demand and spot possible issues with the system.

The platform from HydroIQ, for instance, can forecast the location and time of the peak water demand. With the use of this data, the water distribution system's operations can be modified to guarantee that water is supplied where it is most needed.

The water distribution system's leaks and pipe breaks can be found using HydroIQ's platform. By using this information, the affected area can be swiftly isolated and water supply disruptions can be kept to a minimum.

2019 saw a significant water main break in Phoenix. With the aid of HydroIQ's platform, the city was able to promptly locate the break, isolate the impacted area, and reduce water supply interruptions.

Case Study 3

ML-Based Water Demand Forecasting

The historical data on variables like weather patterns, population growth, and economic development is used to train DWR's machine learning models.

Plans for managing water resources are created using this information to guarantee that there is sufficient water to meet demand and preserve water supplies.

Water demand can be accurately predicted by DWR's machine learning models. Water shortages are prevented and conservation measures are developed using this information.

For instance, California went through a terrible drought in 2021. With precision, DWR's machine learning models were able to forecast how the drought would affect water demand. The state was able to prevent water shortages and develop water conservation measures, like mandatory water restrictions, thanks to the information provided (Emami et al., 2022).

Case Study 4

Applying Machine Learning to Find Water Distribution System Leaks

The data from sensors installed in water distribution systems is used to train Xylem's machine learning models. The sensors gather temperature, flow, and pressure data about the water.

The sensor data can be analysed by Xylem's ML models to find patterns that point to the existence of leaks. A leak may be indicated, for instance, by an abrupt drop in water pressure or a shift in water temperature ().

Traditional methods are unable to detect leaks that are too small, but Xylem's ML models can. Water utilities may be able to save money and reduce water waste as a result.

Water utilities in the United States, for instance, use Xylem's machine learning models to find leaks in their water distribution networks. These utilities have stated that by utilising Xylem's ML models to find and fix leaks, they have been able to cut water waste by as much as 20%.

These are only a handful of the numerous effective ways that machine learning has been used in weather and water management. Machine learning (ML) is a potent instrument that can be used to minimise waste, maximise resource management, and increase prediction accuracy. We may anticipate seeing even more creative and useful uses of ML in weather and water management as the technology advances (Yussif et al., 2023).

The Impact and Outcomes of ML in Water and Weather Management Applications

A wide range of areas could be significantly impacted by the use of ML in weather and water management, including:

- Increased prediction accuracy: Machine learning (ML) models can be applied to increase the precision of meteorological and hydrological forecasts, including hurricane, flood, and drought forecasts. Making better decisions about how to respond to and get ready for these kinds of events can be aided by this.
- Optimised resource management: Machine learning models may be used to optimise the management of water resources by predicting water demand and identifying areas where water conservation is feasible. You can guarantee that there is sufficient water to suit everyone's requirements by taking this action.
- Less waste: By locating leaks in water distribution systems, for example, ML models can be used to detect and minimise water waste. Water and money can be saved by doing this.
- Enhanced resilience: By using ML models, weather and water systems can become more resilient to disruptions such as climate change. This can lessen the effects of extreme weather events on people and property.

Future Directions

Global water resource management could be greatly enhanced by developments in machine learning. In order to facilitate the rapid dissemination and efficient application of advances in water management, machine learning algorithms have been developed. These algorithms enhance system performance and enhance water resource management by analysing data on pressure, water usage, and other variables. Water utilities can benefit from the application of AI and ML in anticipating future water demand and optimising water supply throughout the day to ensure effective water management. ML has also helped to advance weather forecasting significantly. Machine learning-based weather forecasting models have advanced quickly and produced encouraging outcomes. Machine learning models can detect trends and forecast meteorological phenomena like storms, temperature swings, and rainfall by examining past meteorological data. This makes weather forecasts quicker and more precise. Furthermore, the accuracy of future weather forecasts has been further improved by the application of deep learning algorithms to analogue weather forecasting.

Here are some potential advancements and future trends in ML for water management and weather forecasting:

- Creation of more dependable and accurate machine learning models: As ML algorithms advance and more data becomes available, we should anticipate creating ML models that are more dependable and accurate than ever. Better weather and water condition forecasts will result from this, which will enable us to better manage these resources.
- Integration of machine learning (ML) with other technologies: ML is becoming more and more integrated with big data, artificial intelligence (AI), and the Internet of Things (IoT). New and creative approaches to water management and weather forecasting are being developed as a result of this integration. For instance, real-time data on weather and water conditions can be gathered using ML-powered sensors, and this data can be used to enhance the precision of ML models.
- Using ML to support decision-making: By offering advice and insights to human experts, machine learning (ML) can help with decision-making related to weather forecasting and

water management. For instance, ML models can be used to pinpoint areas that are vulnerable to drought or flooding, or they can be used to create plans for efficiently distributing water.

The trends in machine learning for weather forecasting and water management appear promising for the future. Future developments in the field of water management include the creation of smart systems, the use of AI and ML in rainwater harvesting and recycling, and the relationship between water and energy. These developments will enhance efforts to conserve water and transform the water sector. Similar to this, machine learning models will continue to advance in the future, enabling even quicker and more precise weather forecasts. Weather forecasting models will be further enhanced by utilising AI in conjunction with new types of sensors and data sources. All things considered, there is a lot of promise for better resource management and a deeper comprehension of weather patterns in the developments and trends of machine learning for weather forecasting and water management. These technologies have the ability to forecast the weather, forecast system performance, anticipate future problems, and enable more sustainable and effective management of water resources (Jaseena et al., 2022).

Ongoing Research and Emerging Technologies on Weather and Water Management

The goal of current ML research for weather and water management is to create new methods and algorithms that will increase the precision, dependability, and interpretability of ML models. Additionally, scientists are looking into novel approaches to combine machine learning (ML) with other technologies like artificial intelligence (AI), big data, and the Internet of things.

Here are some specific areas of ongoing research in ML for weather and water management:

- Creating new machine learning algorithms for weather and water forecasting: Scientists are creating new machine learning algorithms especially for weather and water forecasting. In the face of uncertainty, these algorithms can learn from intricate data patterns and produce precise predictions.
- Enhancing ML model interpretability: New techniques are being developed by researchers to improve ML model interpretability. This will assist us in figuring out how the models generate predictions and locating any possible biases.
- Integrating machine learning (ML) with other technologies: Scientists are looking into novel approaches to combine ML with AI, big data, and the Internet of Things. New and creative approaches to water management and weather forecasting are being developed as a result of this integration (Ben-Bouallegu et al., 2023).

Here are some emerging technologies in ML for weather and water management:

- Explainable AI (XAI): This area of study looks for ways to provide an explanation for the predictions made by machine learning models. The characteristics and data points that are most crucial to the model's predictions can be found using XAI tools. By using this data, the model's accuracy and dependability can be increased, and any potential biases can be found.
- Federated learning: This machine learning approach enables several devices to work together to jointly train an ML model without exchanging data. This is helpful for weather and water forecasting, where gathering data from numerous devices is crucial, but data privacy must also be maintained.

- Quantum machine learning: This is a relatively new area of study that combines machine learning and quantum computing. When it comes to some tasks, like weather forecasting, quantum machine learning algorithms may perform better than traditional ones.

These are but a few instances of the current studies and cutting-edge machine learning technologies being applied to weather and water management. We may anticipate seeing even more creative and useful ML applications in these fields as ML technology advances.

CONCLUSION

The rapidly evolving field of machine learning (ML) has the potential to transform water management and weather forecasting. By the ML we can optimise the resources and increase the prediction accuracy, decrease the waste, and also predict the weather climatise. To increase the prediction and accuracy of the ML models can provide us various advantages in weather and waste management. ML models are trained by the large datasets to identify the various patterns, this helps the model to predict the weather and water conditions in future very precisely and accurately. The datasets which ML model used to train can be obtained from the radars, satellites and other social media platforms rather than the traditional forecasting system. By this we can improve the accuracy for those projects which are used for the long run. We can use ML models to predict the weather and water condition of the area which needs conservation, this can help us to preserve the water for the seasons when it is dry.

ML models does not only forecast but it can also be used for the waste management and also predicts the upcoming climatic change. We can use the ML models to locate the leaks in water distribution system by which we can save water as well as money. ML models are also used to detect those locations which are very prone for the floods or sea level rise. As now ML models are used everywhere not only in these fields but also in other different fields we can say that there is enormous potential in the ML. If we can use these models correctly wen can manage weather and water management with high prediction and accuracy. If we are able to use these models correctly and incorporate these models into decision-making we can make better decisions which will eventually are beneficial for all the humankind.

REFERENCES

Adamowski, J. F. (2008). Peak Daily Water Demand Forecast Modeling Using Artificial Neural Networks. In Journal of Water Resources Planning and Management (Vol. 134, Issue 2, pp. 119–128). American Society of Civil Engineers (ASCE). doi:10.1061/(ASCE)0733-9496(2008)134:2(119)

Aguilera, H., Guardiola-Albert, C., Naranjo-Fernández, N., & Kohfahl, C. (2019). Towards flexible groundwater-level prediction for adaptive water management: using Facebook's Prophet forecasting approach. In Hydrological Sciences Journal (Vol. 64, Issue 12, pp. 1504–1518). Informa UK Limited. doi:10.1080/02626667.2019.1651933

Alloghani, M., Al-Jumeily, D., Mustafina, J., Hussain, A., & Aljaaf, A. J. (2019). A Systematic Review on Supervised and Unsupervised Machine Learning Algorithms for Data Science. In *Unsupervised and Semi-Supervised Learning* (pp. 3–21). Springer International Publishing. doi:10.1007/978-3-030-22475-2_1

Almikaeel, W., Čubanová, L., & Šoltész, A. (2022). Hydrological Drought Forecasting Using Machine Learning—Gidra River Case Study. In Water (Vol. 14, Issue 3, p. 387). MDPI AG. doi:10.3390/w14030387

Antunes, A., Andrade-Campos, A., Sardinha-Lourenço, A., & Oliveira, M. S. (2018). Short-term water demand forecasting using machine learning techniques. In Journal of Hydroinformatics (Vol. 20, Issue 6, pp. 1343–1366). IWA Publishing. doi:10.2166/hydro.2018.163

Bata, M., Carriveau, R., & Ting, D. S.-K. (2020). Short-term water demand forecasting using hybrid supervised and unsupervised machine learning model. In Smart Water (Vol. 5, Issue 1). Springer Science and Business Media LLC. doi:10.118640713-020-00020-y

Ben-Bouallegue, Z., Clare, M. C. A., Magnusson, L., Gascon, E., Maier-Gerber, M., Janousek, M., Rodwell, M., Pinault, F., Dramsch, J. S., Lang, S. T. K., Raoult, B., Rabier, F., Chevallier, M., Sandu, I., Dueben, P., Chantry, M., & Pappenberger, F. (2023). *The rise of data-driven weather forecasting* (Version 1). arXiv. doi:10.48550/ARXIV.2307.10128

Bougadis, J., Adamowski, K., & Diduch, R. (2005). Short-term municipal water demand forecasting. In Hydrological Processes (Vol. 19, Issue 1, pp. 137–148). Wiley. doi:10.1002/hyp.5763

Broad, K., Pfaff, A., Taddei, R., Sankarasubramanian, A., Lall, U., & de Assis de Souza Filho, F. (2007). Climate, stream flow prediction and water management in northeast Brazil: societal trends and forecast value. In Climatic Change (Vol. 84, Issue 2, pp. 217–239). Springer Science and Business Media LLC. doi:10.100710584-007-9257-0

Chou, J.-S., Ho, C.-C., & Hoang, H.-S. (2018). Determining quality of water in reservoir using machine learning. In *Ecological Informatics* (Vol. 44, pp. 57–75). Elsevier BV. doi:10.1016/j.ecoinf.2018.01.005

Dadhich, S., Pathak, V., Mittal, R., & Doshi, R. (2021). Machine learning for weather forecasting. In *Machine Learning for Sustainable Development* (pp. 161–174). De Gruyter. doi:10.1515/9783110702514-010

Dehghanisanij, H., Emami, H., Emami, S., & Rezaverdinejad, V. (2022). A hybrid machine learning approach for estimating the water-use efficiency and yield in agriculture. In Scientific Reports (Vol. 12, Issue 1). Springer Science and Business Media LLC. doi:10.103841598-022-10844-2

Duerr, I., Merrill, H. R., Wang, C., Bai, R., Boyer, M., Dukes, M. D., & Bliznyuk, N. (2018). Forecasting urban household water demand with statistical and machine learning methods using large space-time data: A Comparative study. In Environmental Modelling & Software (Vol. 102, pp. 29–38). Elsevier BV doi:10.1016/j.envsoft.2018.01.002

Emami, M., Ahmadi, A., Daccache, A., Nazif, S., Mousavi, S.-F., & Karami, H. (2022). County-Level Irrigation Water Demand Estimation Using Machine Learning: Case Study of California. In Water (Vol. 14, Issue 12, p. 1937). MDPI AG. doi:10.3390/w14121937

Gino Sophia, S. G., Ceronmani Sharmila, V., Suchitra, S., Sudalai Muthu, T., & Pavithra, B. (2020). Water management using genetic algorithm-based machine learning. In Soft Computing (Vol. 24, Issue 22, pp. 17153–17165). Springer Science and Business Media LLC. doi:10.100700500-020-05009-0

Hewage, P., Behera, A., Trovati, M., Pereira, E., Ghahremani, M., Palmieri, F., & Liu, Y. (2020). Temporal convolutional neural (TCN) network for an effective weather forecasting using time-series data from the local weather station. In Soft Computing (Vol. 24, Issue 21, pp. 16453–16482). Springer Science and Business Media LLC. doi:10.100700500-020-04954-0

Inoue, J., Yamagata, Y., Chen, Y., Poskitt, C. M., & Sun, J. (2017). Anomaly Detection for a Water Treatment System Using Unsupervised Machine Learning. In *2017 IEEE International Conference on Data Mining Workshops (ICDMW). 2017 IEEE International Conference on Data Mining Workshops (ICDMW).* IEEE. 10.1109/ICDMW.2017.149

Jakaria, A. H. M., Hossain, M. M., & Rahman, M. A. (2020). *Smart Weather Forecasting Using Machine Learning:A Case Study in Tennessee* (Version 1). arXiv. doi:10.48550/ARXIV.2008.10789

Jaseena, K. U., & Kovoor, B. C. (2022). Deterministic weather forecasting models based on intelligent predictors: A survey. In Journal of King Saud University - Computer and Information Sciences (Vol. 34, Issue 6, pp. 3393–3412). Elsevier BV. doi:10.1016/j.jksuci.2020.09.009

Lowe, M., Qin, R., & Mao, X. (2022). A Review on Machine Learning, Artificial Intelligence, and Smart Technology in Water Treatment and Monitoring. In Water (Vol. 14, Issue 9, p. 1384). MDPI AG. doi:10.3390/w14091384

Miro, M. E., Groves, D., Tincher, B., Syme, J., Tanverakul, S., & Catt, D. (2021). Adaptive water management in the face of uncertainty: Integrating machine learning, groundwater modeling and robust decision making. In *Climate Risk Management* (Vol. 34, p. 100383). Elsevier BV. doi:10.1016/j.crm.2021.100383

Rasp, S., Dueben, P. D., Scher, S., Weyn, J. A., Mouatadid, S., & Thuerey, N. (2020). WeatherBench: A Benchmark Data Set for Data-Driven Weather Forecasting. In Journal of Advances in Modeling Earth Systems (Vol. 12, Issue 11). American Geophysical Union (AGU). doi:10.1029/2020MS002203

Ridwan, W. M., Sapitang, M., Aziz, A., Kushiar, K. F., Ahmed, A. N., & El-Shafie, A. (2021). Rainfall forecasting model using machine learning methods: Case study Terengganu, Malaysia. In Ain Shams Engineering Journal (Vol. 12, Issue 2, pp. 1651–1663). Elsevier BV. doi:10.1016/j.asej.2020.09.011

Rozos, E. (2019). Machine Learning, Urban Water Resources Management and Operating Policy. In Resources (Vol. 8, Issue 4, p. 173). MDPI AG. doi:10.3390/resources8040173

Sun, A. Y., & Scanlon, B. R. (2019). How can Big Data and machine learning benefit environment and water management: a survey of methods, applications, and future directions. In Environmental Research Letters (Vol. 14, Issue 7, p. 073001). IOP Publishing. doi:10.1088/1748-9326/ab1b7d

Wee, W. J., Zaini, N. B., Ahmed, A. N., & El-Shafie, A. (2021). A review of models for water level forecasting based on machine learning. In Earth Science Informatics (Vol. 14, Issue 4, pp. 1707–1728). Springer Science and Business Media LLC. doi:10.100712145-021-00664-9

Yussif, A.-M., Sadeghi, H., & Zayed, T. (2023). Application of Machine Learning for Leak Localization in Water Supply Networks. In Buildings (Vol. 13, Issue 4, p. 849). MDPI AG. doi:10.3390/buildings13040849

Zhu, M., Wang, J., Yang, X., Zhang, Y., Zhang, L., Ren, H., Wu, B., & Ye, L. (2022). A review of the application of machine learning in water quality evaluation. In Eco-Environment & Health (Vol. 1, Issue 2, pp. 107–116). Elsevier BV. doi:10.1016/j.eehl.2022.06.001

Chapter 5
Optimizing Water Resources With IoT and ML: A Water Management System

Rakhi Chauhan
Chitkara University, India

Neera Batra
Department of CSE, Maratha Mandal Engineering College, Maharishi Markandeshwar University (Deemed), Mullana, India

Sonali Goyal
Department of CSE, Maratha Mandal Engineering College, Maharishi Markandeshwar University (Deemed), Mullana, India

Amandeep Kaur
https://orcid.org/0000-0002-8418-7954
Department of CSE, Maratha Mandal Engineering College, Maharishi Markandeshwar University (Deemed), Mullana, India

ABSTRACT

The necessity for effective water management systems has increased recently due to the rising demand for water resources and the negative effects of climate change. This chapter gives a thorough investigation into the installation of a water management system (WMS) using machine learning (ML) and the internet of things (IoT) technologies to address these issues. The proposed WMS makes use of internet of things (IoT)-enabled sensors placed throughout various water infrastructure sites, including reservoirs, water supply networks, and pipelines to gather real-time information on critical variables like weather conditions, water quality, etc. The acquired data is then subjected to sophisticated ML algorithms to optimize water use and distribution. The WMS described in this chapter serves as an example of how ML and IoT have the potential to fundamentally alter current approaches to water management.

DOI: 10.4018/979-8-3693-1194-3.ch005

INTRODUCTION

Water is an essential requirement for the sustenance of human life on Earth. The presence of water is essential for the survival of a wide range of biological species. Growing concerns are emerging in relation to water scarcity due to the escalating water use resulting from the expanding human population (Ma.T et al. 2020). In addition to overarching concerns regarding the paucity of freshwater resources for drinking purposes, there is a growing apprehension around the constrained availability of water, specifically in relation to its accessibility for agricultural applications (Rosa, L et.al. 2020, Vallino, E et al. 2020). In order to address the challenges associated with water scarcity, it is crucial to implement a complete water management system. The real-time monitoring of water level and quality is of great significance in the field of water management. The utilization of real-time water level monitoring offers a practical approach to efficiently address the issue of water wastage caused by tank overflow. By implementing a methodical and ongoing process of monitoring water levels at regular intervals throughout the day, the water management system exhibits the capacity to aid in the detection of water leaks within an intelligent residential environment. The establishment of an intelligent water management system holds significant importance in facilitating the advancement of a more intelligent and sustainable global environment. The agricultural and farming industries exhibit a substantial reliance on the cyclical patterns of seasons and the prevailing climatic conditions.

The temperature has a crucial role in the cultivation of several organic crops, vegetables, and pulses. At present, there is a lack of thorough understanding regarding weather forecasting, leading to ranchers relying on projected forecasts to carry out their obligations. However, there are situations in which humans experience disasters as a result of the lack of precision in weather forecasts. The emergence of technical improvements and the accessibility of advanced weather forecasting technologies have enabled ranchers to obtain instantaneous updates on their mobile devices. The indispensability of training in this specific field is indisputable; nonetheless, it is crucial to precisely delineate the particular group of ranchers in the area who will be utilizing the meteorological predictions. Weather forecasting involves the prediction of atmospheric conditions by the analysis of several aspects, including geographical location and temporal variables. The identification of prevailing weather conditions in each zone and technology will empower ranchers to make well-informed judgments and implement suitable measures. The correlation between climatic conditions and agricultural operations has consequently led to a need for accurate weather prediction. The aforementioned requirement stems from the farmers' inclination to make well-informed decisions in order to limit potential risks and prevent unfavorable results. The agricultural productivity of crops is substantially impacted by factors such as temperature, duration of daylight precipitation, ensuring appropriate thermal conditions along with enough hydration and nourishment, holds paramount significance in the context of domesticated animals. The water system pertains to a conceptual framework for the utilization of water resources in terrestrial environments, specifically designed to facilitate the spread and cultivation of agricultural activities. The variability in meteorological patterns exerts a substantial impact on the hydrological system and agricultural output.

Timing and evapotranspiration are essential variables that exert a substantial influence on meteorological phenomena. It is crucial for individuals involved in ranching activities to possess the requisite preparation to effectively respond to and adjust to changes in the surrounding ecological conditions. One of the noteworthy consequences within the water system framework is the advent of the dry season, which is defined by extended durations of arid conditions. Therefore, if their sound assessment is correct, the potential risks of a disaster are significantly reduced in comparison to the initial estimates. The dry

Figure 1. Overview of a water management system

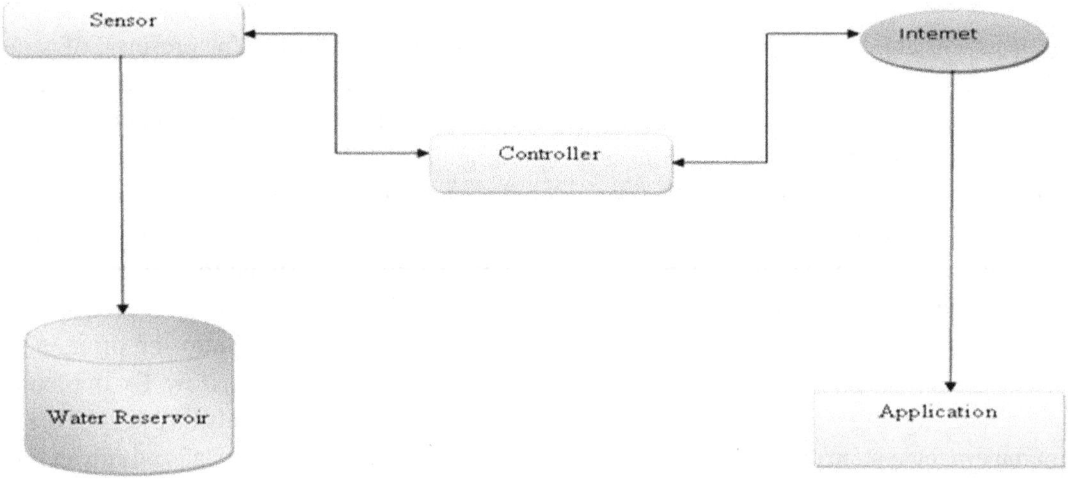

season, defined by decreased humidity levels and increased temperatures, has the potential to increase agricultural water demand. Under situations of high temperature, irrigators are obligated to carefully monitor the usage of water for both daily irrigation and intermittent harvesting activities. Moreover, it is anticipated that they would utilize pioneering methodologies and strategies that facilitate significant enhancements in agricultural output. The limited adoption of smart water management technologies can be attributed to the significant financial burden they impose. The expense linked to the deployment of intelligent urban areas has seen a significant decline in recent times, primarily attributed to the rise of the Internet of Things (IoT). The phrase "Internet of Things" (IoT) refers to a network including interconnected devices that possess the capability to exchange data. In contemporary times, there has been an increasing prevalence of cost-effective water management systems that are reliant on the Internet of Things. IoT devices possess the capability to autonomously transmit data, hence eliminating the need for human involvement. In Figure 1 with the help of IOT principal's water management explanations can be seen. The following system can be made up of data visualization application, a variety of sensor and control unit. Water treatment control systems employ sensors. They are in charge of converting physical parameters from equipment and processes into digital data that robots and humans can use. Simply put, without sensors, water treatment plant control is impossible. A water sensor is a device that measures water levels in a range of applications. Water sensors come in a number of designs, such as ultrasonic sensors, pressure transducers, bubblers, and float sensors.

Water level controllers that adjust the water level automatically save energy. They accomplish this by turning off the motor automatically when the tank is full and when there is no water flow to the tank. This means that controlling a water supply requires less water and energy. Water management controllers are very useful tools for improving the performance and efficiency of water management systems in a variety of applications. These advanced devices improve system performance, reduce energy consumption, and extend the life of your pump by intelligently regulating speed, pressure, and flow. The term "controller" refers to a device or system that is responsible for managing and regulating. Water is a vital substance for sustaining life on Earth. It is a transparent, odorless. A reservoir is a man-made or natural body of water that is used to store and supply.

An application can be defined as a software program that is deployed on the internet, designed to receive data from a controller and present it in a user interface. The incorporation of Internet of Things (IoT) technology in water management systems offers a financially efficient strategy that enables swift scalability within a constrained duration. The simplicity of water quality testing for the detection of several contaminants is facilitated by the utilization of affordable sensors. The ease of integration of regularly utilized communication technologies into pre-existing systems is mostly attributed to their widespread availability. The implementation of Internet of Things (IoT) platforms facilitates convenient remote monitoring and control.

PRELIMINARY ASPECTS FOR WATER MANAGEMENT USING MACHINE LEARNING

Machine learning techniques are utilized to develop predictions about the occurrence of rainfall. The exploitation of time series data is a critical aspect in the application of machine learning. The time series data can be assessed using the following algorithms.

The moving average model: The utilization of time series models is commonly regarded as a direct and uncomplicated method. Despite its apparent simplicity, this model exhibits unexpected levels of acceptability and serves as a robust foundation. The application of the moving average facilitates the detection of significant patterns within the dataset. The utilization of a windowing methodology can enhance the application of the moving average model to proficiently attenuate noise and catch various patterns within a time series. The moving-average model can be conceptualized as a filter that possesses a finite impulse response. It is commonly employed for the purpose of processing white noise. Additionally, this model accommodates other perspectives. There exist two fundamental differentiations in the role of random shocks between the moving average (MA) model and the autoregressive (AR) model.

Exponential smoothing: It is a widely used forecasting technique in time series analysis. It is based on the principle of assigning exponentially decreasing weights to the statistical technique employed for predicting time series data is known as time series forecasting. The technique described is extensively employed across various disciplines, including economics, finance, and operations research. Exponential smoothing utilizes a comparable idea to that of a moving average. However, in this instance, individual perceptron are assigned separate diminishing weights. Over time, there has been a diminishing emphasis on the significance attributed to individual perceptron.

Double exponential smoothing: The technique discussed above is utilized to analyze and make predictions on data sets that have a temporal structure. The method under consideration is an extension of the fundamental exponential smoothing technique, which takes into account both. This method is commonly referred to as Holt's method. The aforementioned technique is frequently employed in the study of time series data for the purpose of forecasting. Double exponential smoothing is a method that builds upon simple exponential smoothing. While simple exponential smoothing only takes into account the most recent observation, double exponential smoothing is used when there is a noticeable pattern present in the time series data. Given the consideration of pertinent aspects, our approach entails the iterative utilization of exponential smoothing methodology.

RELATED WORK

Up to 70% of all freshwater consumption worldwide is used by agriculture (Kolencik & Simansky, 2016), which highlights the need for strong water regulation to maintain the security of food and water for the entire population. A considerable role is played by field application techniques and water system frameworks for the development of yields. While attempting to avoid efficiency loss caused by water concerns, Ranchers who over irrigate their land are therefore assessing the impact on profitability. Both water and energy are being squandered. On the contrary, the precision of the water system can enhance the efficiency and profitability of water utilization by maintaining a prudent separation from both the sub aquatic and super aquatic systems. It is crucial to manage water correctly for the precision water system in agriculture in order to boost crop output, lower expenses, and enhance ecological sustainability. Even while the (IoT) (Atzori et al. 2010) lacks fully evolved fusion of various technologies required running seamlessly realistically speaking, it is quickly becoming the go-to solution for innovative boardroom applications. The notable rise of the Internet of Things (IoT) can be ascribed to multiple factors, encompassing the proliferation of compact devices and energy-efficient remote technologies, the availability of cloud server farms for scalability and strategic purposes, the establishment of managerial frameworks to handle unstructured data from social networks, the utilization of high-performance computing resources in software development phases, and the deployment of computational intelligence algorithms to efficiently manage the extensive volume of data implicated. There are now a few obstacles that must be overcome in order for IoT to be widely adopted for the precision water system. The writing of code is still not totally automated for Internet of Things (IoT)-based smart applications, such as irrigation systems for farming (Kamienski et al. 2019). Additionally, it has been observed that there is a scarcity of sophisticated programming frameworks designed specifically for the Internet of Things (IoT). A streamlined transmission mechanism has been developed to improve the efficiency of pilot application submissions for intelligent water managers, automate a step in the process, and automate mist figuring, these frameworks would incorporate innovations like IoT, big data analysis, distributed computing, and mist figuring. Third, the integration of numerous and diverse sensors necessitates the use of appropriate data models and principles. The construction and evaluation of an intelligent water platform based on Internet of Things (IoT) technologies was undertaken by the SWAMP project. The primary objective of this platform was to enhance the precision of water management in agricultural settings. The accomplishment of this endeavor was achieved by the utilization of a pragmatic approach, involving the implementation of four preliminary inquiries conducted in Brazil, Italy, and Spain (Nagajayanthi, 2021). The SWAMP platform demonstrates the potential for customization and implementation in a variety of ways, allowing for adjustments to meet the specific needs and expectations of different domains. The aforementioned attribute, which includes many elements such as environments, countries, climate, soils, and crops, requires a significant degree of adaptability to suit diverse organizational structures that incorporate a dynamic range of enhancements.

This article provides a full overview of the SWAMP project, which is a methodology that aims to improve inferred frameworks by employing a situation-based approach. This article provides a complete overview of the SWAMP project, which is a methodology focused on situational contexts with the aim of improving inferred frameworks. The paper presents a comprehensive examination of the engineering elements, stages, and preliminary endeavors associated with the project. The SWAMP layered engineering paradigm comprises three distinct categories of administrative tasks that are essential for achieving successful replication and adaptability. The functioning of artificial intelligence (AI), information analysis, Internet of Things (IoT), and storage services is facilitated by highly replicable systems. The board offers comprehensive mobile water information management services that incorporate a diverse array of academic expertise and specialized methodologies specifically designed for different water systems and distribution networks.

Application-specific services require a greater level of work to evolve due to their focus on specific domains or industries. The SWAMP Platform incorporates the fundamental components of FIWARE (Roffia et al. 2018), together with the semantic enhancements provided by a SPARQL-based inference engine (Ahanger & Aljumah, 2019). The stage can be deployed in many configurations to address segment situations in cloud or fog environments, incorporating the utilization of Internet of Things (IoT) communication advancements and advanced cloud-based computations and analysis. Additionally, intelligent decisions can be made based on data located on the farm premises. The purpose of this study is to examine various potential outcomes for the SWAMP Platform and enhance our comprehension of the replicabilty and situational adaptability of its component elements. The process of finding the most efficient technique to handle such a stage can be sped up by looking at numerous possibilities regarding various sending settings.

The importance of adaptability in IoT applications underscores the need for focused research on the coding components of the FIWARE-controlled platform in order to tailor the presentation of each pilot project to its individual context. Based on the empirical evidence, it can be inferred by researchers that this specific stage exhibits the potential to fulfill the requirements of pilots, albeit with certain notable limitations. The research findings revealed that specific components of FIWARE required tweaks to improve their functionality, while others required extensive overhauls to boost their adaptability and lower their use of computing resources.

Moreover, previous research has confirmed that MongoDB was recognized as a potential bottleneck in the development of FIWARE, posing a risk of system breakdowns. The feasibility of establishing a direct connection to SWAMP is unlikely given the lack of a publicly available Internet of Things (IoT) network that is specifically designed for precise agricultural irrigation, according to existing research. The Internet of Things (IoT) involves several security requirements, which include but are not limited to ensuring protection, preserving privacy, assuring integrity, authenticating entities, authorizing access, and conducting auditing (Ullah et al. 2021). Existing research (Kamilaris et al. 2016) has shown that security issues provide a hindrance to the completion of IoT stages.

The current open source frameworks pertaining to precise water systems within the Internet of Things (IoT) area exhibit a deficiency in the validation of concepts. The approaches listed above demonstrate either a significant adherence to established norms or an excessive emphasis on comprehensiveness. However, both approaches lack a succinct and well-structured framework that effectively enhances reliability and facilitates the implementation of innovative systems. In the realm of promoting hydration among executives, there are numerous independent initiatives that are not directly associated with current stages and models.

The FIGARO project, for instance, does not utilize the Internet of Things (IoT) in a direct manner. Nevertheless, the primary aim of this initiative is to improve water management practices and maximize water efficiency by implementing a precise water system at the organizational level (Brewester et al. 2017). In their study, Brewester et al. (2013) present a comprehensive examination of a well constructed, albeit constrained, Internet of Things (IoT)-enabled platform employed for the purpose of gathering data in specific domains focused on precise agricultural and environmental monitoring. The framework known as Agri-IoT, as outlined in reference (Atlam et al. 2018), employs the Internet of Things (IoT) to improve agricultural processes. This technology exhibits certain similarities to SWAMP in terms of its capability for information analysis and continuous management. In recent years, there has been much research and examination into the possible uses of the Internet of Things (IoT) when integrated with cloud-based services and thorough data analysis. The worry in Europe currently revolves around the problems and extensive implications of IoT in the context of large-scale studies for intelligent agriculture. The study conducted by Brewster et al. (year) investigates the architectural framework of extensive Internet of Things (IoT) initiatives within the agricultural industry. The authors' primary emphasis is on examining innovations and configurations that have the potential to be used across several agrifood sectors, such as dairy, organic products, arable yields, and meat and vegetable supply chains (Morabito et al. 2015). Several Internet of things (IoT)-based applications utilized FIWARE as a registration platform within the context of smart farming.

In a study conducted by Cheng et al. (2018), a brief writing survey was administered and the Agricolus stage was presented as a precision agricultural instrument. The study conducted focused on the development of an application utilizing FIWARE, a technology platform, with the aim of improving the accuracy of a water system utilized in agricultural practices in the southern part of Spain. It is crucial to recognize that the implementation mostly revolves around a specific use case (Drakaki et al. 2022). This document presents a comprehensive examination of the operational capabilities of various equipment and apparatus, with an analysis of the functioning principles of water systems.

Using FIWARE, engineering, and a platform, this study demonstrates four notable use-cases for business frameworks. The growth of IoT-based frameworks will result in an explosion of data, necessitating the adoption of a novel paradigm known as hazy processing (Sattar et al. 2023). Another cutting-edge strategy (Park et al. 2022) for dealing with the haze is compartment-based virtualization, which is a lightweight alternative to conventional hypervisors. Docker holders for FIWARE Generic Enablers are also available for the SWAMP mist figuring approach. Fog Flow is a technique for creating Internet of Things (IoT) applications for densely populated urban areas with inhabitants spread throughout fog and haze on the system's periphery. Despite the inclusion of Fog Flow within FIWARE, the SWAMP project optimally utilizes FIWARE's components by directly utilizing them, augmenting them where necessary with additional segments designed expressly for the SWAMP's unique agricultural settings.

Annually, a substantial number of research publications convey the achievements, progress and gaps in the vast domain of water resource management (WRM). In the field of hydrology, numerous inquiries make use of recorded data, encompassing measures of stream flow, precipitation, and temperature. Research methodologies that use processed data, such as global atmospheric data, are an option that obtained from gauge and satellite observations in relation to prognostications within the given domain. Within the domain of data analysis, it has been noted that data-driven models demonstrate greater performance in comparison to statistical models. This can be attributed to their heightened ability to navigate complex situations.

Table 1: Table summarizes recent uses of machine learning (ML) for forecasting in the field of water resource management (WRM). The glossary provided below presents meanings for the essential terminology utilized within the table.

Table 1. Recent uses of machine learning (ML) for forecasting in the field of water resource management (WRM)

Research Field	Algorithm	Goals	Ref.
Hydropower	DNN	This study aims to investigate the viability of utilizing historical energy production data as a predictive tool, as well as the potential benefits of forecasting future energy inputs to model subsequent energy output.	(Sun. J et al. 2022)
Hydropower	ANN, ARIMA, SVM	This study set out to evaluate several ML algorithms' prognostic accuracy in estimating the electricity output of a Chinese reservoir. Dates from 1979 to 2016 were used in the analysis.	(Sharghi et al. 2022)
Water Levels	LSTM, GRU	This study aims to construct a data model that is both efficient and exact in predicting water levels characterized by significant temporal fluctuations.	(Liu, J et al. 2022)
Stream flow	CNN-BAT	This study aims to evaluate the predictive performance of a Convolutional Neural Network (CNN) by employing the BAT met heuristic Algorithm.	(Cho & Kim, 2022)
Ground water	Using a BP-ANN (back propagation ANN), an ARIMA model, an LSTM, and an AR model	The goal of this study is to evaluate the performance of three commonly employed data-driven models in making monthly and daily predictions of the global water level (GWL). These models are the autoregressive integrated moving average (ARIMA), the back-propagation artificial neural network (BP-ANN), and the long short-term memory (LSTM) network.	(Apaydin et al. 2021)
Ground water	Methods that work as an ensemble include LASTM, FFNN, and ANFIS.	The goal here is to provide forward-looking predictions about the groundwater level (GWL) at the centroid of each cluster's piezo meters.	(Ghobadi & Kang, 2023)
Stream flow	LSTM	This study aims to develop a system for hourly stream flow forecasting by utilizing machine learning techniques in conjunction with weather forecasts, simulations of land surface hydrological models, and other factors.	(Peng et al. 2022)
Precipitation	LSTM	This study suggests integrating the Long Short-Term Memory (LSTM) network with the Weather Research and Forecasting Hydrological Modeling System (WRF-Hydro) to boost the accuracy of stream flow forecasts.	(Feng et al. 2022)
Stream flow	ANN, CNN, LSTM	The purpose of this research is to compare the accuracy of neural network methods for monthly stream flow forecasting while using one of three preprocessing procedures: singular spectral analysis (SSA), seasonal-trend decomposition exploiting loess (STL), and attribute selection.	(Nguyen et al. 2022)
Stream flow	CONV LSTM, LST Net, 3D-CNN, TD-CNN, transformer	In order to improve the precision of multi-step forward forecasting, this study employs attention-based deep neural networks (DNNs) and meso scale hydro climatic data as booster predictors.	(Liu et al. 2022)
Water quality	Deep transfer learning based on transformer (TLT)	This study proposes a transfer learning strategy to enhance the accuracy of water quality forecasts in low-data settings.	(Adnan et al. 2021)
Stream flow	ANN, ELM, SVM, EMD, EEMD	In order to improve hydrological forecasting, this research will combine the parallel cooperation search algorithm (PCSA) with the ELM method.	(Adnan et al. 2021)
Stream flow	BART	This study presents a new hybrid model termed GA-BART to address the difficulty of hourly stream flow forecasting. The integration of GA and BART is an attempt to enhance the reliability and accuracy of forecasts by means of a hybrid model.	(Ikram et al. 2022)

ISSUES AND CHALLENGES IN WATER MANAGEMENT

Numerous serious issues that can be broken down into several classes arise when deep learning methods are applied to water management. The subject under consideration revolves around water management systems that employ deep learning methodologies, focusing especially on difficulties associated with data quality and availability. Modern water management and conservation systems incorporate deep learning networks to efficiently recognize large amounts of test data or real-time data collected from sensors or photographs. For training purposes, these networks require enormous data sets. Consideration of the necessary facts is also crucial. The collecting of data from scientific and commercial sectors is subject to limits due to their high sensitivity within the application domain and potential for conferring competitive advantages. Obtaining crucial information regarding government entities for the sake of research and development poses significant challenges as a result of diverse political and legal constraints. The limitations linked to demography are likewise applicable to data. There is an increasing apprehension regarding the quality of real-time large-scale data in light of the substantial need for its application in the training of deep learning models. The task of evaluating the accuracy and reliability of individual data points gets increasingly complex when dealing with the collection of a significant amount of data for the purpose of training. We can also explain some challenges in water management using following diagram.

Figure 2. Challenges in water management

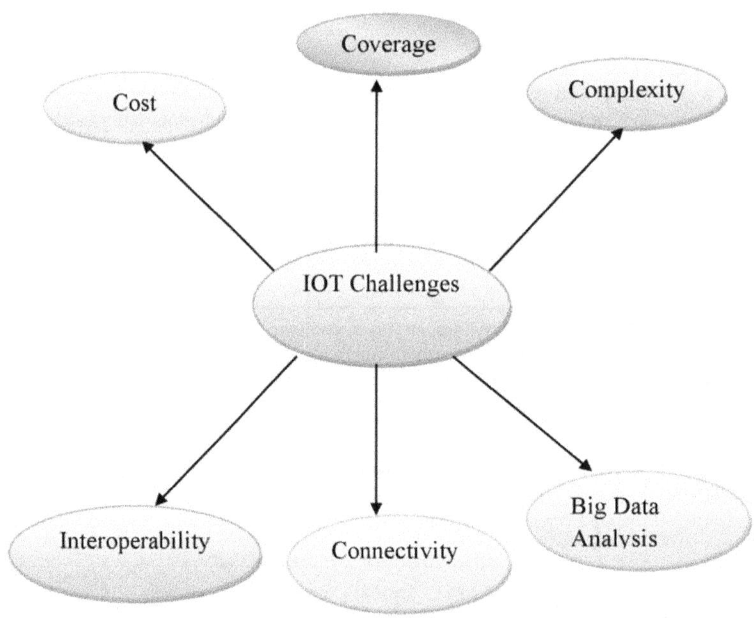

The challenges with the help of water management can be seen like cost, complexity, connectivity, coverage, interoperability, big data analysis etc.

Therefore, it is not conceivable to claim that the trained model was constructed using data of greater quality. If the pre-processing step is not performed accurately, there is a potential for the trained system to display vulnerability to errors due to the presence of outliers or noisy data. Deep learning networks utilized in the realm of water management systems heavily depend on comprehensive datasets, a substantial amount of which is publically available or may be accessed by many stakeholders within the academic or industry sectors. The alterations made to the incoming data might be seen as a manifestation of the system's behavior. Potential adversaries possess the motivation to abuse the system for this particular objective. As an illustration, due to the possession of a competitive advantage, an attacker has the capacity to manipulate an irrigation management system, resulting in an excessive discharge of water that jeopardizes the entirety of the agricultural production. A deep learning model supported by AI is crucial for usage in water management, smart agriculture, and smart farming. However, the AI will be useless if the final system can be easily hacked. Therefore, it is imperative to incorporate cyber security rules and ensure data access integrity in order to effectively integrate deep learning-based water management systems, akin to smart systems. In actuality, deep learning is not limited to a particular method, but instead emphasizes the construction of an architecture or model. As technological systems advance to successive iterations, it becomes imperative to make adaptations to the algorithms. In the case of a significant technical advancement to the model, the level of retraining necessary for a deep-learning network will consistently be insufficient intelligent sensors could be used to evaluate water quality by selectively analyzing its qualities.

SMART SOLUTIONS FOR WATER MANAGEMENT

Despite the myriad of concerns and challenges associated with water management, there are some potential solutions that might be implemented. The following possibilities are delineated below.

The efficient and secure management of water resources by both end-users and workers is made possible by intelligent sensor devices, which play a pivotal role in the successful implementation of policies. These devices provide the prompt acquisition of data, together with the timely dissemination of notifications and the execution of suitable measures to proactively resolve any concerns. The degree of automation, real-time alarm production, and insights acquisition may vary based on the infrastructure, ultimately resulting in the attainment of enhanced levels of efficiency.

The integration of devices and sensors into water industrial infrastructure presents several benefits for utilities and their personnel, as it enables efficient monitoring of physical infrastructure and ensures the safety and welfare of utility staff. As a result of the escalating quantity of data, service teams have developed the capability to efficiently manage remote infrastructure and reduce the necessity for on-site physical repairs. In the event of any potential dangers to the infrastructure, the pertinent service teams are expeditiously informed. Furthermore, the utilization of automated methods is utilized to proactively execute the necessary modifications, hence obviating the necessity for support workers to physically address the difficulties in person. The current subject of inquiry pertains to the various factors involved in the management, storage, and distribution of water.

The allocation of water resources exerts a substantial influence on the economic advancement of nations on a worldwide scale. Nevertheless, the incidence of water loss, predominantly attributed to leakage and various other contributing reasons, presents a significant peril of environmental pollution. This phenomenon exerts a significant influence on the physical and mental well-being, as well as the total welfare, of individuals. The application with technology can be utilized to effectively accomplish the objectives of preserving water quality and reducing waste. The potential benefits of combining energy harvesting with Internet of Things (IoT) technologies offer a promising opportunity to tackle the energy conservation issues associated with water delivery systems. In addition to the graphical user interface, the smart water system can benefit greatly from the usage of solar cells, piezoelectric harvesting, electromagnetic harvesting, and thermoelectric harvesting. X-ray photoelectron spectroscopy (XPS) analysis and additional research into micro-electromechanical systems (MEMS), such as lead zirconate- titanate (PZT) films, piezoelectric nano wires, multi-parameter sensors, and iridium oxide films, may aid in better water quality management.

The prioritization of network and device security is of paramount significance, and this procedure commences with the identification and recognition of assets, afterwards succeeded by the formulation of a complete security strategy. The failure to recognize a single overlooked device has the capacity to make the entire network vulnerable to security threats. Hence, it is imperative to classify both obsolete and current equipment into distinct categories based on their presence or necessity. The incorporation of network segmentation into discrete zones facilitates the mitigation of assaults across the entire infrastructure. Threat detection techniques and real-time warning systems are deployed with the purpose of promptly notifying IT personnel, hence facilitating their prompt response and mitigation of recognized risks. The concept of edge data processing entails the execution of data processing tasks in immediate vicinity to the source of the data, as opposed to depending exclusively on centralized computer infrastructures. The integration of developing (IoT) devices into the current infrastructure will lead to a significant increase in data that requires both gathering and processing. The process of transferring data to cloud servers can pose notable difficulties and result in considerable costs, so hindering the timely processing and dissemination of critical data required for swift mitigation of potential risks. To maximize cost-effectiveness and improve the efficiency of data processing, it is recommended to incorporate intelligence throughout the data collecting phase. The topic under investigation is popularly known as edge computing, which is a conceptual framework facilitating various automated procedures within a system. Utilities possess the capability to augment their operating capabilities by leveraging real-time data analysis and programming of devices or infrastructure. This allows individuals to effectively implement established policies. This facilitates individuals to participate in particular activities in accordance with the data that has been generated. The incorporation of remote infrastructure and devices plays a vital role in facilitating the execution of preventive maintenance protocols, hence enabling the prompt identification of problems and timely intervention. The application of this particular automation methodology holds significant potential in offering a diverse array of benefits to utilities operating within expansive territory. Furthermore, it enables the acquisition of extensive knowledge to bolster continuous improvements in infrastructure and the extension of infrastructure into previously uncharted areas. The inadequate implementation of edge applications has the potential to lead to disorder and unpredictable consequences. The remote monitoring of device performance, timely distribution of updates, and execution of necessary patches are routine operational procedures that may provide potential challenges and burdens. One conceivable obstacle that utility companies could face due to the inappropriate implementation of edge computing is the requirement for local decision-making to be carried out at the central control center, leading to a delay in addressing local concerns. In the context of data minimization, the process of transmitting large amounts of data across a communication network carries the inherent potential of exceeding the system's capacity or incurring unnecessary costs. The successful resolution of storage difficulties faced by wireless sensor networks has been achieved by the merging of Internet of Things (IoT) and cloud computing. Various techniques were utilized to enhance the efficiency of storing and processing sensor data in cloud computing systems. The strategies stated above encompassed the utilization of secure protocols, continuous monitoring in real-time, and the implementation of cluster technologies. The transmission of sensor data to cloud-based systems can be accomplished by employing advanced algorithms and protocols, ensuring the reliability of the network. The effective incorporation of a novel network necessitates the implementation of strategies to guarantee adherence to established norms and enforcement procedures. The implementation of a strategy that places emphasis on compliance and proactive approaches would yield a more thorough evaluation of the team's abilities, along with the recognition of any shortcomings and the requisite steps to rectify them. The pragmatic approach involves the implementation of several strategies such as asset identification, network partitioning, threat detection and mitigation, and the integration of security measures for operational technologies in the field of information technology. When incorporating edge computing into a system, it is imperative to deliberately implement it in areas where localized decision-making is necessary, regardless of the presence of connectivity to the central control center or the need for substantial data reduction. The potential for improved communication between a central hub and a remote site can be realized through the use of peripheral data reduction techniques. The achievement of cost reduction and the exploitation of other data sources can be realized using the same networks.

Sensors with the capability to identify device faults or water floods in specific areas have the capacity to autonomously divert excess supply within the network. In addition, these sensors possess the capability to detect deviations from the pre-established water level, thereby alerting designated employees to promptly attend to any emerging concerns. By keeping tabs on and relaying real-time threshold and location data for each team member, the Internet of Things (IoT) enhances safety protocols in the working environment. The potential advantages of incorporating Internet of Things (IoT) technology into utility services include the ability to promptly identify and alert providers about potential risks arising from the introduction of contaminants or chemicals into water sources due to infrastructure degradation.

FUTURE SCOPE

In the context of potential future system advancements, it will be crucial to undertake the process of retraining the deep neural networks responsible for supervising the intelligent sensors. The acquisition of new skills and knowledge will be crucial in order to effectively adjust to the probable integration of additional variables. The salient characteristics of this phenomenon reside in its inherent unpredictability and inherent challenges with regards to enhancement. When applied to real-time data-dependent smart or IoT devices, the re-training method presents substantial hurdles. The fundamental cause of this problem is the unpredictability and lack of contextual comprehension displayed by neural networks after training. Training's usefulness: Deep neural networks are used in real-time systems so that they can quickly adjust to new circumstances or improvements. However, it is important to acknowledge that the enhanced approaches do not provide a conclusive guarantee of achieving the same level of accuracy or optimization as the current procedure in the same context. This is due to the inherent uncertainties associated with the deep learning network. The alterations applied to the sensors possess the capability to elicit modifications inside the neural network, hence potentially impacting the behavior of the current model. Hence, it is imperative that deep neural networks undergo comprehensive training to achieve maximum performance and gain the requisite level of accuracy, particularly when substantial adjustments are made to the overall system. In the coming years, forthcoming systems are expected to demonstrate characteristics of automation and construction methodologies that use intelligent technology rooted in artificial intelligence (AI). In order to effectively include these technologies as viable solutions from both a commercial and technological standpoint, it is imperative to develop systems that possess a substantial degree of context-awareness. In order to achieve successful implementation of these systems, it is imperative to establish rigorous standards for data quality and provide extensive training on quality assurance. Systems that demonstrate both technological and economic viability must possess the capability to effectively address the aforementioned challenges within a real-time environment.

CONCLUSION

The agricultural industry holds significant importance in the Indian economy. The (IoT) is of paramount importance in augmenting the effectiveness of agricultural techniques and raising the overall welfare of farmers. To optimize the efficiency and caliber of decision-making procedures in the agricultural domain, it is crucial to furnish farmers with instructional materials that include visual cues. The evaluation of data points can be conducted by utilizing a range of interconnected devices.

There exists a prevailing knowledge deficit among Indian farmers pertaining to soil factors and the optimal selection of crops based on specific soil features. Increased crop yields can be achieved through the use of IoT devices in agriculture. These gadgets exemplify the capacity to effectively collect and assess crucial data pertaining to soil composition, fertilizer needs, and water levels that are of importance to agricultural areas. The application of visual alerts in the native language of agricultural workers showcases its notable efficacy in meteorological prediction and theft deterrence, thereby providing an extra benefit.

This literature review proposes the use of reinforcement learning (RL) as a viable method to tackle the multifaceted research related to integrated water resources management (WRM), especially in light of the complex interplay between water resources and their socio-economic and environmental aspects, as well as the challenges posed by water crises and resource distribution. It is anticipated that reinforcement learning agents will direct their attention towards this field in the near future, with the goals of improving our understanding of interconnected hydrological systems, crafting more comprehensive policies, and settling on more efficient management structures.

The increasing adoption of intelligent control and monitoring systems in the water resource management (WRM) field can be ascribed to the progress made in machine learning (ML) approaches.

REFERENCES

Adnan, R. M., Mostafa, R. R., Elbeltagi, A., Yaseen, Z. M., Shahid, S., & Kisi, O. (2021). Development of new machine learning model for streamflow prediction: case studies in Pakistan. In Stochastic Environmental Research and Risk Assessment (Vol. 36, Issue 4, pp. 999–1033). Springer Science and Business Media LLC. doi:10.100700477-021-02111-z

Adnan, R. M. R., Mostafa, R., Kisi, O., Yaseen, Z. M., Shahid, S., & Zounemat-Kermani, M. (2021). Improving streamflow prediction using a new hybrid ELM model combined with hybrid particle swarm optimization and grey wolf optimization. In *Knowledge-Based Systems* (Vol. 230, p. 107379). Elsevier BV. doi:10.1016/j.knosys.2021.107379

Ahanger, T. A., & Aljumah, A. (2019). Internet of Things: A Comprehensive Study of Security Issues and Defense Mechanisms. In IEEE Access (Vol. 7, pp. 11020–11028). Institute of Electrical and Electronics Engineers (IEEE). doi:10.1109/ACCESS.2018.2876939

Apaydin, H., Taghi Sattari, M., Falsafian, K., & Prasad, R. (2021). Artificial intelligence modelling integrated with Singular Spectral analysis and Seasonal-Trend decomposition using Loess approaches for streamflow predictions. In *Journal of Hydrology* (Vol. 600, p. 126506). Elsevier BV. doi:10.1016/j.jhydrol.2021.126506

Atlam, H., Walters, R., & Wills, G. (2018). Fog Computing and the Internet of Things: A Review. In Big Data and Cognitive Computing (Vol. 2, Issue 2, p. 10). MDPI AG. doi:10.3390/bdcc2020010

Atzori, L., Iera, A., & Morabito, G. (2010). The Internet of Things: A survey. In Computer Networks (Vol. 54, Issue 15, pp. 2787–2805). Elsevier BV. doi:10.1016/j.comnet.2010.05.010

Brewster, C., Roussaki, I., Kalatzis, N., Doolin, K., & Ellis, K. (2017). IoT in Agriculture: Designing a Europe-Wide Large-Scale Pilot. In IEEE Communications Magazine (Vol. 55, Issue 9, pp. 26–33). Institute of Electrical and Electronics Engineers (IEEE). doi:10.1109/MCOM.2017.1600528

Cheng, B., Solmaz, G., Cirillo, F., Kovacs, E., Terasawa, K., & Kitazawa, A. (2018). Easy Programming of IoT Services Over Cloud and Edges for Smart Cities. In IEEE Internet of Things Journal (Vol. 5, Issue 2, pp. 696–707). Institute of Electrical and Electronics Engineers (IEEE). doi:10.1109/JIOT.2017.2747214

Cho, K., & Kim, Y. (2022). Improving streamflow prediction in the WRF-Hydro model with LSTM networks. In *Journal of Hydrology* (Vol. 605, p. 127297). Elsevier BV. doi:10.1016/j.jhydrol.2021.127297

Drakaki, K.-K., Sakki, G.-K., Tsoukalas, I., Kossieris, P., & Efstratiadis, A. (2022). Day-ahead energy production in small hydropower plants: uncertainty-aware forecasts through effective coupling of knowledge and data. In *Advances in Geosciences* (Vol. 56, pp. 155–162). Copernicus GmbH. doi:10.5194/adgeo-56-155-2022

Feng, Z., Shi, P., Yang, T., Niu, W., Zhou, J., & Cheng, C. (2022). Parallel cooperation search algorithm and artificial intelligence method for streamflow time series forecasting. In *Journal of Hydrology* (Vol. 606, p. 127434). Elsevier BV. doi:10.1016/j.jhydrol.2022.127434

Ghobadi, F., & Kang, D. (2023). Application of Machine Learning in Water Resources Management: A Systematic Literature Review. In Water (Vol. 15, Issue 4, p. 620). MDPI AG. doi:10.3390/w15040620

Ikram, R. M. A., Ewees, A. A., Parmar, K. S., Yaseen, Z. M., Shahid, S., & Kisi, O. (2022). The viability of extended marine predators algorithm-based artificial neural networks for streamflow prediction. In *Applied Soft Computing* (Vol. 131, p. 109739). Elsevier BV. doi:10.1016/j.asoc.2022.109739

Kamienski, C., Soininen, J.-P., Taumberger, M., Dantas, R., Toscano, A., Salmon Cinotti, T., Filev Maia, R., & Torre Neto, A. (2019). Smart Water Management Platform: IoT-Based Precision Irrigation for Agriculture. In Sensors (Vol. 19, Issue 2, p. 276). MDPI AG. doi:10.339019020276

Khosravi, K., Golkarian, A., & Tiefenbacher, J. P. (2022). Using Optimized Deep Learning to Predict Daily Streamflow: A Comparison to Common Machine Learning Algorithms. In Water Resources Management (Vol. 36, Issue 2, pp. 699–716). Springer Science and Business Media LLC. doi:10.100711269-021-03051-7

Kolenčík, M., & Šimanský, V. (2016). Application of various methodological approaches for assessment of soil micromorphology due to VESTA program applicable to prediction of the soil structures formation. *Acta Fytotechnica et Zootechnica*, *19*(2), 68–73. doi:10.15414/afz.2016.19.02.68-73

Liu, J., Yuan, X., Zeng, J., Jiao, Y., Li, Y., Zhong, L., & Yao, L. (2022). Ensemble streamflow forecasting over a cascade reservoir catchment with integrated hydrometeorological modeling and machine learning. In Hydrology and Earth System Sciences (Vol. 26, Issue 2, pp. 265–278). Copernicus GmbH. doi:10.5194/hess-26-265-2022

Liu, Y., Hou, G., Huang, F., Qin, H., Wang, B., & Yi, L. (2022). Directed graph deep neural network for multi-step daily streamflow forecasting. In *Journal of Hydrology* (Vol. 607, p. 127515). Elsevier BV. doi:10.1016/j.jhydrol.2022.127515

López-Riquelme, J. A., Pavón-Pulido, N., Navarro-Hellín, H., Soto-Valles, F., & Torres-Sánchez, R. (2017). A software architecture based on FIWARE cloud for Precision Agriculture. In *Agricultural Water Management* (Vol. 183, pp. 123–135). Elsevier BV. doi:10.1016/j.agwat.2016.10.020

Ma, T., Sun, S., Fu, G., Hall, J. W., Ni, Y., He, L., Yi, J., Zhao, N., Du, Y., Pei, T., Cheng, W., Song, C., Fang, C., & Zhou, C. (2020). Pollution exacerbates China's water scarcity and its regional inequality. *Nature Communications*, *11*(1), 650. doi:10.103841467-020-14532-5 PMID:32005847

Morabito, R., Kjallman, J., & Komu, M. (2015). Hypervisors vs. Lightweight Virtualization: A Performance Comparison. In *2015 IEEE International Conference on Cloud Engineering. 2015 IEEE International Conference on Cloud Engineering (IC2E).* IEEE. 10.1109/IC2E.2015.74

Nagajayanthi, B. (2021). Decades of Internet of Things Towards Twenty-first Century: A Research-Based Introspective. In Wireless Personal Communications (Vol. 123, Issue 4, pp. 3661–3697). Springer Science and Business Media LLC. doi:10.100711277-021-09308-z

Nguyen, D. H., Le, X. H., Anh, D. T., Kim, S.-H., & Bae, D.-H. (2022). Hourly streamflow forecasting using a Bayesian additive regression tree model hybridized with a genetic algorithm. In *Journal of Hydrology* (Vol. 606, p. 127445). Elsevier BV. doi:10.1016/j.jhydrol.2022.127445

Park, K., Jung, Y., Seong, Y., & Lee, S. (2022). Development of Deep Learning Models to Improve the Accuracy of Water Levels Time Series Prediction through Multivariate Hydrological Data. In Water (Vol. 14, Issue 3, p. 469). MDPI AG. doi:10.3390/w14030469

Peng, L., Wu, H., Gao, M., Yi, H., Xiong, Q., Yang, L., & Cheng, S. (2022). TLT: Recurrent fine-tuning transfer learning for water quality long-term prediction. In Water Research (Vol. 225, p. 119171). Elsevier BV. doi:10.1016/j.watres.2022.119171

Roffia, L., Azzoni, P., Aguzzi, C., Viola, F., Antoniazzi, F., & Salmon Cinotti, T. (2018). Dynamic Linked Data: A SPARQL Event Processing Architecture. In Future Internet (Vol. 10, Issue 4, p. 36). MDPI AG. doi:10.3390/fi10040036

Rosa, L., Chiarelli, D. D., Rulli, M. C., Dell'Angelo, J., & D'Odorico, P. (2020). Global agricultural economic water scarcity. In Science Advances (Vol. 6, Issue 18). American Association for the Advancement of Science (AAAS). doi:10.1126ciadv.aaz6031

Sattar Hanoon, M., Najah Ahmed, A., Razzaq, A., Oudah, A. Y., Alkhayyat, A., Huang, F. Y., Kumar, P., & El-Shafie, A. (2023). Prediction of hydropower generation via machine learning algorithms at three Gorges Dam, China. In Ain Shams Engineering Journal (Vol. 14, Issue 4, p. 101919). Elsevier BV. doi:10.1016/j.asej.2022.101919

Sharghi, E., Nourani, V., Zhang, Y., & Ghaneei, P. (2022). Conjunction of cluster ensemble-model ensemble techniques for spatiotemporal assessment of groundwater depletion in semi-arid plains. In *Journal of Hydrology* (Vol. 610, p. 127984). Elsevier BV. doi:10.1016/j.jhydrol.2022.127984

Sun, J., Hu, L., Li, D., Sun, K., & Yang, Z. (2022). Data-driven models for accurate groundwater level prediction and their practical significance in groundwater management. In *Journal of Hydrology* (Vol. 608, p. 127630). Elsevier BV. doi:10.1016/j.jhydrol.2022.127630

Ullah, R., Abbas, A. W., Ullah, M., Khan, R. U., Khan, I. U., Aslam, N., & Aljameel, S. S. (2021). EEWMP: An IoT-Based Energy-Efficient Water Management Platform for Smart Irrigation. In S. Nazir (Ed.), *Scientific Programming* (Vol. 2021, pp. 1–9). Hindawi Limited. doi:10.1155/2021/5536884

Vallino, E., Ridolfi, L., & Laio, F. (2020). Measuring economic water scarcity in agriculture: a cross-country empirical investigation. In Environmental Science & Policy (Vol. 114, pp. 73–85). Elsevier BV. doi:10.1016/j.envsci.2020.07.017

Chapter 6
Utilizing Machine Learning for Enhanced Weather Forecasting and Sustainable Water Resource Management

Risha Dhargalkar
Government College Sanquelim, Goa University, India

Viosha Cruz
Carmel College of Arts Science and Commerce, India

Abdullah Alzahrani
Oakland University, USA

ABSTRACT

Every phase of human life is influenced by nature; therefore, weather forecasting and water management are challenging tasks as they work according to environmental changes. The traditional weather forecasting model was done using historical data in a physics model, which leads to unsteady results. With machine learning and artificial intelligence advancement, weather forecasting and water management have undergone revolutions to predict future data analysis. This chapter provides an overview of essential weather forecasting attributes and different data acquisition and preprocessing elements in water management. The chapter's subsequent sections detail the many stages needed for weather forecasting and the various machine-learning algorithms that may be used to forecast weather conditions by recognizing patterns and then analyzing them. In addition to this, the chapter also highlights applications of water resource management. Since water is a vital resource, automation and controlling allocation and distribution are crucial tasks, which are also outlined.

DOI: 10.4018/979-8-3693-1194-3.ch006

INTRODUCTION

Climate change is the global phenomenon that is most talked about in this generation. It is known to cause natural calamities like storms, cyclones, and floods. From this fact stems the need to have accurate systems to help us tackle these issues. The advancement of technology can provide real solutions to such problems. This chapter focuses on two significant issues. The first is weather forecasting, and the second is water management. Both these topics are closely related since they deal with monitoring and using natural resources.

As weather conditions keep changing, predicting weather forecasts around the globe is very important. According to the study conducted by Salcedo-Sanz et al. (2023), traditional numerical weather prediction methods involve using historical data in a physics model that simulates the atmosphere's behavior based on thermodynamics and fluid dynamics principles, leading to unsteady results. These models use a system of partial differential equations to represent the physical processes governing the atmosphere's motion and thermodynamics, and they aim to predict how the atmosphere will evolve. However, these systems, which include partial differential equations for a given physical model that are unstable and incurred uncertainties in the initial measurements of the atmospheric conditions, restricted accurate weather forecasting only for a few days, after which weather forecasts became significantly unreliable. To get beyond the limitations above and enhance forecasting for precise results, meteorologists have combined Machine Learning (ML) algorithms with traditional methodologies that use historical data to train models, and then these models will be used to anticipate future data analysis.

However, the current generation's most pressing problem is the unprecedented pressure on water resources. Furthermore, unabating water scarcity and extreme weather conditions like floods and dry spells are perceived as the biggest threats to global prosperity and stability. There is an increased level of awareness in acknowledging the fact that there is a need to preserve and use water resources more efficiently to reduce wastage and avoid drought situations. Several researchers have predicted that the world is bound to face extreme water scarcity. A study conducted by Parwal (2015) discusses the water management issues faced in India. The author emphasizes the high demand for water sources for agriculture, industries, cooling, and extraction. The author also sheds light on the outcome of various reports published by world-renowned bodies that mention the country's dire state of water resources. It is, therefore, imperative to maintain the integrity of the planet's natural reserves.

Additionally, a report published by ADRI (India Water Facts, 2017) highlights the major concerns faced by India about water management. It states that the country faces a problem of water depletion. It mentions that rainfall, occurring annually, is a significant source of replenishment for water sources, contributing about 58% of the total replenishment sources. It also states that 70% of the water sources are polluted with contaminants like arsenic and fluoride.

Furthermore, in another study conducted by Cosgrove and Loucks (2015), the paper discusses the challenges and possible future directions for effective water management. It may seem at times that a particular area has an adequate supply of water; however, the spatial and temporal distribution of the water resource is inadequate, and the regions located away from the source will most definitely face challenges in getting access to the water and also in distributing it efficiently. The authors also elaborate on the failure of traditional methods, as previously erroneous estimates and incomplete studies led to problems like disruption and overallocation of resources. Overuse of resources has also been the cause of pollution of water sources. The tremendous increase in population has led to high per capita water consumption. With the population still on the rise, the water sources are also bound to increase consumption. The study also mentions some of the possible solutions to combat the problem of water source depletion by stressing that the entire methodology and production techniques will have to be changed to see any improvement in water source sustainability. The latest technologies used for the same include artificial intelligence, cybernetics, nanotechnology, and biotechnology.

The paper discusses the use of computer-based optimization and simulation models. Additionally, decision support models can provide an excellent solution for efficient decision-making on subjects like reallocating water resources for better utility.

Targeting the issue of water management has to be done holistically. More is needed to suggest improvements in technology to solve this problem. There is a need to focus on the implementation of suggested technology-based improvisations of existing systems. Additionally, it is of critical importance to keep in mind that for the technological solutions to be effective, intensive monitoring of the areas of interest must be done. Government support and backing are crucial to making such strategies work. As mentioned in a study C. P. Kumar (2018), schemes such as "Har Khet Ko Pani" have aided in managing water resource usage for agriculture. The concerned government authorities need to monitor the contamination of resources; for example, the river Ganga is the most utilized water source in that region of the country, and this has caused heavy contamination of the river. The author suggests integrated assessments of surface water and groundwater to help authorities develop effective plans. The paper also mentions model-assisted planning of groundwater development as one of the methods to make effective decisions on the redistribution of water. Another possible method is using predictive numerical software solutions for recharge estimation. Such techniques have proven to be effective in water management. This research focuses on using ML methodologies to provide feasible and proficient solutions to the water management problem. This would include solutions to problems like agriculture during dry seasons, drought and flood management, and water reallocation. The following section discusses studies that effectively use ML to manage and sustain water resources.

LITERATURE REVIEW

The ability of ML to analyze big datasets, find patterns, and make predictions based on past data has made it a crucial tool in weather forecasting. Some of the ways ML can be applied in weather forecasting are weather prediction, weather pattern recognition, data assimilation, extreme weather events, climate modeling, and nowcasting. Many researchers have conducted research in the above domains, and a few of them are highlighted below.

Singh, Kaushik, Gupta, and Malviya (2019) proposed techniques to study and analyze three alternative ML models for weather prediction: Support Vector Machines (SVM), Artificial Neural Networks (ANN), and time series-based Recurrent Neural Networks (RNN). Based on the root mean squared error between predicted and actual values, these models are compared. The information is gathered from the airport weather stations of various cities, considering various weather-related factors, including air temperature, atmospheric pressure, relative humidity, wind direction, and cloud cover. The tools used for data analysis are Pandas,

Keras, NumPy, Git, Matplotlib, Tensorflow, and Anaconda. It has been observed that time-series-based RNN outperforms any other algorithm. According to Lakshmi, Ajimunnisa, Prasanna, Yugasravani, and raviteja (2021), a model for forecasting the weather has been presented in which data is gathered and examined using a back propagation neural network, which lowers the error rate by considering many qualities and producing precise results.

Regression and classification neural network designs are used with frameworks like Tensorflow and Keras to enhance algorithm performance. According to Jakaria, Tennessee, and Rahman (2020), ML models can be built by mixing historical data from adjacent cities with data from a specific city to anticipate its weather for the upcoming few days.

The case study is based on the American city of Nashville, Tennessee, where the weather is constantly changing. All weather observations from all cities at a specific timestamp are compiled to form a record. Each record includes several characteristics for all the cities, such as temperature, humidity, wind direction, atmospheric pressure, condition, etc. Temperature is the goal variable at the same timestamp as the following day. The proposed model showed that a simple model can provide a reliable weather forecast for this city.

Different machine methods have been described in J. S. Kumar (2022) discussion on weather forecast prediction. Historical data is used as the input to algorithms that train a model to predict the weather more accurately than the standard physics model. The accuracy of various algorithms is determined after they have been trained on data, and this information is then used to demonstrate the algorithm's performance. A ML-based weather forecasting model was proposed by Oshodi (2022) using four classifier algorithms that were trained using a dataset from Kaggle and whose performance was evaluated, proving that Gaussian Naive Bayes outperformed the remaining algorithms. Raut (2021) studied different ML and Deep Learning (DL) algorithms by considering different features such as temperature, humidity, rainfall, air pressure, and so on, and it was proved that hybrid ML and DL strategies can give the best accuracy.

ML IN WEATHER FORECASTING PARAMETERS OF WEATHER FORECAST

- Predicting the future state of the atmosphere based on present and past meteorological data involves various vital attributes. A few of them are listed below.
- Temperature includes different atmospheric aspects such as air pressure, humidity, and cloud formation. Since temperatures constantly change, monitoring them for weather forecasts to get accurate results is very important.
- Air Pressure, also known as atmospheric pressure, it indicates how much of the Earth's surface is being pressed by air to indicate changes in weather systems.
- Humidity refers to the amount of moisture in the air. High levels of humidity lead to the formation of clouds and precipitation, while low levels result in dry and clear conditions.
- The movement of weather systems that have the potential for cyclones and storms can be determined by meteorologists by understanding speed and wind direction.
- Precipitation includes various forms of water that fall from the atmosphere, such as rain, snow, sleet, and hail. Predicting the type and amount of precipitation is vital for public safety, transportation, and agriculture.

Table 1. Mumbai weather dataset

Date Time	Temp	Dew	Humidity	Sea Level Pressure	Winder	Solar Radiation	Windspeed
01-01-	28.4	11.9	37.8	10164	147.5	216.1	16.6
2016							
02-01-	26.8	13	44.8	1017.2	110.3	215.7	16.6
2016							
03-01-	25.5	14.6	52.8	1015.7	145.3	221.1	18.4
2016							
04-01-	26.4	13.1	46.6	1015.3	126.9	216.2	16.6
2016							
05-01-	27.1	13.5	44.9	1014.4	125.5	208.1	14.8
2016							
06-01-	26.9	14.3	48.2	1015.1	110.2	200.8	15.8
2016							
07-01-	26.1	17	58	1015.4	149.4	189.9	13
2016							
08-01-	26.6	16.5	55.2	1013.8	157.7	185	16.6
2016							

Dataset for Weather Forecasting

Various sources of datasets are available for weather forecasting that consider parameters like temperature, pressure, humidity, rainfall, etc. One example of a dataset from Kaggle is taken in this chapter to better understand how a dataset looks and works.

Table 1 shows an example of the Mumbai Weather Dataset. The dataset is downloaded from the Kaggel website and is available in CSV format. Kaggel is the most significant ML and data science community, which consists of numerous datasets.

Table 1 illustrates the dataset which contains day-by-day data and weather variables that can be used as predictors in models for predicting rainfall, such as temperature, dewpoint, humidity, wind speed, and air pressure.

Process of Weather Forecasting

The following steps are considered to predict future weather conditions using a machine-learning approach.

- The weather prediction dataset is taken as an input, which consists of different parameters like temperature, pressure, humidity, etc.
- Preprocessing and cleaning of data are done to remove unwanted data
- Weather prediction attributes are selected among multiple attributes or parameters available in the dataset.

- Then, input and output label pairs are identified from multiple available attributes, including input parameters taken from the dataset and output parameters for attributes you want to predict.
- The ML model is trained using recognized pairs using techniques that divide the dataset into a training and testing model, with 90% of the data being used for the trained model and the remaining 10% for accuracy testing.
- ML models with algorithms are tested, and the model is hosted or launched
- Once the model is launched, the prediction and visualization of the data are one Figure 1 describes an overview of the ML process for Weather Prediction.

Figure 1. Overview of the ML process for weather prediction

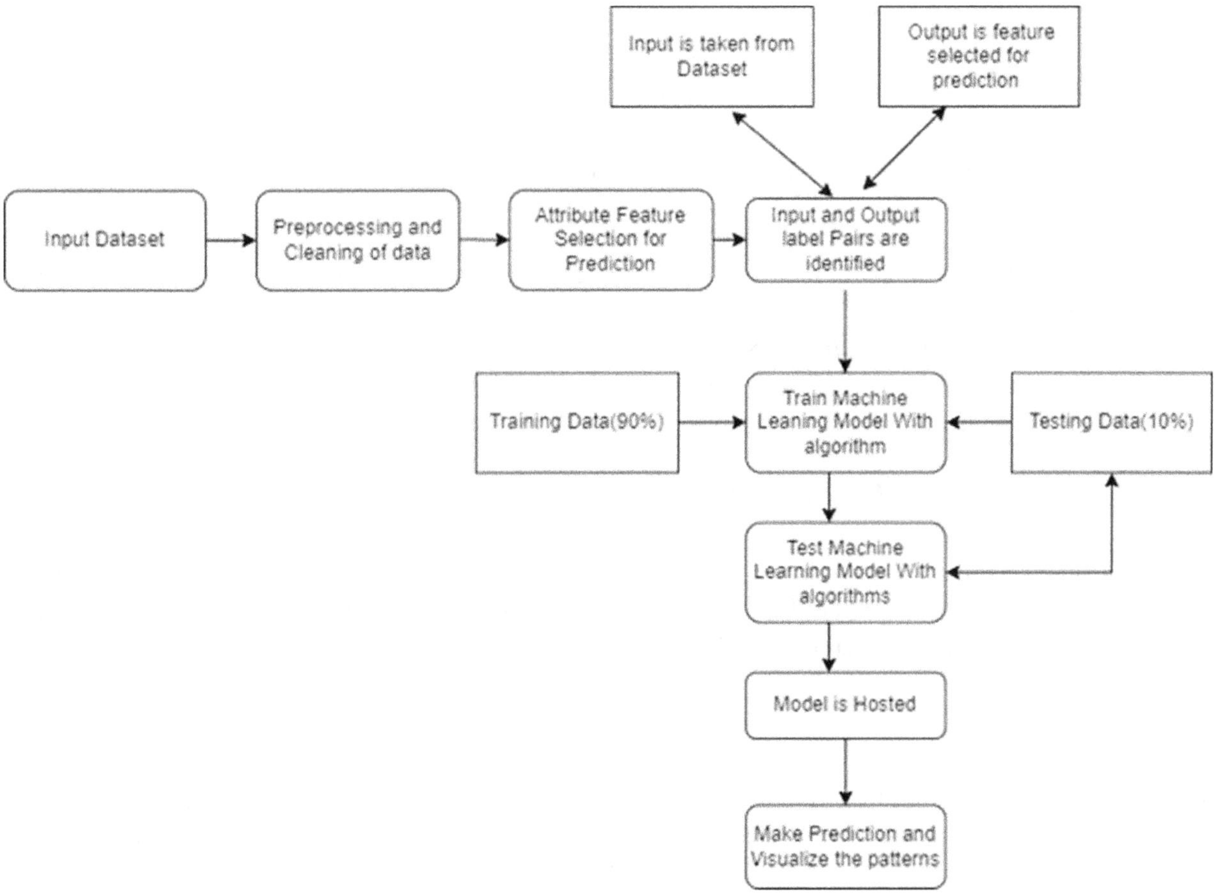

Applications of Weather Forecasting

Multiple weather forecasting applications are available across various fields; some are listed below:

- ***Agriculture:*** A weather forecast helps farmers decide when to plan for plantation, irrigation, and harvesting. This helps them optimize crop management and minimize losses.

- ***Transportation:*** To plan routes, avoid bad weather, and protect the safety of passengers and cargo, airlines, shipping firms, and other transportation providers employ weather forecasts.
- ***Emergency Management:*** Predicting and planning for natural catastrophes like hurricanes, tornadoes, floods, and wildfires requires using weather predictions. Authorities can better allocate resources and evacuate high-risk locations using this information.
- ***Energy Industry:*** Weather conditions impact energy generation and delivery, particularly for renewable sources like wind and solar. Utility companies use accurate projections to balance supply and demand and schedule repairs.
- ***Retail:*** Businesses in the retail industry, particularly those that offer seasonal goods like apparel, rely on weather forecasts to predict demand trends and modify their inventory accordingly.
- ***Climate Research:*** Weather forecasts are used by climate scientists to comprehend short-term variations in climate patterns, which can yield helpful information for long-term climate studies.
- ***Health and Safety:*** Certain weather conditions, such as intense heat or cold, may impact public health and safety. Forecasts aid in preparing medical facilities, emergency personnel, and local governments.
- ***Air Quality Monitoring:*** Air quality can be impacted by weather. The ability to monitor and control air pollution, essential for public health, is made possible by accurate forecasts.
- ***Renewable Energy Planning:*** Weather pattern forecasting aids utilities in maximizing the use of renewable energy sources while minimizing dependency on fossil fuels, particularly wind and solar energy.
- ***Tourism and Outdoor Activities:*** Travelers must consider the weather when organizing outdoor activities. Knowing the weather in advance is essential for any activity, whether a day at the beach, a hike, or a sporting event.

Different Techniques Used for Weather Forecasting

A complicated endeavor, weather forecasting uses meteorological data, physical models, and computing methods. The accuracy of weather forecasts has recently been improved through ML. Here are several popular ML algorithms and methods for forecasting the weather:

Regression Models

By examining historical weather data and finding patterns that can be utilized to forecast future weather conditions, regression models can be helpful for weather forecasting. Depending on the unique requirements and characteristics of the meteorological data, many regression models can be used for this purpose. Several typical regression models for weather forecasting are listed below:

- ***Linear Regression:*** This is the simplest form of regression method where goal variables(for example, the temperature the following day) and input features (such as temperature, humidity, pressure, etc.) are assumed to have a linear relationship. Although simple to interpret, linear regression may miss complicated nonlinear correlations in the data.
- ***Multiple Linear Regression:*** This is a multiple-input feature-supported extension of linear regression. When predicting future weather conditions, you may consider characteristics like historical temperature, pressure, humidity, wind speed, etc.

- ***Polynomial Regression:*** Regression of this kind uses polynomial functions to capture nonlinear relationships. It might be more appropriate when the relationship between the variables is curvilinear.

Time Series Analysis

Time series analysis is a powerful technique for forecasting many data types, including weather data. Predicting future weather patterns using data from the past and a variety of meteorological variables is the aim of weather forecasting. An outline of how time series analysis might be used to improve weather forecasting is given below:

- ***Data Collection:*** Obtain historical weather data, which frequently contains information on temperature, humidity, wind speed, precipitation, air pressure, and more. The data should cover a sufficiently long period to allow for the identification of seasonal and long-term trends.
- ***Data Preprocessing:*** Clean the data by handling missing numbers, outliers, and conflicts. Transform the data into a time series format with a consistent time interval (hourly or daily) for analysis.
- ***Exploratory Data Analysis (EDA):*** Analyze the data initially to determine its patterns, trends, and seasonality. Line plots, scatter plots, and autocorrelation plots are a few visualization approaches that can be useful.
- ***Time Series Decomposition:*** The time series should be broken down into its constituent parts, which are usually the trend, seasonality, and residual (random) components. The underlying patterns are easier to understand because of this dissection.
- ***Model Selection:*** Select a suitable time series forecasting model depending on the data's features. In addition to more complex techniques like Prophet or DL models (for example, LSTM for sequence prediction), standard models include ARIMA (AutoRegressive et al.), SARIMA (Seasonal ARIMA), and others.
- ***Model Training:*** Create training and validation sets from the data. The chosen model should be trained using the training data, its parameters adjusted, and its performance verified using the validation data.
- ***Model Evaluation:*** Assess the model's performance using the appropriate metrics, such as mean absolute error (MAE), root mean squared error (RMSE), or mean absolute percentage error (MAPE).
- ***Forecasting:*** Use the model to predict future weather after it has been trained and assessed. The model can produce point forecasts (single values) or probabilistic forecasts (ranges of values) depending on the methodology chosen.
- ***Ensemble Methods:*** To improve overall forecast accuracy, random forests, and gradient boosting are employed to aggregate the predictions of many models.

Random Forest

The often employed ML method. The weather may be predicted using a random forest. It uses an ensemble learning technique to integrate the output of various decision trees to increase the precision of each prediction. Random Forest is a versatile solution for various weather forecasting issues because it can be used for classification (predicting categories or classes) and regression (predicting continuous values). Here is how to use Random Forest to predict the weather:

- ***Data Preparation:*** As previously mentioned, gather historical weather data, including features (input variables) and the goal variable you wish to predict (such as temperature, precipitation, etc.).
- ***Data Preprocessing:*** Data cleaning and preparation, management of missing values, scaling of features, and sometimes engineering of new features
- ***Feature Selection/Engineering:*** Identify the feature that can be used to identify weather predictions
- ***Model Selection:*** Select Random Forest as the algorithm for your weather prediction task. Random Forest is renowned for managing intricate data interactions and minimizing overfitting.
- ***Data Splitting:*** You produce training and testing (or validation) sets from your data. As a result, you can use one subset of the data to train the model and another unused subset to test the model's performance.
- ***Model Training:*** Utilize the training data to train the Random Forest model. Using bootstrapping, the algorithm creates numerous decision trees on various subsets of the data and then combines their predictions to get the final forecast.
- ***Model Evaluation:*** Use the appropriate metrics to evaluate the model's performance on the testing and validation data, such as mean absolute error (MAE), root mean squared error (RMSE), or others relevant to your specific forecasting task.
- ***Hyperparameter Tuning:*** In Random Forest, various hyperparameters can be changed to improve the model's performance. Methods like a grid or randomized search can be used to locate the ideal collection of hyperparameters.
- ***Forecasting:*** Once the model has been trained and evaluated, you may use it to give future weather forecasts by providing the required input features.

Gradient Boosting

Algorithms that use gradient boosting, such as XGBoost and LightGBM, have produced encouraging results in weather forecasting. They work well with big datasets and can be applied to various weather prediction jobs. The steps for gradient boosting are as follows:

- ***Data Collection:*** Obtain past weather information for the desired area. Temperature, humidity, wind speed, atmospheric pressure, and any other pertinent characteristics should all be included in this data. You also require information on the target variable you want to forecast, such as the upcoming temperature.
- ***Feature Engineering:*** The data should be processed and ready for training. Missing values, encoding categorical variables, and developing lag features (which use historical weather data as predictors) are all included in this. Gradient boosting models work best when features are engineered.
- ***Model Selection:*** Select a method that uses gradient boosting, such as XGBoost, LightGBM, or CatBoost. These packages offer practical gradient-boosting implementations and can manage big datasets.

- ***Training the Model:*** Create training and validation sets from your data. The validation set is used to fine-tune hyperparameters (such as learning rate, maximum tree depth, and number of trees) after the gradient boosting model has been trained on the training set. While enhancing model performance, overfitting should be avoided.
- ***Prediction:*** You can use the model to anticipate future weather conditions once trained and fine-tuned. This entails feeding the model the most recent information available and letting it forecast the important weather variables.
- ***Model Evaluation:*** The model's effectiveness can be evaluated using a separate test dataset or other appropriate metrics, such as the mean squared error (MSE) for temperature prediction or accuracy for categorical weather events (such as rain or no rain).
- ***Continuous Improvement:*** The model should be updated frequently

Support Vector Machines (SVM)

SVMs can be used in weather forecasting for classification tasks such as determining whether it will rain based on various input features. Here is a general outline of how SVM could be used in a weather forecasting context, particularly for specific sub-tasks:

- ***Binary Classification:*** Weather forecasting activities that need binary classification can be completed using SVM. For instance, based on past weather information and other pertinent variables, project whether it will rain (class 1) or not (class 0).
- ***Extreme Weather Event Detection:*** It is possible to identify extreme weather events using SVMs. Detecting days with extremely high or low temperatures, severe winds, or significant fluctuations in air pressure, for instance.
- ***Anomaly Detection:*** SVMs can be used to find weather data abnormalities. Doing so makes it possible to spot odd weather patterns or data points that point to sensor problems or other anomalies.
- ***Feature Selection:*** The forecasting task's most crucial features (variables) can be determined using SVMs as a feature selection method.
- ***Ensemble Methods:*** The ensemble methods that combine the predictions can include SVMs
- ***Neural Networks:*** DL models like convolutional and recurrent neural networks are suitable for weather prediction problems because they can handle massive amounts of meteorological data for pattern recognition.
- ***Cluster A nalysis:*** To classify places with comparable weather conditions for forecasting, clustering algorithms can be used to find similar weather patterns.
- ***Long Short-Term Memory (LSTM):*** LSTM, an RNN variant, excels at processing data sequences, making time-dependent weather data a good fit.
- ***Data Assimilation:*** Observational data and model predictions are combined using Kalman filters and other data assimilation techniques to reduce forecasting errors.
- ***Gaussian Processes:*** These probabilistic models can be applied to weather forecast uncertainty estimation

Notably, the selection of an algorithm is influenced by the particular weather prediction task, the data at hand, and the desired level of accuracy. To get the best results, several weather forecasting models combine these methods. Additionally, the performance of these algorithms is significantly enhanced by domain expertise, feature engineering, and data pretreatment.

ML IN WATER MANAGEMENT

Methodologies Used in Water Management

This section outlines the various methodologies already being used in designing water management systems that perform well in real-life situations. The main water management problem can be divided into further classes: water use in agriculture, flood management, drought management, intelligent urban use, environment-related water management, and groundwater redistribution. This section is focused on some of the critical classes of problems and the key attributes of each category.

Smart Water Management Systems Using ML

Such a system aims to maximize information gain so people can make better decisions.

A study by Jenny et al. Liakos, Busato, Moshou, Pearson, and Bochtis (2018) discusses the principles of artificial intelligence used in the context of water utilities. Artificial intelligence is defined as the simulation of intelligent human-like behavior in machines. The study suggests using a vital parameter called the "unaccounted-for water" parameter. This attribute is used to determine the efficiency of a water source. The gap between the volume of water delivered in a network and legitimate consumption, metered and unmetered, is called unaccounted-for water. Utilities must determine how much water is lost since it affects their profitability because utilities, i.e., users, must pay for unused water.

Furthermore, water is being squandered unnecessarily despite being a precious resource. Traditionally, various methods like water leak detection techniques, acoustic sensors, and gas tracers are used to monitor the possible causes of failures in the water management system. Most systems are designed on the principle of mass and momentum conservation. Recently, human effort in the interpretation of data has been dramatically reduced by using artificially intelligent acoustic correlators. Additionally, parameters like pressure flow and node consumption are inputs to the monitoring systems. ML algorithms can then detect spatial and temporal patterns in the flow and pressure parameters.

Hydraulic modeling is the practice of simulating a pipe network using physical properties and mathematics. The network might be any sort that transports liquid or gas through pipes or open channels. A pressure differential or gravity conveys the medium across the network in a hydraulic model. The study suggests two main approaches described below:

- ***Physically-based methods:*** These methods use hydraulic models for state estimation techniques and pressure sensitivity analysis. Such methods can only function well if a hydraulic model is defined and used.
- ***Data-driven methods:*** These methods use artificial neural networks, support vector machines, classification trees, and other classification techniques. The advantage of such methods over physically based methods is that they can find patterns without relying on network equations that may be too complex to formulate. The following steps have been defined in the process of the digital transformation of a water management system:

- Defining targets and key performance indicators
- Defining and describing a hydraulic model to be able to obtain data from
- Measure and monitor critical parameters
- Make use of leakage and burst detection mechanisms
- Make use of calculated error values as feedback for the improvisation of the ML algorithms used. Optimization algorithms can also be used for the same purpose.

ML in Agriculture

A study conducted by Parwal (2015) provides insight into the role of water management in agriculture. As discussed in the study, water management in agricultural production necessitates substantial effort and plays a vital role in hydrological, climatological, and agronomical balance. The estimation of evapotranspiration is crucial for crop production. Evapotranspiration is the sum of all processes that transport water from the ground surface to the atmosphere through evaporation and transpiration. Evapotranspiration involves water evaporation into the atmosphere from the soil surface, evaporation from the groundwater table's capillary fringe, and evaporation from land-based water bodies. The flow of water from the soil to the atmosphere by plants is also included in evapotranspiration. On the other hand, the daily dew point temperature is also an essential factor in determining anticipated weather conditions. Thus providing help in predicting adverse weather warnings.

Neural networks, also known as artificial neural networks (ANNs) or simulated neural networks (SNNs), are a subset of ML that provides the foundation for DL techniques. They work on the same principle as the human brain. They are designed to work like biological neurons that pass and process perceived information to make inferences used by the brain to make decisions.

Artificial neural networks (ANNs) comprise node layers, each with an input layer, one or more hidden layers, and an output layer. Each node, or artificial neuron, is linked to another node, and each node has an assigned weight and threshold. If the output of any particular node exceeds the given threshold value, that node is activated and begins transferring data to the network's next tier. Otherwise, no data is sent to the next layer in the hierarchy of layers. The data is input into the system, and the network learns from the input data to provide valuable inferences and predict and forecast accurately.

ML for Environment and Water Management

In another study by Sun and Scanlon (2019), the author shares the rapid improvements in IOT, cloud computing, and remote sensing. The latest big data platforms, like Hadoop, are used to work with the extensive collections of geographic data available. Algorithms like the MapReduce algorithm have performed well with substantially large datasets. Apache SPARK processes data in memory and can be much quicker than MapReduce. The developers use the help of programming support libraries like Spark, Mlib, and Mahout, the latter of which focuses on distributed ML, to design effective solutions. To work with massive datasets, PIG and HIVE provide scripting capability. Hive is designed to have a SQL-like interface. The study also suggests that a data standard like WaterML can be used for modeling hydrological time series.

ML for Water Quality Evaluation

Water management deals with not only monitoring the system and redistribution but also checking for water quality. This is important, as the decision to commercially use a water source or not entirely depends on the quality of the water. If the water is contaminated, the source cannot be used for redistribution. In a study by Zhu et al. (2022), water quality and water pollution control are issues that ML can help with. Various methods can measure water quality, including the Water Quality Index (WQI). The individual index values of some or all of the parameters within the five categories of water quality parameters are averaged to create the WQI.

- Water clarity is measured by the depth of light in the water column. Units used are turbidity (NTU*) and Secchi disk depth (meters or feet);
- Dissolved oxygen is the amount of oxygen available to organisms living under the water. The oxygen is available in a dissolved state. Dissolved oxygen concentration (mg/l);
- The number of oxidizable substances in the water measures oxygen demand. These can lower the dissolved oxygen content by getting oxidized.biochemical oxygen demand (mg/l), chemical oxygen demand (mg/l), and/or total organic carbon (mg/l);
- Nutrients: total nitrogen (mg/l and/or total phosphorus (mg/l)
- Bacteria: total coliform (# per mg/l and/or fecal coliform (# per mg/l)

ML has been used as the basis for several applications for wastewater management systems. The image below briefly explains what neural networks are used for different types of water sources. The primary cause of the declining water quality in metropolitan areas is human activity's municipal and industrial wastewater. The authors concluded that because real-world water treatment and management systems might have incredibly complicated conditions, the present methods may only be used with specific systems, which limits the scope of ML methods.

ML for Freshwater Management

It has been observed that water control may affect the aquatic environment, which has resulted in regulatory restrictions on the frequency and amount of freshwater withdrawals becoming increasingly complex. In a research endeavor, Murgatroyd and W Hall (2021) propose a combination of modeling and simulation to suggest and test reductions to regulatory limits on river water withdrawals. The Lotic-Invertebrate Index for Flow Evaluation (LIFE) is related to prior flow statistics observed in the Lee watershed, England, using multilevel linear regression for its computation. The system is modeled by a series of nodes and arcs in the network flow model, WATHNET-5. Multilevel linear regression modeling was used to understand the impact of future changes in flow on macroinvertebrate ecology in the area under study. For each macroinvertebrate sample, seasonal hydrological indicators (HI) were calculated. The indicators considered indicated many aspects of the flow regime, such as severe flows, low flows, high flows, and rates of change in flow, and they were ecologically relevant. Multilevel models were fitted using features from the winter and summer seasons preceding a sample from the summer macroinvertebrate. Furthermore, R packages like Glmulti and lme4 were used for processing data. Their study is an excellent example of the empirically driven multilevel linear modeling used that works by incorporating macroinvertebrate responses in water resource system simulation.

ML for Water Management in Drought-Prone Areas

A research venture by Shihu (2011) investigates the climate change and water resource management difficulties water suppliers face in drought-prone regions, focusing on the American West, where agencies must balance the management of imported and local water resources in the face of many future uncertainties. The authors apply robust decision-making (RDM) to the San Bernardino Valley Municipal Water District's (Valley District's) water management planning. They explore the performance of a machine. Learning-based implementation of the system with the help of random forests, ANNs, and RNNs (recursive neural networks). MODFlow modeling software is used for creating a model that represents the water basin flow and generates a simulation. Due to time and resource constraints, the research team failed to run the MODFLOW model for many iterations, thus restricting the number of input and output datasets needed for training and testing ML algorithms.

Data Acquisition

The data used for water flow modeling is geographic. This kind of data can be obtained in several ways. The two main methods are discussed depending on the nature of the data under consideration.

- ***Passive Data:*** This type of data is obtained from online search terms on social media. Phone GPS systems are also a source that provides such data. Data collected from several online users is also called crowdsourced data. This type of data usually needs to be more structured.
- ***Active Data:*** This type of data is obtained from planned surveys, field campaigns, and remote sensors strategically placed in the field of interest. This type of data is usually structured.
  Some familiar sources of data are listed below:
- ***Data from Earth Observation Systems:*** NASA already has a widely used Earth Observation System (EOS) in place (NASA's Earth Observing System, 2023). NASA's Earth Observing System (EOS) is a system of polar-orbiting and low-inclination satellites designed to provide long-term worldwide observations of the Earth's surface, the biosphere, the atmosphere, and the seas. Data from such a system helps ML algorithms train well and provide accurate results.
- ***Large Scale Datasets:*** Several datasets are available for researchers to work with. The detection of global surface water in very-high-resolution (VHR) satellite data can directly be used in critical applications such as enhanced flood mapping and water resource assessment. Similarly, surface soil moisture (0–5 cm vertical average),
  root-zone soil moisture (0–100 cm vertical average), and additional research applications that compute surface meteorological forcing variables, soil temperature, evapotranspiration, and net radiation are all included in the SMAP Level-4 (L4) Soil Moisture dataset.
- ***Large-scale Earth System Models:*** Earth system models aid in understanding and providing critical information on water supply, drought, climate and temperature extremes, ice sheets and sea levels, and land-use change. For example, the Earth System Modeling Framework (ESMF) is free and open-source software for creating climate, numerical weather prediction, data assimilation, and other related software applications. These are computationally intensive applications that are typically executed on supercomputers.
- ***Multi-sensor Data:*** Several sensors can gather information about flow, pressure, etc. In a study by Shihu (2011), the author discusses using sensors to measure parameters like chlorine level,

turbidity, PH conductivity, and sample temperature. Data of a geographic nature obtained from the above-discussed sources usually has a typical format. The most commonly used formats are as follows:

- GRIB is an acronym for "General Regularly Distributed Information in Binary Form" and is a WMO (World Meteorological Organization) standard format for preserving and distributing gridded data.
- NetCDF is a file format that stores complex structured scientific data (variables) such as temperature, humidity, pressure, wind speed, and direction. Each variable is defined as a dimension (for example, time) in ArcGIS by constructing a layer or table view from the netCDF file.
- The National Center for Supercomputing Applications (NCSA) created the Hierarchical Data Format (HDF) to aid users in storing and processing scientific data across several operating systems and scientific equipment.

Figure 2 describes Apache Big Data Ecosystem.

Figure 2. Apache big data ecosystem

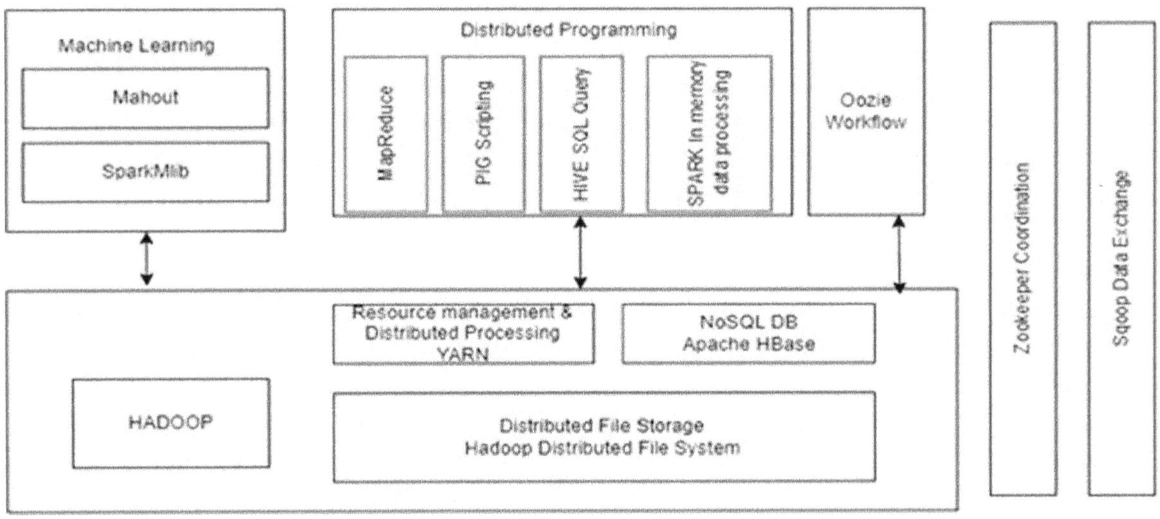

MODELS AND ALGORITHMS FOR WATER MANAGEMENT

Researchers popularly use several models to represent water management systems.

These models are used to study the water management problem by collecting data and generating accurate simulations of water flows. Two of the models used have been briefly described below.

- ***Wathnet5:*** This software is a simulation engine used to simulate water resource systems. It works by optimizing attributes while maintaining set constraints to discover new solutions to water management problems. It provides accurate decision support for high-level decision-making. It is a modeling suite that is robust. It aids in all stages of water management: building, calibration, and optimization.

- ***EPANet:*** This is a software tool that is used for the simulation of water distribution networks all over the world. It was created to understand the movement and distribution of drinking water elements within various networks. However, it can also be used for various distribution system-related analytic applications C. P. Kumar (2018).
- **GSFlow:** This model combines groundwater and surface water flows. The model was designed to represent connected groundwater and surface water flows in one or more watersheds. It provides a simulation of the fluid flow throughout the land surface, subsurface, saturated and unsaturated materials, streams, and lakes all at the same time. A GSFLOW simulation is driven by climate data such as actual or predicted precipitation, air temperature, and solar radiation, as well as groundwater stressors (such as withdrawals) and boundary conditions.
- ***MODFLOW:*** This is an object-oriented application and framework designed to allow many models and models of different sorts within the same simulation. MODFLOW uses the finite-difference method for groundwater flow modeling. It was developed by the United States Geological Survey (USGS). It allows the user to create a numerical representation (i.e., a groundwater model) of the hydrogeologic environment at a field site. Using the finite-difference approach, it divides the groundwater flow model's domain into rows, columns, and layers. It defines a unique set of grid blocks (i.e., model cells) to reflect the distribution of hydrogeologic features and hydrologic boundaries within the model domain. Once the model has been set up for use, the data collected from the model is fed as input to ML algorithms designed to work on neural networks. Several variations of these algorithms, like ANNs, CNNs for image classification, and RNNs for spatio-temporal data, are used in water management systems. Tensorflow and Keras provide support for implementing neural networks. Various libraries like WaterML and SparkMlib also help developers design and write effective solutions.

CONCLUSION

This chapter focuses on ML in weather forecasting and water management. This chapter offers an overview of the ML procedure for predicting the weather and an example of a dataset used to produce a prediction, which serves as a summary of the entire chapter. This research concludes by describing various approaches or ML algorithms that can be used in weather forecasting.

Furthermore, ML in water management focuses on how water resources are under unprecedented stress for the current generation, with chronic scarcity, hydrological unpredictability, and catastrophic weather events posing severe threats to the stability and growth of the global economy. India needs help managing its water resources because annual rainfall only makes up 58% of its total replenishment supplies and is 70% contaminated with contaminants like arsenic and fluoride. With 70% of the country's water sources poisoned, there is a water shortage.

The study emphasizes the application of ML in designing systems for groundwater redistribution, flood warning systems, drought management, and water quality measuring systems, to name a few.It has been observed that various models, including GSFLow, Modflow, etc., are employed to simulate the actual issue of water management. Following that, neural network frameworks like ANNs, CNNS, and RNNs use the model data as input. The data gathered from these models follows conventional patterns that make it easier for developers to separate crucial aspects for classification and prediction. Such models frequently use optimization algorithms to improve their performance, and the simulation of such models depends on computing power and support libraries, both of which are receiving more and more attention from academics. The study shows how water management systems use current models and ML to create simulations of the water flow, spot abnormalities, and make more precise predictions.

REFERENCES

Cosgrove, W. J., & Loucks, D. P. (2015). Water management: Current and future challenges and research directions. *Water Resources Research, 51*(6), 51. doi:10.1002/2014WR016869

Jakaria, A. H. M., Tennessee, T. M. H., & Rahman, M. A. (2020). *Smart weather forecasting using machine learning: a case study in*. Research Gate.

Kumar, C. P. (2018). Water Resources Issues and Management in India. *The Journal of Scientific and Engineering Research, 5*(9), 2394–2630.

Kumar, J. S. (2022). *Smart Weather Prediction Using Machine Learning*. Jibendu Kumar Mantri.

Lakshmi, N. S., Ajimunnisa, P., Prasanna, V. L., Yugasravani, T., & RaviTeja, M. (2021). Prediction of weather forecasting by using machine learning. *International Journal of Innovative Research in Computer Science & Technology, 9*(4). Advance online publication. doi:10.21276/ijircst.2021.9.4.7

Liakos, K. G., Busato, P., Moshou, D., & Pearson, S., & Bochtis, D. (2018). Machine Learning in Agriculture: A Review. *Sensors (Basel)*, 18–18. PMID:30110960

Parwal, M. (2015). A Review Paper on Water Resource Management. *International Journal of New Technology and Research, 1*(2), 2454–4116.

Raut, J. (2021). A Review on Weather Forecasting using Machine Learning and Deep Learning Techniques. *International Advanced Research Journal in Science, Engineering and Technology, 8*, 5–5.

Salcedo-Sanz, S., Pérez-Aracil, J., Ascenso, G., Ser, J. D., Casillas-Pérez, D., Kadow, C., ... Castelletti, A. (2023, August). Analysis, characterization, prediction, and attribution of extreme atmospheric events with machine learning and deep learning techniques: A review. *Theoretical and Applied Climatology*. Advance online publication. doi:10.100700704-023-04571-5

Shihu, S. (2011). Multi-sensor Remote Sensing Technologies in Water System Management. *3rd International Conference on Environmental Science and Information Application Technology*, 152–157. 10.1016/j.proenv.2011.09.027

Singh, S., Kaushik, M., Gupta, A., & Malviya, A. K. (2019). Weather Forecasting using Machine Learning Techniques. SSRN *Electronic Journal*.

Chapter 7
AI-Based Smart Water Quality Monitoring and Wastewater Management:
An Integrated Bio-Computational Approach

Dipankar Ghosh
JIS University, India

Sayan Adhikary
JIS University, India

Srijaa Sau
JIS University, India

ABSTRACT

Water is unambiguously susceptible to contamination, as it is able to dissolve a broader spectrum of substances than any other liquid on Earth. Increasing population and urbanization have been imposed to monitor water quality and wastewater management in the current global scenario. Conventional water quality monitoring involves water sampling, testing, and investigation, which are usually performed manually and are not dependable. Rapid economic prosperity generates a larger quantity of wastewater enriched with a broad range of pollutants that pose serious threats to the environment and human health. Advancements in artificial intelligence and machine learning approaches have shown breakthrough potential toward large dataset capture and analysis of large datasets to attain complex large-scale water quality monitoring and wastewater management systems. The current chapter summarizes prospects and potentials of AI technologies for the amelioration of water quality monitoring and wastewater management to establish an integrated sustainable biocomputation platform in the near future.

DOI: 10.4018/979-8-3693-1194-3.ch007

INTRODUCTION

One of the most important things for leading a healthy lifestyle is to stay hydrated; for that reason, water is very beneficial for human beings. Despite the fact that water covers approximately 71% of the Earth's surface, obtaining clean water for drinking and other essential uses remains a challenge for human civilization (Fishman, 2011). There are many reasons, such as population, industrialization, and water pollution, that are to blame for the escalating daily water crisis. Along with the other causes, water wastage and water distribution also play an important role in this water crisis (Sivakumar, 2011; Jury & Vaux, 2007). From the time of evolution of the human civilization, people had started to use the water more than the requirement, and for that reason, the formation of the wastewater reached up to the usual level, and the level of the fresh ground water has also reached below the acceptance level (Lofrano & Brown, 2010). Consequently, as the global population increases, so does the need for fresh subterranean water (Contreras et al., 2017). The water crisis indirectly indicates an increase in the amount of waste water, which leads to many health deaths, such as malaria and cholera (Marshall, 2011). Therefore, to address all these types of difficulties, critical thinking and progressive approaches for wastewater treatment are among the most important steps to follow (Bhargava, 2016). With this perspective, currently, different biocomputational platforms have taken an important role in different water quality management systems, such as groundwater monitoring, surface water monitoring, and wastewater management (Malviya & Jaspal, 2021; Zaresefat & Derakhshani, 2023; Oruganti et al., 2023). Artificial intelligence is a new technology that is used to perform different types of analytical, statistical, and industrial work without any mistakes at a rapid speed as a replacement for human power (Dwivedi et al., 2021). In water monitoring and management, artificial intelligence (AI) is primarily used to control and monitor several parameters, including biological oxygen demand (BOD), chemical oxygen demand (COD), and concentrations of nitrogen and sulfur (Malviya & Jaspal, 2021). Artificial intelligence (AI) algorithms, including support vector machines (SVMs), artificial neural networks (ANNs), adaptive neurofuzzy inference systems (ANFISs), and deep neural networks (DNNs), have been broadly employed in the monitoring and management of wastewater and various water bodies (Hussain & Naaz, 2020).

The primary data processing models used by artificial neural networks (ANNs) are derived from the way organic nervous systems, such as the brain, handle information. In contrast to SVM, which can solve small samples in terms of nonlinear, high-dimensional, localized minima and other partial elements, artificial neural network (ANN) solutions are applicable to both linear and nonlinear problem types. ANNs build several processing units based on interconnected connections, mimicking the way the human brain analyzes and processes data (Dastres & Soori, 2021; Walczak, 2019). Additionally, it has a modular architecture that permits component design implementation to be performed independently (Yahya et al., 2019). Along with the mentioned models, RF, DNN and many other algorithms have also been proven to have proper functionality in terms of predicting water quality parameters, automating water treatment plants and many other water management processes. Therefore, effective use of AI will be successful in wastewater management and water quality monitoring.

Conventional Approaches for Water Quality Monitoring

Water quality monitoring has come into main focus since the last 19th century for the sudden and enormous appearance of water-borne diseases (İçağa, 2007; Neary et al., 2009; Behmel et al., 2016). There are a number of common ways to monitor water quality, including technology-based approaches and conventional approaches (Thiyagarajan et al., 2017; Chen et al., 2021). Conventional methods for water quality monitoring are the most well-known traditional methods for water quality assessment. The conventional techniques are totally dependent on the large amount of dataset or information, which needs to be analyzed or examined for the proper result (Ahmed et al., 2019; Geetha & Gouthami, 2016; U. Ahmed et al., 2019). The accurate identification of many water quality indices, including pH, turbidity, COD, BOD, oxygen density, and microbiological presence, is essential for water quality monitoring (AlMetwally et al., 2020). Conventional water quality monitoring techniques comprise some stages to collect and analyze data, including water sampling, sample testing and investigative analysis (Thiyagarajan et al., 2017; Rahman et al., 2020). As the mentioned stages deal with a large amount of data, these should be performed by the experts and with proper concentration (Von Haefen et al., 2023). Recently, technology has been introduced in water quality monitoring systems to avoid any mistakes and save more time (Adu-Manu et al., 2017).

However, it is crucial to have a thorough understanding of the conventional and traditional methods of water quality monitoring before learning about sophisticated systems. As stated before, this approach has 3 steps: water sampling, testing samples and investigative analysis (Figure 1).

Figure 1. Comparative overview of conventional approaches toward water quality monitoring and wastewater treatment

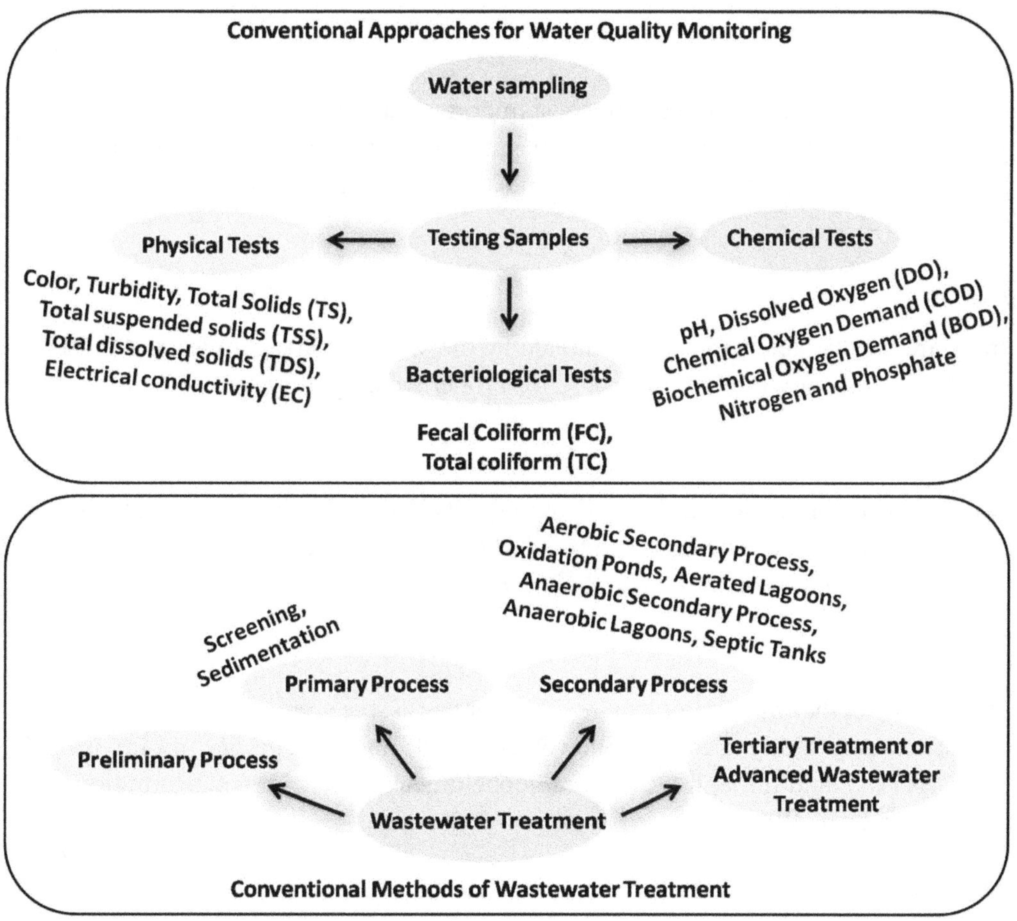

Water Sampling

Water sampling, as the name implies, is the process of gathering water samples to evaluate water quality. It is critical to gather water samples correctly to conduct an accurate water quality examination. Water sampling procedures can only be conducted by ISO-certified laboratories (Thiyagarajan et al., 2017). There are different criteria for collecting water samples, such as:

Sample Collection Containers

Selection of the perfect sampler is very important. The container must be clean, small, and transparent.

Types of Water Sample

For the water quality test, three main types of water samples were collected: blank, spike and replicate. Blanks and spikes are mainly used to estimate biased or detection limits. Replicate samples are mainly used to identify the variability between two identical samples (Foreman et al., 2021; Queensland, G., 2019; LaTour et al., 2006).

There are also a few more criteria, such as location for sample collection, proper laboratory facilities, and sample preservation, that are very important for proper water sampling.

Testing Samples

After the collection of the water samples, the most important part of the quality analysis, sample testing, needs to be performed. There are three main types of testing procedures performed for each sample. The three parameters include physical tests, chemical tests, and bacteriological tests (Thiyagarajan et al., 2017; Patil, 2012).

Physical Tests

Physical traits are the characters that are examined by the senses (Thiyagarajan et al., 2017). There are many physical traits that are tested or measured for proper water quality analysis.

Color

The state of the water can be ascertained by looking at its hue. For instance, by looking at the color of the water, the zooplankton presence can be observed (Wissel et al., 2003). Heavy metal contamination also changes the color of the water (Jain et al., 2020).

Turbidity

A gauge of the purity of the water is its turbidity; in other words, the amount of light that can flow through the water indicates its turbidity (Davies-Colley & Smith, 2001). The unit of turbidity is nephelometric turbidity units (NTUs), and through the help of nephelometer or turbidimeter turbidity measurements (Collado-Fernández et al., 2000).

Total Solids (TS)

It is one of the oldest analyses of water and represents the number of total solid materials present in water, such as sulfur, phosphorus, and calcium (Bhandari & Nayal, 2008; Ali & Qamar, 2013).

Total suspended solids (TSS)

To put it simply, the total amount of suspended solids comprises both organic and inorganic solid materials that are still suspended in water (U. Ahmed et al., 2019). The turbidity of water is completely dependent on the TSS (Nasrabadi et al., 2016).

Total Dissolved Solids (TDS)

The total amount of dissolved organic and inorganic solid material in the water is known as the total suspended solids (U. Ahmed et al., 2019). For estimation of the groundwater quality, TDS plays a very important role and helps to understand the effect of seawater intrusion (Rusydi, 2018).

Electrical Conductivity (EC)

EC means the potential of the water to pass the electric current (Khatoon, 2013). Water's electrical conductivity is temperature dependent. Therefore, it is necessary to record the electrical conductivity value at different temperatures (Hayashi, 2004). The total dissolved solids in the water are indicated by this EC value (Thirumalini & Joseph, 2009).

Chemical Tests

After collecting data about the physical traits, the water quality analysis was run through the chemical test. Chemical test results are the main indication of whether the water is actually suitable for drinking. The water quality index (WQI) can be measured to carry out these procedures (Lumb et al., 2011). The chemical traits of water are

pH

One of the most crucial aspects of water is its pH, or hydrogen potential. Fresh water has a large amount of dissolved salt and organic molecules, which can be used to measure the water's alkalinity and acidity (Baldisserotto, 2011). While the pH of pure water is 7, surface and ground water often have pH ranges of 6.5 to 8.5 and 6 to 8.5, respectively. Drinking water should have a pH between 6.5 and 8.5. (Dirisu et al., 2016; Whelton et al., 2007).

Dissolved Oxygen (DO)

Aquatic creatures flourish in water because of the amount of dissolved oxygen. For instance, a decreased concentration of dissolved oxygen suggests that aquatic creatures are growing more slowly. 5 mg/ml is the minimum dissolved oxygen criterion for the healthy survival of aquatic habitats (Tomasso, 1997; Bozorg-Haddad et al., 2021).

Chemical Oxygen Demand (COD)

The amount of oxygen that water consumes during the oxidation of inorganic matter and the breakdown of organic matter is known as COD (Samudro & Mangkoedihardjo, 2010). COD is known as the visual representation of the contamination of water caused by organic water bodies (Li et al., 2017).

Biochemical Oxygen Demand (BOD)

The BOD of water is the quantity of oxygen needed by microorganisms during organic matter decomposition (Nayar, 2020). A higher number of BOD represents an increased amount of dissolved oxygen (Dey et al., 2021).

Nitrogen and Phosphate

The increasing amount of nitrogen in water is the result of the overuse of fossil fuels and fertilizers, and it is very much responsible for creating disruptions in aquatic lives (Grizzetti et al., 2011); Assessment, 2005). Phosphorous is mainly responsible for eutrophication in water (Hao et al., 2019). Only orthophosphate is good for aquatic environments, while some inorganic phosphate, pyrophosphate, phosphodiesters, and organic phosphateesters can be chemically or enzymatically transformed to orthophosphate in an aqueous environment (Correll, 1998; Su et al., 2013).

Bacteriological Tests

Bacteriological tests refer to the analysis of bacterial contamination in the water or, in simple words, the existence of microorganisms in water (Akhtar et al., 2019). The two types of tests that are performed to analyze the microbial presence in water are total fecal coliform bacteria (TFCB) and total plate count (TPC) tests (Khan et al., 2023).

Fecal Coliform (FC)

According to reports from the WHO, nearly 2 billion people are affected by contaminated drinking water, and fecal coliforms are one of the reasons for this water contamination (Pras & Mamane, 2023). The fecal coliform test uses *Escherichia coli* (*E. coli*) as an indicator to reveal additional fecal coliform contaminations in water, including Salmonella, Giardia, norovirus, Campylobacter, and Cryptosporidium (Eregno et al., 2016; Orr et al., 2023).

Total Coliform (TC)

Total coliform counts are the sum of the fecal and other types of nonfecal bacteria counts (Hamadou et al., 2023). The normal ground water is free from coliform, so its total coliform count is 0 (Cabral & Marques, 2006). The membrane filtration process is widely used to count total coliform contamination in water samples (Kemper et al., 2023).

Applications of Artificial Intelligence (AI) in Water Quality Monitoring

The quality of water worldwide is negatively impacted by the excessive use of pesticides and fertilizers brought on by a growing population and industrialization. Consequently, checking water quality has become crucial to maintaining global equilibrium and health. The process of establishing methods of hand collecting water samples from different areas and analyzing different parameters of those samples to predict water quality becomes extremely laborious and time-consuming (Zainab et al., 2023). Because artificial intelligence (AI) can handle enormous amounts of data, researchers are turning to AI as a substitute tool for sophisticated nonlinear hydrological modeling due to the different limitations associated with conventional techniques for water quality monitoring (Remesan & Mathew, 2014). Thus, compared to other traditional statistical techniques, artificial intelligence (AI) has become more beneficial in terms of reducing the amount of time needed for data sampling and obtaining high accuracy and reliability to discover nonlinear patterns of input and output (Ay & Özyıldırım, 2018).

Applications of AI in Groundwater Quality Monitoring

Groundwater is the world's largest freshwater reserve, making up thirty percent of all freshwater found worldwide (Masood et al., 2022). The demand for freshwater is continuously increasing with the growing population in many developing countries. An increasing number of anthropogenic factors—such as mining, xenobiotic compound overuse, and industrial pollutants—as well as environmental factors—such as seawater intrusion and interactions with rock–water leaching—provide a significant risk to the management of water quality by changing the quality of groundwater and influencing consumption patterns (Tirkey et al., 2017; Nourbakhsh et al., 2015).

Artificial neural networks (ANNs), support vector machines (SVMs), random forests (RFs), ordinary decision trees (ODTs), and adaptive network-based fuzzy inference systems (ANFISs) are a few machine learning algorithms that have been applied recently for the assessment of groundwater quality (Han et al., 2010; Jeihouni et al., 2019; El-Magd et al., 2023; Kılıçaslan et al., 2014).

Nathan et al. compared and applied multilinear regression (MLR) and ANN models to groundwater quality using the Canadian Water Quality Index (CWQI). A total of 1065 water samples were gathered, and 17 different bacteriological and physiochemical parameters were assessed; however, the scope of the investigation was restricted to the ten physiochemical parameters (EC, TDS, TH, HCO3-, Cl-, SO42-, Ca2+, Mg2+, Na+, and K+) that were related to pollution. Using the ten parameters listed above, hierarchical cluster analysis (HCA) produced three distinct clusters. With high R2, MAE, and RMSE values, the ANN and MLR models both performed well in predicting the CWQI in clusters 1 and 2, but only the ANN model performed well in cluster 3 (Nathan et al., 2017).

Sakizadeh started an experiment to evaluate the accuracy of three different ANN algorithms for predicting the ground water quality index: ANNs with early stopping, an ensemble of ANNs, and ANNs with Bayesian regularization. The outcomes demonstrated that the ensemble averaging method and the Bayesian regularization method had both reached the minimum level of generalization ability (Sakizadeh, 2015). Mohapatra et al. predicted seasonal ground water levels by employing three powerful machine learning techniques, ANFIS, SVM and DNN (deep neural network). In most agro-ecological zones, the DNN model outperformed ANFIS in terms of seasonal ground water level prediction (Mohapatra et al., 2021).

The ANFIS method was utilized by Ashiyani et al. to forecast the groundwater quality index in Matar Taluka and Nadiad Taluka (Ashiyani, 2015). Tutmez et al. used an ANFIS model to demonstrate the relationship between the electrical conductivity of groundwater and its major cations. In regard to statistical prediction, the ANFIS model outperformed more conventional techniques for simulating electrical conductivity based on TDS in water (Tütmez et al., 2006). In one study, groundwater quality was simulated using a coactive neuro-fuzzy inference system (CANFIS), and the spatial variation in groundwater quality was displayed using a geographic information system (GIS). The incorporation of CANFIS and GIS resulted in enhanced efficiency in assessing groundwater quality (Gholami et al., 2016).

Norouzi et al. employed RF algorithms to evaluate the quality of groundwater because of their high precision in forecasting, aptitude for understanding nonlinear relationships, and capacity to identify the critical variables in the prediction. The study demonstrated the reliability of RF techniques in groundwater vulnerability assessment (Norouzi & Moghaddam, 2020). Madani et al. presented a model that uses RF algorithms and multivariate logistic regression (MLR) to forecast groundwater contamination. At 93% accuracy, the random forest model outperformed the MLR model when assessing contamination, but when the "SOLVER" and "C" parameters were modified, the accuracy of the MLR model increased from 70 to 83% (Madani et al., 2022).

Applications of AI in Surface Water Quality Monitoring

Only one percent of the Earth's surface is covered with freshwater, although 75% of it is covered by water. The world population is expanding at an accelerated rate, and industrialization is making the shortage of fresh water resources even worse (Aani et al., 2019). The accumulation of municipal and industrial wastewater in surface waters as a result of numerous human activities is the main factor contributing to the deterioration of urban water quality. The low quality of surface water endangers plant and animal life, ecosystems and human health (Mohammadpour et al., 2014). Traditional methods are often applied to assess surface water quality, but they are limited by realistic difficulties. As a result, it is essential to create fresh methods for assessing and tracking water quality.

Recently, two main types of models, namely, mechanism and nonmechanism-oriented models, have been available for water quality modeling and forecasting. Due to its relative sophistication, any water body can use the mechanism model, which simulates water quality using sophisticated system structure data. To predict WQC, Aldhyani et al. used ANN models, specifically the long short-term memory (LSTM) and nonlinear autoregressive neural network (NARNET), as well as SVM, K-nearest neighbor (K-NN), and naive Bayes. In predicting WQI, the NARNET model outperformed the LSTM model by a small margin. Furthermore, the SVM algorithm outperformed the naive Bayes and K-NN models in assessing WQC (Aldhyani et al., 2020).

The presence of algal blooms causing eutrophication is a major concern in surface water quality. Quang et al. applied eight different machine learning algorithms (GRU, RNN, LSTM, LR, SARIMAX, DTR, SVR and ANFIS) to predict algal blooms considering the most significant elements, such as BOD, DTN, DTP, DO, pH, precipitation and temperature, bringing to algal blooms. The ANFIS model performed the best in terms of quantitative and qualitative prediction and confirmation that the complex interactions between nutrients, organic contaminants, and environmental factors were the cause of eutrophication and the proliferation of algal blooms (Ly et al., 2021).

In a study, TSS, TS, and DS in surface water were predicted using a variety of models, including ANNs, SNMs, and GPR models. The ANN model appeared to be the best model outperforming the other models that exhibited the characteristics of overfitting (Rizal et al., 2022). However, due to the instability of the neural network in terms of model parameter optimization, nonlinear disturbance has a notable effect on the ANN model's accuracy. Therefore, in some cases, the SVM model may show higher generalization ability and make a prediction with higher accuracy than the ANN (Balabin & Lomakina, 2011).

Among all the parameters, dissolved oxygen (DO), which is thought to be a crucial indicator for assessing surface water quality, is important for controlling biological activity (W. Li et al., 2020). Tomić created a nonlinear polynomial neural network (PNN) model to forecast the DO concentration in surface water, and the PNN model was used to ascertain the correlation between DO and other parameters for water quality assessment. The variables influencing DO content were found to be temperature, pH, BOD, and phosphorus concentration; however, in severe circumstances, the significance of bicarbonates and alkalinity outweighed pH and BOD (Tomić et al., 2018). The relationship between DO and factors such as specific conductance, pH, and water temperature were analyzed by three algorithms, namely, the multiverse optimizer (MVO), SCE and BHA, through the principle of a multilayer perceptron network. While networks based on BHA and SCE could both anticipate DO levels with a fairly straightforward configuration, MVO was able to obtain a deeper comprehension of the relationship (Yang et al., 2021).

Conventional Methods of Wastewater Treatment

Water is the most important supplement to human living. Living entities require fresh water for other necessities in addition to using subsurface water as a source of drinking water (Gedda et al., 2021). Therefore, the freshwater crisis has recently become an important threat to human civilization. This threat is becoming a major matter to be looked out for the increasing amount of waste water and lack of waste water management systems. Wastewater can be formed from places such as kitchens and factories. The addition of different chemical compounds, organic compounds, solid wastes, plastics, etc., can also pollute water. Even a concentration of heavy metals in the water greater than its concentration range is responsible for water pollution. Because of the increasing amount of untreated wastewater, the physical health of human beings is adversely affected as well (Kesari et al., 2021; Sonune & Ghate, 2004). By seeing the current situation where the fresh water and waste water balance is completely imbalanced, it is evident that our future generation will face an enormous fresh water crisis (Burkhard et al., 2000). To solve this problem, the treatment and recycling of waste water is one of the best methods. Currently, various types of methods are introduced to process wastewater, and the main motives of these techniques are to remove/decrease inorganic, organic, and toxic materials and kill pathogenic microorganisms. Among all other methods, conventional wastewater treatments are one of the most known and used wastewater treatments and comprise four steps, namely, preliminary treatment, primary treatment, secondary treatment and tertiary treatment (Figure 1) (Gedda et al., 2021; Gupta et al., 2012; Tchobanoglous et al., 2003; Topare et al., 2011).

Preliminary Process

This step's primary goal is to remove large materials, plastics, coarse solids, and other debris so that maintenance and operational issues with treatment operations, processes, and auxiliary systems will be less difficult (Sonune & Ghate, 2004; Topare et al., 2011). Although this treatment is primarily related to sewage or municipal wastewater, it is also crucial for industrial wastewater, sometimes known as "trade effluent". Preliminary treatment is also important in the removal of oil/grease, which are present in sewage water (Sidwick, 1991). One of the most important features of this step is that it helps to reduce BOD by approximately 15-30%.

Primary Process

The primary wastewater process is mainly used to remove solids, inorganic and organic compounds, floating compounds, plastics, and compounds that are readily settled by gravity. This method is very important for highly polluted water, as it contains various steps, such as screening, filtration, centrifugation, sedimentation, and coagulation (Gupta et al., 2012). This process reduces BOD by approximately 25-50%, total suspended solids by 50-70% and floating wastes by 65% (Sonune & Ghate, 2004).

Screening

First, the water is gone through the screening process, in which solid wastes such as wood, plastic, cloth, and metal products are separated from the water (Gupta et al., 2012). In this process, a screen made up of long, closely spaced, narrow metal bars is present, and in the time of wastewater processing, the large solid waste particles become stuck in the screen bars and separate.

Sedimentation

In this process, wastes are separated by the help of gravity. The water is left undisturbed, and with time, the solid waste settles down at the bottom.

Floatation and coagulation, in addition to screening and sedimentation, are crucial for the removal of TSS, oil, grease, etc. (Gupta et al., 2012).

Secondary Process

This is the only step of the wastewater management system in which microbial influence has an important role. Microbes help to reduce the organic and inorganic pollution of water (Gupta et al., 2012). With the help of secondary treatment, 5-10% of BOD is reduced, and among different wastes, approximately 30% suspended, 6% colloidal and approximately 65% dissolved solids are separated from the wastewater. Two types of secondary treatment are available, namely, aerobic and anaerobic processes (Sonune & Ghate, 2004; Gupta et al., 2012; Topare et al., 2011).

Aerobic Secondary Process

The aerobic secondary process is the method by which aerobic microbes breakdown organic matter. This can be done by different processes, such as oxidation ponds and aerated lagoons (Topare et al., 2011).

Oxidation Ponds

There are three types of wastewater stabilization ponds (WSPs). Among them, one of the simple and effective ponds is oxidation ponds. At an approximate depth of 1–1.90 meters, oxidation ponds carry out their function with the aid of algae and bacteria. However, the drawbacks of these ponds are partially dependent on the environment, meaning that the function of the oxidation ponds changes with changes in the climate (A. E. Ali et al., 2020).

Aerated Lagoons

Aerated lagoons are a type of wastewater treatment medium where the operator can control the operating parameter to achieve or predict the goal (Godini et al., 2021). Diversity in microbial density is the main functional operator of aerated lagoons (Nunes et al., 2021).

Anaerobic Secondary Process

Anaerobic microbes degrade and stabilize organic matter during this process. This process can also be performed by different processes, such as anaerobic lagoons and septic tanks (Topare et al., 2011).

Anaerobic Lagoons

Anaerobic lagoons are a low-rate anaerobic digestion technology that are badly influenced by the environment, such as temperature (Schmidt et al., 2019). To maintain an anaerobic environment and ensure a low loss of heat, a layer of scum occasionally accumulates on the lagoon's surface (Musa & Idrus, 2021).

Septic Tanks

Septic tanks are the most globally used method to process wastewater with complicated biodegradation mechanisms (Shaw & Dorea, 2021). With the help of sedimentation and anaerobic fermentation, septic tanks remove suspended organic matter from rural domestic sewage (Liu et al., 2022).

Tertiary Treatment or Advanced Wastewater Treatment

Tertiary treatment involves numerous kinds of procedures. In short, it can be said that this is an extremely important step for the formation of high-quality water. This process includes many processes, such as evaporation, crystallization, solvent extraction, ion exchange, advanced oxidation, micro- and ultrafiltration, precipitation, and reverse osmosis (Gupta et al., 2012). To lessen pathogens, turbidity, other dissolved ions, colloids, and other contaminants—including complex refractory organic matter—tertiary treatment is needed (Ghangrekar, 2022).

Applications of Artificial Intelligence in Wastewater Management

Large-scale hazardous contaminants can seriously impair the quality of various water resources when they are present in commercial and industrial effluent. Therefore, it is essential to efficiently monitor and simulate all wastewater quality characteristics prior to discharge into the mainstream when planning water resources and creating water treatment methods. However, measuring all the characteristics takes time and necessitates difficult tests, as indicated in the standard methods for the inspection of water and wastewater. As a result, AI systems have demonstrated their ability to manage intricate, dynamic, and interactive wastewater treatment issues. It can be used for estimating effluent quality, forecasting how well water treatment will work, optimizing water quality parameters, designing treatment units, and automating the system (Fu et al., 2018; Baird et al., 2017; Sheng & Xu, 2018).

Water Quality Parameter Monitoring

Nourani et al. conducted a comparison of three distinct models that assess the performance of wastewater treatment plants by taking into account variables including BOD, COD, and TSS. The ensemble neural network model outperformed other models in terms of dependability when predicting BOD, COD, and total nitrogen effluent (Nourani & Elkiran, 2018). Wireless sensor networks enhance current centralized systems and traditional manual approaches by decentralizing smart water quality monitoring systems that adapt to the diverse and ever-changing water distribution infrastructure of cities. An integrated Internet of Things (IoT) system was presented by Martinez et al. for tracking water quality and tailoring it for initial validation in use cases, including wastewater treatment plants (Martínez et al., 2020). Qin et al. employed a boosting-iterative predictor weighting-partial least squares (boosting-IPW-PLS) method to monitor the TSS, COD and oil and grease contents of the effluents in addition to a multisensor water quality monitoring system that included a turbidimeter and UV/Vis spectrometer. As a result, during the calibration model building process, they were able to control the noise and information instability in the fused UV/Vis spectra and turbidity measurements. By assigning small weights to quality-irrelevant variables, the Boosting-IPW-PLS method uses IPW to muffle them. Based on the weighted variables, models are created for wastewater quality prediction, resulting in a high correlation coefficient between the actual value and the predicted value and moderate performance in terms of monitoring water quality (Qin et al., 2012). Two nonlinear models (ANN and SVM) were used in a comparison study to forecast the total nitrogen concentration of wastewater treatment plant effluent at a 1-day interval. The ANN model's capacity to generate predictions with a high degree of accuracy was superior (Guo et al., 2015).

Contaminant and Substance Removal

Cu(II) and Cr(VI) elimination processes were predicted by Rebouh et al. using an ANFIS technique in conjunction with a traditional mathematical isotherm model such as the Langmuir, Redlich-Peterson, and Freundlich models. ANFIS was the best model for estimating the adsorption of both Cu(II) and Cr(VI), according to an analysis of both interpolated and extrapolated data (Rebouh et al., 2015). In a study by Dolatabadi et al., it was shown how well the ANN and ANFIS models predicted sawdust's capacity for adsorption in the elimination of heavy metal ions such as Cu(II) and a cationic dye simultaneously (Dolatabadi et al., 2018).

Antwi and colleagues developed a three-layered feedforward backpropagation artificial neural network (BPANN) model to assess the removal of COD from wastewater treated by industrial starch processing using an upflow anaerobic sludge blanket (UASB) reactor.

The model performed significantly well, with an R2 of 87% (Antwi et al., 2018). Using SVM and ANFIS models, Manu et al. assessed the nitrogen removal efficiency from aerobic biological wastewater treatment plants while taking into account input variables including COD, pH, TS, free ammonia, ammonia nitrogen, and influent nitrogen. While the Gbell MF MODEL demonstrated efficiency in modeling the nonlinear time series, the SVM model performed better in producing a dependable prediction result compared to ANFIS (Manu & Thalla, 2017).

A three-layer backpropagation ANN model was created by Jing et al. to forecast the elimination of PAHs (naphthalene) from marine oily effluent by using UV irradiation. The outcomes demonstrated that ANN could accurately replicate the breakdown of naphthalene, offering insights into the mechanisms and characteristics of the photoinduced degradation process of PAHs (Jing et al., 2014). To forecast

2-chlorophenol's adsorptive removal from aqueous solution, Singh et al. built five nonlinear models: the multilayer perceptron network (MLPN), SVM, generalized regression neural network, gene expression programming models and radial basis function network (RBFN). They then compared the results of these models. Compared to the other models, the RBFM and MLPN demonstrated superior generalization and predictive ability (Singh et al., 2012).

Treatment Plant Efficiency and Automation

Wen et al. developed a PID controller and an RBF network-based feedforward controller that together form a real-time DO intelligent control system. The feedforward controller can set ideal input values for a feedback control system, potentially enabling an intelligent control system for waste water treatment plants. The feedback controlling system can measure stable qualified effluent by precisely controlling the DO concentration in real time (Wen et al., 2017). Bongard et al. used artificial intelligence (AI) to optimize the processing of sewerage systems and concluded that automation and AI systems are more cost-effective than traditional engineering techniques (Bongards et al., 2014). Mamandipoor et al. presented techniques for deep neural network-based autonomous wastewater treatment facility defect detection. By integrating with environmental decision support systems, these techniques help WWTPs maintain high performance and low estimations even in the face of adversity with minimal human oversight (Mamandipoor et al., 2020; Ghosh et al., 2022).

The primary limitation of wastewater treatment plant automation control is its incapacity to adapt appropriately to variations in influent load or flow. Artificial intelligence has been widely employed to monitor treatment plants around the clock to solve this issue and help plant operators with their everyday tasks. Hernández-del-Olmo et al. suggested that the complete wastewater treatment plant be automated (D. Ghosh et al., 2023; Hernández-Del-Olmo et al., 2012). Furthermore, complex and dynamic problems involving the design, automation, quality/energy aspect, and sustainability of WWTPs could be resolved with the help of decision support systems (DSSs) that can integrate multiple data sources and multicriteria viewpoints to produce more trustworthy results (Mannina et al., 2019; Ghorai & Ghosh, 2022).

CONCLUSION

As the entire text makes clear, wastewater management has emerged as the sole practical solution to one of the most important problems the world is currently facing: water contamination. A large amount of wastewater has been recycled for management purposes. There are many different kinds of traditional techniques for controlling the treatment of water and keeping an eye on its quality from different reservoirs. Based on the findings of the quality monitoring, future water management takes place. However, the conventional water monitoring process takes much time, and there is an extreme amount of risk for incorrect results. As a solution, AI currently comes into play, where all analyses are automated and based on computerized AI algorithms. Artificial neural networks (ANNs), support vector machines (SVMs), random forests (RFs), ordinary decision trees (ODTs), and adaptive network-based fuzzy inference systems (ANFISs) are just a few examples of the many different machine learning techniques available. Using these algorithms, the process of water monitoring is simple, less time consuming and more efficient. Following the observation of the water quality, wastewater needs to be treated before being made available to the general public. For a long time, conventional approaches have been used to treat wastewater

Figure 2. Comparative overview of conventional and AI-based water monitoring systems and waste water management platforms

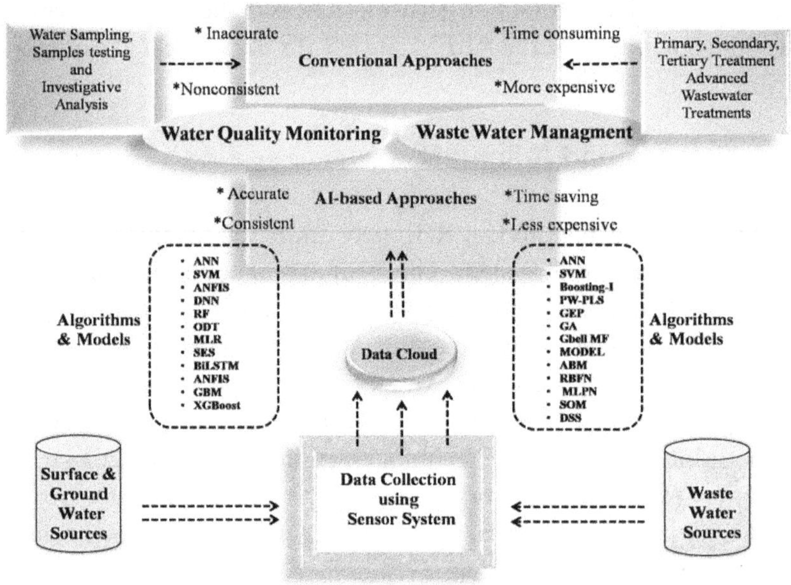

by primary, secondary and tertiary processes. Through these processes, solid wastes and gaseous waste are removed, BOD and COD are controlled, and many other water parameters are controlled. Nevertheless, traditional approaches to managing water quality are likewise labor intensive and ineffective, much as traditional approaches to monitoring water quality. For that reason, a wide range of companies are trying to use AI and machine learning models as water quality management tools. With the help of AI, wastewater quality monitoring and management tend to be performed properly. The use of AI makes these steps easier, saves time and correct (Figure 2). However, AI has a limited boundary in water quality monitoring and management. At present, it only focuses on emphasizing either targeted exposure or effect detection. However, in the future, AI might be used at a broader level where automated sampling processes and analysis processes could be introduced. Additionally, in vitro, in vivo and in situ bioanalytical methods might be introduced for monitoring water quality. AI may bring an enormous improvement in the physio-chemical water testing fields. Artificial intelligence may open many more new doors for water quality monitoring and management and serve this field with a new level of improvement.

Diagrammatic presentation depicts the impact of conventional and AI-based accelerated approach usage for water quality monitoring and waste water treatment concerning accuracy, time bounds, economic sustainability and consistency. Figure 2 also depicts the name of the algorithms and models that are in use for water quality monitoring on the left-hand dotted rectangular box, and the right-hand dotted rectangular box indicates waste water management. Water quality monitoring platforms are needed to maintain the quality of surface and groundwater sources, while wastewater management systems are in use to recycle wastewater from diverse ranges of sources or reservoirs.

REFERENCES

Aani, S. A., Bonny, T., Hasan, S. W., & Hilal, N. (2019). Can machine language and artificial intelligence revolutionize process automation for water treatment and desalination? *Desalination*, *458*, 84–96. doi:10.1016/j.desal.2019.02.005

Adu-Manu, K. S., Tapparello, C., Heinzelman, W., Katsriku, F. A., & Abdulai, J. (2017). Water quality monitoring using wireless sensor networks. *ACM Transactions on Sensor Networks*, *13*(1), 1–41. doi:10.1145/3005719

Ahmed, A. N., Othman, F., Afan, H. A., Ibrahim, R. K., Fai, C. M., Hossain, S., Ehteram, M., & El-Shafie, A. (2019). Machine learning methods for better water quality prediction. *Journal of Hydrology (Amsterdam)*, *578*, 124084. doi:10.1016/j.jhydrol.2019.124084

Ahmed, U., Mumtaz, R., Anwar, H., Mumtaz, S., & Qamar, A. M. (2019). Water quality monitoring: From conventional to emerging technologies. *Water Science and Technology: Water Supply*, *20*(1), 28–45. doi:10.2166/ws.2019.144

Akhtar, S., Fatima, R., Soomro, Z. A., Hussain, M., Ahmad, S. R., & Ramzan, H. S. (2019). Bacteriological quality assessment of water supply schemes (WSS) of Mianwali, Punjab, Pakistan. *Environmental Earth Sciences*, *78*(15), 458. Advance online publication. doi:10.100712665-019-8455-1

Aldhyani, T. H. H., Al-Yaari, M., Alkahtani, H., & Maashi, M. (2020). Water quality prediction using artificial intelligence algorithms. *Applied Bionics and Biomechanics*, *2020*, 1–12. doi:10.1155/2020/6659314 PMID:33456498

Ali, A. E., Salem, W. M., Younes, S. M., & Kaid, M. (2020). Modeling climatic effect on physiochemical parameters and microorganisms of Stabilization Pond Performance. *Heliyon*, *6*(5), e04005. doi:10.1016/j.heliyon.2020.e04005 PMID:32478191

Ali, M., & Qamar, A. M. (2013). Data analysis, quality indexing and prediction of water quality for the management of rawal watershed in Pakistan. In *Eighth International Conference on Digital Information Management (ICDIM 2013)* (pp. 108-113). IEEE. 10.1109/ICDIM.2013.6694009

AlMetwally, S. H., Hassan, M. K., & Mourad, M. (2020). Real time Internet of Things (IoT) based water quality management system. *Procedia CIRP*, *91*, 478–485. doi:10.1016/j.procir.2020.03.107

Antwi, P., Li, J., Meng, J., Deng, K., Quashie, F. K., Li, J., & Boadi, P. O. (2018). Feedforward neural network model estimating pollutant removal process within mesophilic upflow anaerobic sludge blanket bioreactor treating industrial starch processing wastewater. *Bioresource Technology*, *257*, 102–112. doi:10.1016/j.biortech.2018.02.071 PMID:29486407

Ashiyani, N. (2015). *Adaptive Neuro Fuzzy Inference System (ANFIS) for prediction of groundwater quality index in Matar Taluka and Nadiad Taluka*. https://api.semanticscholar.org/CorpusID:40125977

Assessment, M. E. (2005). *Ecosystems and human Well-Being: Multiscale Assessments: Findings of the Sub-Global Assessments Working Group*. Island Press.

Ay, M., & Özyıldırım, S. (2018). Artificial intelligence (AI) studies in water resources. *Natural and Engineering Sciences*, *3*(2), 187–195. doi:10.28978/nesciences.424674

Baird, R. B., Rice, E. W., & Eaton, A. (2017). *Standard Methods for the Examination of Water and Wastewater.*http://dspace.uniten.edu.my/handle/123456789/14241

Balabin, R. M., & Lomakina, E. I. (2011). Support vector machine regression (SVR/LS-SVM)—An alternative to neural networks (ANN) for analytical chemistry? Comparison of nonlinear methods on near infrared (NIR) spectroscopy data. *Analyst*, *136*(8), 1703. doi:10.1039/c0an00387e PMID:21350755

Baldisserotto, B. (2011). Water pH and hardness affect growth of freshwater teleosts. *Brazilian Journal of Animal Science*, *40*(1), 138–144.

Behmel, S., Damour, M., Ludwig, R., & Rodriguez, M. J. (2016). Water quality monitoring strategies — A review and future perspectives. *The Science of the Total Environment*, *571*, 1312–1329. doi:10.1016/j.scitotenv.2016.06.235 PMID:27396312

Bhandari, N. S., & Nayal, K. (2008). Correlation Study on Physico-Chemical Parameters and Quality Assessment of Kosi River Water, Uttarakhand. *E-Journal of Chemistry*, *5*(2), 342–346. doi:10.1155/2008/140986

Bhargava, A. (2016). Physico-Chemical Waste Water Treatment Technologies: An Overview. *International Journal of Scientific Research and Education*. https://doi.org/ doi:10.18535/ijsre/v4i05.05

Bongards, M., Gaida, D., Trauer, O., & Wolf, C. (2014). Intelligent automation and IT for the optimization of renewable energy and wastewater treatment processes. *Energy, Sustainability and Society*, *4*(1), 19. Advance online publication. doi:10.118613705-014-0019-3

Bozorg-Haddad, O., Delpasand, M., & Loáiciga, H. A. (2021). Water quality, hygiene, and health. In Elsevier eBooks (pp. 217–257). doi:10.1016/B978-0-323-90567-1.00008-5

Burkhard, R., Deletić, A., & Craig, T. (2000). Techniques for water and wastewater management: A review of techniques and their integration in planning. *Urban Water*, *2*(3), 197–221. doi:10.1016/S1462-0758(00)00056-X

Cabral, J. P., & Marques, C. J. (2006). Fecal Coliform Bacteria in Febros River (Northwest Portugal): Temporal Variation, Correlation with Water Parameters, and Species Identification. *Environmental Monitoring and Assessment*, *118*(1–3), 21–36. doi:10.100710661-006-0771-8 PMID:16897531

Chen, B., Mu, X., Chen, P., Wang, B., Choi, J., Park, H., Xu, S., Wu, Y., & Yang, H. (2021). Machine learning-based inversion of water quality parameters in typical reach of the urban river by UAV multispectral data. *Ecological Indicators*, *133*, 108434. doi:10.1016/j.ecolind.2021.108434

Collado-Fernández, M., González-SanJosé, M. L., & Pino-Navarro, R. (2000). Evaluation of turbidity: Correlation between Kerstez turbidimeter and nephelometric turbidimeter. *Food Chemistry*, *71*(4), 563–566. doi:10.1016/S0308-8146(00)00212-0

Contreras, J. D., Meza, R., Siebe, C., Rodríguez-Dozál, S., López-Vidal, Y., Castillo-Rojas, G., Amieva, R. I., Solano-Gálvez, S. G., Mazarí-Hiriart, M., Silva-Magaña, M. A., Vázquez-Salvador, N., Rosas-Pérez, I., & Romero, L. M. (2017). Health risks from exposure to untreated wastewater used for irrigation in the Mezquital Valley, Mexico: A 25-year update. *Water Research, 123*, 834–850. doi:10.1016/j.watres.2017.06.058 PMID:28755783

Correll, D. L. (1998). The role of phosphorus in the eutrophication of receiving waters: A review. *Journal of Environmental Quality, 27*(2), 261–266. doi:10.2134/jeq1998.00472425002700020004x

Dastres, R., & Soori, M. (2021). Artificial neural network systems. *International Journal of Imaging and Robotics, 21*(2), 13-25. https://www.researchgate.net/publication/350486076_Artificial_Neural_Network_Systems

Davies-Colley, R. J., & Smith, D. G. (2001). Turbidity suspended sediment, and water clarity: A review. *Journal of the American Water Resources Association, 37*(5), 1085–1101. doi:10.1111/j.1752-1688.2001.tb03624.x

Dey, S., Botta, S., Kallam, R., Angadala, R., & Andugala, J. (2021). Seasonal variation in water quality parameters of Gudlavalleru Engineering College pond. *Current Research in Green and Sustainable Chemistry, 4*, 100058. doi:10.1016/j.crgsc.2021.100058

Dirisu, C., Mafiana, M., Dirisu, G. B., & Amodu, R. (2016). Level of Ph in drinking water of an oil and gas producing community and perceived biological and health. *ResearchGate*. https://www.researchgate.net/publication/332012834_LEVEL_OF_pH_IN_DRINKING_WATER_OF_AN_OIL_AND_GAS_PRODUCING_COMMUNITY_AND_PERCEIVED_BIOLOGICAL_AND_HEALTH_IMPLICATIONS

Dolatabadi, M., Mehrabpour, M., Esfandyari, M., Hossein, A., & Davoudi, M. (2018). Modeling of simultaneous adsorption of dye and metal ion by sawdust from aqueous solution using of ANN and ANFIS. *Chemometrics and Intelligent Laboratory Systems, 181*, 72–78. doi:10.1016/j.chemolab.2018.07.012

Dwivedi, Y. K., Hughes, L., Ismagilova, E., Aarts, G., Coombs, C., Crick, T., Duan, Y., Dwivedi, R., Edwards, J. S., Eirug, A., Galanos, V., Ilavarasan, P. V., Janssen, M., Jones, P., Kar, A. K., Kizgin, H., Kronemann, B., Lal, B., Lucini, B., ... Williams, M. D. (2021). Artificial Intelligence (AI): Multidisciplinary perspectives on emerging challenges, opportunities, and agenda for research, practice and policy. *International Journal of Information Management, 57*, 101994. doi:10.1016/j.ijinfomgt.2019.08.002

El-Magd, S. A., Ismael, I. S., El-Sabri, M. S., Abdo, M. S., & Farhat, H. I. (2023). Integrated machine learning–based model and WQI for groundwater quality assessment: ML, geospatial, and hydroindex approaches. *Environmental Science and Pollution Research International, 30*(18), 53862–53875. doi:10.100711356-023-25938-1 PMID:36864333

Eregno, F. E., Tryland, I., Tjomsland, T., Myrmel, M., Robertson, L. J., & Heistad, A. (2016). Quantitative microbial risk assessment combined with hydrodynamic modeling to estimate the public health risk associated with bathing after rainfall events. *The Science of the Total Environment, 548–549*, 270–279. doi:10.1016/j.scitotenv.2016.01.034 PMID:26802355

Fishman, C. (2011). *The big Thirst: The Secret Life and Turbulent Future of Water*. Simon and Schuster.

Foreman, W. T., Williams, T. L., Furlong, E. T., Hemmerle, D. M., Stetson, S., Jha, V. K., Noriega, M. C., Decess, J. A., Reed-Parker, C., & Sandstrom, M. W. (2021). Comparison of detection limits estimated using single- and multiconcentration spike-based and blank-based procedures. *Talanta, 228*, 122139. doi:10.1016/j.talanta.2021.122139 PMID:33773706

Fu, Z., Cheng, J., Yang, M., & Batista, J. R. (2018). Prediction of industrial wastewater quality parameters based on wavelet denoised ANFIS model. In *2018 IEEE 8th Annual Computing and Communication Workshop and Conference (CCWC)* (pp. 301-306). IEEE. 10.1109/CCWC.2018.8301761

Gedda, G., Balakrishnan, K., Devi, R. U., Shah, K. J., Gandhi, V., Gandh, V., & Shah, K. L. (2021). Introduction to conventional wastewater treatment technologies: Limitations and recent advances. *Mater. Res. Found, 91*, 1–36. doi:10.21741/9781644901151-1

Geetha, S., & Gouthami, S. (2016). Internet of things enabled real time water quality monitoring system. *Smart Water, 2*(1), 1. Advance online publication. doi:10.118640713-017-0005-y

Ghangrekar, M. M. (2022). Aerobic Wastewater Treatment Systems. In *Wastewater to Water: Principles, Technologies and Engineering Design* (pp. 395–474). Springer Nature Singapore. doi:10.1007/978-981-19-4048-4_10

Gholami, V., Khaleghi, M. R., & Sebghati, M. (2016). A method of groundwater quality assessment based on fuzzy network-CANFIS and geographic information system (GIS). *Applied Water Science, 7*(7), 3633–3647. doi:10.100713201-016-0508-y

Ghorai, P., & Ghosh, D. (2022). Sustainable Approach for Insoluble Phosphate Recycling from Wastewater Effluents. In Springer eBooks (pp. 77–86). doi:10.1007/978-3-030-94148-2_7

Ghosh, D., Chaudhary, S., & Dhara, S. (2023). Prospects and Potentials of Microbial Applications on Heavy-Metal Removal from Wastewater. *Metal Organic Frameworks for Wastewater Contaminant Removal,* 177–201. doi:10.1002/9783527841523.ch8

Ghosh, D., Debnath, S., & Das, S. (2022). Microbial electrochemical platform: A sustainable workhorse for improving wastewater treatment and desalination. In Elsevier eBooks (pp. 239–268). doi:10.1016/B978-0-323-90765-1.00014-9

Godini, K., Azarian, G., Kimiaei, A., Drăgoi, E. N., & Curteanu, S. (2021). Modeling of a real industrial wastewater treatment plant based on aerated lagoon using a neuro-evolutive technique. *Process Safety and Environmental Protection, 148*, 114–124. doi:10.1016/j.psep.2020.09.057

Grizzetti, B., Bouraoui, F., Billen, G., Van Grinsven, H., Cardoso, A. C., Thieu, V., Garnier, J., Curtis, C., Howarth, R. W., & Johnes, P. J. (2011). Nitrogen as a threat to European water quality. In Cambridge University Press eBooks (pp. 379–404). doi:10.1017/CBO9780511976988.020

Guo, H., Jeong, K., Lim, J., Jo, J., Kim, Y. M., Park, J., Kim, J. H., & Cho, K. H. (2015). Prediction of effluent concentration in a wastewater treatment plant using machine learning models. *Journal of Environmental Sciences (China), 32*, 90–101. doi:10.1016/j.jes.2015.01.007 PMID:26040735

Gupta, V. K., Ali, I., Saleh, T. A., Nayak, A., & Agarwal, S. (2012). Chemical treatment technologies for waste-water recycling—An overview. *RSC Advances, 2*(16), 6380. doi:10.1039/c2ra20340e

Hamadou, W. S., Sulieman, A. M. E., Alshammari, N., Snoussi, M., Alanazi, N. A., Alshammary, A., & Al-Azmi, M. (2023). Water quality assessment of the surface and groundwater from Wadi Al-Adairey, Hail, Saudi Arabia. *Sustainable Water Resources Management*, *9*(5), 144. Advance online publication. doi:10.100740899-023-00923-1

Han, Y., Zou, Z., & Wang, H. (2010). Adaptive neuro fuzzy inference system for classification of water quality status. *Journal of Environmental Sciences (China)*, *22*(12), 1891–1896. doi:10.1016/S1001-0742(09)60335-1 PMID:21462706

Hao, H., Wang, Y., & Shi, B. (2019). NaLa(CO3)2 hybridized with Fe3O4 for efficient phosphate removal: Synthesis and adsorption mechanistic study. *Water Research*, *155*, 1–11. doi:10.1016/j.watres.2019.01.049 PMID:30826591

Hayashi, M. (2004). Temperature-Electrical conductivity relation of water for environmental monitoring and geophysical data inversion. *Environmental Monitoring and Assessment*, *96*(1–3), 119–128. doi:10.1023/B:EMAS.0000031719.83065.68 PMID:15327152

Hernández-Del-Olmo, F., Gaudioso, E., & Nevado, A. (2012). Autonomous Adaptive and Active Tuning Up of the Dissolved Oxygen Setpoint in a Wastewater Treatment Plant Using Reinforcement Learning. *IEEE Transactions on Systems, Man, and Cybernetics. Part C, Applications and Reviews*, *42*(5), 768–774. doi:10.1109/TSMCC.2011.2162401

Hussain, A., & Naaz, S. (2020). Prediction of diabetes mellitus: Comparative study of various machine learning models. In Advances in intelligent systems and computing (pp. 103–115). doi:10.1007/978-981-15-5148-2_10

İçağa, Y. (2007). Fuzzy evaluation of water quality classification. *Ecological Indicators*, *7*(3), 710–718. doi:10.1016/j.ecolind.2006.08.002

Jain, R., Thakur, A., Kaur, P., Kim, K., & Devi, P. (2020). Advances in imaging-assisted sensing techniques for heavy metals in water: Trends, challenges, and opportunities. *Trends in Analytical Chemistry*, *123*, 115758. doi:10.1016/j.trac.2019.115758

Jeihouni, M., Toomanian, A., & Mansourian, A. (2019). Decision Tree-Based Data Mining and Rule Induction for Identifying High Quality Groundwater Zones to Water Supply Management: A Novel Hybrid Use of Data Mining and GIS. *Water Resources Management*, *34*(1), 139–154. doi:10.100711269-019-02447-w

Jing, L., Chen, B., & Zhang, B. (2014). Modeling of UV-Induced photodegradation of naphthalene in marine oily wastewater by artificial neural networks. *Water, Air, and Soil Pollution*, *225*(4), 1906. Advance online publication. doi:10.100711270-014-1906-0

Jury, W. A., & Vaux, H. J. (2007). The emerging global water crisis: managing scarcity and conflict between water users. In Advances in Agronomy (pp. 1–76). doi:10.1016/S0065-2113(07)95001-4

Kemper, M., Veenman, C., Blaak, H., & Schets, F. M. (2023). A membrane filtration method for the enumeration of *Escherichia coli* in bathing water and other waters with high levels of background bacteria. *Journal of Water and Health*, *21*(8), 995–1003. doi:10.2166/wh.2023.004 PMID:37632376

Kesari, K. K., Soni, R., Jamal, Q. M. S., Tripathi, P., Lal, J. A., Jha, N. K., Siddiqui, M. H., Kumar, P., Tripathi, V., & Ruokolainen, J. (2021). Wastewater Treatment and Reuse: A Review of its Applications and Health Implications. *Water, Air, and Soil Pollution*, *232*(5), 208. Advance online publication. doi:10.100711270-021-05154-8

Khan, S., Saeed, S., Khan, S. A., & Abrar, M. (2023). Impact of Bacteriological Water Quality on Water Borne Diseases and its Health Costs among Students of the Institutions. *Journal of Social Sciences Review*, *3*(1), 510–518. doi:10.54183/jssr.v3i1.164

Khatoon, N. (2013). Correlation Study For the Assessment of Water Quality and Its Parameters of Ganga River, Kanpur, Uttar Pradesh, India. *IOSR Journal of Applied Chemistry*, *5*(3), 80–90. doi:10.9790/5736-0538090

Kılıçaslan, Y., Tuna, G., Gezer, G., Gülez, K., Arkoç, O., & Potirakis, S. M. (2014). ANN-Based estimation of groundwater quality using a wireless water quality network. *International Journal of Distributed Sensor Networks*, *10*(4), 458329. doi:10.1155/2014/458329

LaTour, J., Weldon, E., Dupré, D. H., & Halfar, T. M. (2006). *Water Resources Data for Illinois - Water Year 2005 (Includes historical data)*. doi:10.3133/wdrIL051

Li, J., Luo, G., He, L., Jing, X., & Lyu, J. (2017). Analytical Approaches for Determining Chemical oxygen Demand in water bodies: A review. *Critical Reviews in Analytical Chemistry*, *48*(1), 47–65. doi:10.1080/10408347.2017.1370670 PMID:28857621

Li, W., Fang, H., Qin, G., Tan, X., Huang, Z., Zeng, F., Du, H., & Li, S. (2020). Concentration estimation of dissolved oxygen in Pearl River Basin using input variable selection and machine learning techniques. *The Science of the Total Environment*, *731*, 139099. doi:10.1016/j.scitotenv.2020.139099 PMID:32434098

Liu, N., Cheng, S., Wang, X., Li, Z., Zheng, L., Lyu, Y., Ao, X., & Wu, H. (2022). Characterization of microplastics in the septic tank via laser direct infrared spectroscopy. *Water Research*, *226*, 119293. doi:10.1016/j.watres.2022.119293 PMID:36323216

Lofrano, G., & Brown, J. (2010). Wastewater management through the ages: A history of mankind. *The Science of the Total Environment*, *408*(22), 5254–5264. doi:10.1016/j.scitotenv.2010.07.062 PMID:20817263

Lumb, A., Sharma, T. C., & Bibeault, J. (2011). A review of genesis and evolution of Water Quality Index (WQI) and some future directions. *Water Quality, Exposure, and Health*, *3*(1), 11–24. doi:10.100712403-011-0040-0

Ly, Q. V., Nguyen, X. C., Lê, N. C., Truong, T., Hoang, H. H., Park, T. J., Maqbool, T., Pyo, J., Cho, K. H., Lee, K., & Hur, J. (2021). Application of Machine Learning for eutrophication analysis and algal bloom prediction in an urban river: A 10-year study of the Han River, South Korea. *The Science of the Total Environment*, *797*, 149040. doi:10.1016/j.scitotenv.2021.149040 PMID:34311376

Madani, A., Hagage, M., & Elbeih, S. F. (2022). Random Forest and Logistic Regression algorithms for prediction of groundwater contamination using ammonia concentration. *Arabian Journal of Geosciences*, *15*(20), 1619. Advance online publication. doi:10.100712517-022-10872-2

Malviya, A., & Jaspal, D. (2021). Artificial intelligence as an upcoming technology in wastewater treatment: A comprehensive review. *Environmental Technology Reviews*, *10*(1), 177–187. doi:10.1080/21622515.2021.1913242

Mamandipoor, B., Majd, M., Sheikhalishahi, S., Modena, C., & Osmani, V. (2020). Monitoring and detecting faults in wastewater treatment plants using deep learning. *Environmental Monitoring and Assessment*, *192*(2), 148. Advance online publication. doi:10.100710661-020-8064-1 PMID:31997006

Mannina, G., Rebouças, T. F., Cosenza, A., Sànchez–Marrè, M., & Gibert, K. (2019). Decision support systems (DSS) for wastewater treatment plants – A review of the state of the art. *Bioresource Technology*, *290*, 121814. doi:10.1016/j.biortech.2019.121814 PMID:31351688

Manu, D. S., & Thalla, A. K. (2017). Artificial intelligence models for predicting the performance of biological wastewater treatment plant in the removal of Kjeldahl Nitrogen from wastewater. *Applied Water Science*, *7*(7), 3783–3791. doi:10.100713201-017-0526-4

Marshall, S. (2011). The water crisis in Kenya: Causes, effects and solutions. *Global Majority E-Journal*, *2*(1), 31–45.

Martínez, R. F., Vela, N., Aatik, A. E., Murray, E., Roche, P. C., & Navarro, J. M. (2020). On the Use of an IoT Integrated System for Water Quality Monitoring and Management in Wastewater Treatment Plants. *Water (Basel)*, *12*(4), 1096. doi:10.3390/w12041096

Masood, A., Tariq, M. U. R., Hashmi, M. Z. U. R., Waseem, M., Sarwar, M. K., Ali, W., Farooq, R., Almazroui, M., & Ng, A. W. M. (2022). An Overview of Groundwater Monitoring through Point-to Satellite-Based Techniques. *Water (Basel)*, *14*(4), 565. doi:10.3390/w14040565

Mohammadpour, R., Shaharuddin, S., Chang, C. K., Zakaria, N. A., Ghani, A. A., & Chan, N. W. (2014). Prediction of water quality index in constructed wetlands using support vector machine. *Environmental Science and Pollution Research International*, *22*(8), 6208–6219. doi:10.100711356-014-3806-7 PMID:25408070

Mohapatra, J. B., Jha, P., Jha, M. K., & Biswal, S. (2021). Efficacy of machine learning techniques in predicting groundwater fluctuations in agro-ecological zones of India. *The Science of the Total Environment*, *785*, 147319. doi:10.1016/j.scitotenv.2021.147319 PMID:33957597

Musa, M. H., & Idrus, S. (2021). Physical and Biological Treatment Technologies of slaughterhouse Wastewater: A review. *Sustainability (Basel)*, *13*(9), 4656. doi:10.3390u13094656

Nasrabadi, T., Ruegner, H., Sirdari, Z. Z., Schwientek, M., & Grathwohl, P. (2016). Using total suspended solids (TSS) and turbidity as proxies for evaluation of metal transport in river water. *Applied Geochemistry*, *68*, 1–9. doi:10.1016/j.apgeochem.2016.03.003

Nathan, N. S., Saravanane, R., & Sundararajan, T. (2017). Application of ANN and MLR models on groundwater quality using CWQI at Lawspet, Puducherry in India. *Journal of Geoscience and Environment Protection*, *05*(03), 99–124. doi:10.4236/gep.2017.53008

Nayar, R. (2020). Assessment of Water Quality Index and Monitoring of Pollutants by Physico-Chemical Analysis in Water Bodies: A review. *International Journal of Engineering Research & Technology (Ahmedabad), V9*(01). Advance online publication. doi:10.17577/IJERTV9IS010046

Neary, D. G., Ice, G. G., & Jackson, C. R. (2009). Linkages between forest soils and water quality and quantity. *Forest Ecology and Management, 258*(10), 2269–2281. doi:10.1016/j.foreco.2009.05.027

Norouzi, H., & Moghaddam, A. A. (2020). Groundwater quality assessment using random forest method based on groundwater quality indices (case study: Miandoab plain aquifer, NW of Iran). *Arabian Journal of Geosciences, 13*(18), 912. Advance online publication. doi:10.100712517-020-05904-8

Nourani, V., Elkiran, G., & Abba, S. I. (2018). Wastewater treatment plant performance analysis using artificial intelligence – an ensemble approach. *Water Science and Technology, 78*(10), 2064–2076. doi:10.2166/wst.2018.477 PMID:30629534

Nourbakhsh, Z., Mehrdadi, N., Moharamnejad, N., Hassani, A. H., & Yousefi, H. (2015). Evaluating the suitability of different parameters for qualitative analysis of groundwater based on analytical hierarchy process. *Desalination and Water Treatment, 57*(28), 13175–13182. doi:10.1080/19443994.2015.1056837

Nunes, J. V., Da Silva, M. W. B., Couto, G. H., Bordin, E. R., Ramsdorf, W. A., Flôr, I. C., Vicente, V. A., De Almeida, J. D., Celinski, F., & Xavier, C. R. (2021). Microbiological diversity in an aerated lagoon treating kraft effluent. *BioResources, 16*(3), 5203–5219. doi:10.15376/biores.16.3.5203-5219

Orr, I., Mazari, K., Shukle, J. T., Li, R., & Filippelli, G. M. (2023). The impact of combined sewer outflows on urban water quality: Spatiotemporal patterns of fecal coliform in indianapolis. *Environmental Pollution, 327*, 121531. doi:10.1016/j.envpol.2023.121531 PMID:37004861

Oruganti, R. K., Biji, A. P., Lanuyanger, T., Show, P. L., Sriariyanun, M., Upadhyayula, V. K., Gadhamshetty, V., & Bhattacharyya, D. (2023). Artificial intelligence and machine learning tools for high-performance microalgal wastewater treatment and algal biorefinery: A critical review. *The Science of the Total Environment, 876*, 162797. doi:10.1016/j.scitotenv.2023.162797 PMID:36907394

Patil, P. N. (2012). Physico-chemical parameters for testing of water -A review. *ResearchGate*. https://www.researchgate.net/publication/344323551_Physico-chemical_parameters_for_testing_of_water_-A_review

Pras, A., & Mamane, H. (2023). Nowcasting of fecal coliform presence using an artificial neural network. *Environmental Pollution, 326*, 121484. doi:10.1016/j.envpol.2023.121484 PMID:36958657

Qin, X., Gao, F., & Chen, G. (2012). Wastewater quality monitoring system using sensor fusion and machine learning techniques. *Water Research, 46*(4), 1133–1144. doi:10.1016/j.watres.2011.12.005 PMID:22200261

Queensland, G. (2019). *Environmental protection (water) policy 2009-monitoring and sampling manual physical and chemical assessment*. Academic Press.

Rahman, M., Bepery, C., Hossain, M. J., & Islam, M. M. (2020). Internet of Things (IoT) based water quality monitoring system. *ResearchGate*. https://www.researchgate.net/publication/344167317_Internet_of_Things_IoT_Based_Water_Quality_Monitoring_System

Rebouh, S., Bouhedda, M., & Hanini, S. (2015). Neuro-fuzzy modeling of Cu(II) and Cr(VI) adsorption from aqueous solution by wheat straw. *Desalination and Water Treatment*, *57*(14), 6515–6530. doi:10.1080/19443994.2015.1009171

Remesan, R., & Mathew, J. (2014). Machine Learning and Artificial Intelligence-Based approaches. In Springer eBooks (pp. 71–110). doi:10.1007/978-3-319-09235-5_4

Rizal, N. N. M., Hayder, G., Mnzool, M., Elnaim, B. M. E., Mohammed, A. O. Y., & Khayyat, M. M. (2022). Comparison between Regression Models, Support Vector Machine (SVM), and Artificial Neural Network (ANN) in River Water Quality Prediction. *Processes (Basel, Switzerland)*, *10*(8), 1652. doi:10.3390/pr10081652

Rusydi, A. F. (2018). Correlation between conductivity and total dissolved solid in various type of water: A review. *IOP Conference Series*, *118*, 012019. 10.1088/1755-1315/118/1/012019

Sakizadeh, M. (2015). Artificial intelligence for the prediction of water quality index in groundwater systems. *Modeling Earth Systems and Environment*, *2*(1), 8. Advance online publication. doi:10.100740808-015-0063-9

Samudro, G., & Mangkoedihardjo, S. (2010). Review on BOD, COD and BOD/COD ratio: a triangle zone for toxic, biodegradable and stable levels. *ResearchGate*. https://www.researchgate.net/publication/228497615_Review_on_BOD_COD_and_BODCOD_ratio_a_triangle_zone_for_toxic_biodegradable_and_stable_levels

Schmidt, T., Harris, P., Lee, S., & McCabe, B. K. (2019). Investigating the impact of seasonal temperature variation on biogas production from covered anaerobic lagoons treating slaughterhouse wastewater using lab scale studies. *Journal of Environmental Chemical Engineering*, *7*(3), 103077. doi:10.1016/j.jece.2019.103077

Shaw, K., & Dorea, C. C. (2021). Biodegradation mechanisms and functional microbiology in conventional septic tanks: A systematic review and meta-analysis. *Environmental Science. Water Research & Technology*, *7*(1), 144–155. doi:10.1039/D0EW00795A

Sheng, S., & Xu, G. (2018). Novel performance prediction model of a biofilm system treating domestic wastewater based on stacked denoising autoencoders deep learning network. *Chemical Engineering Journal*, *347*, 280–290. doi:10.1016/j.cej.2018.04.087

Sidwick, J. M. (1991). The preliminary treatment of wastewater. *Journal of Chemical Technology and Biotechnology*, *52*(3), 291–300. doi:10.1002/jctb.280520302

Singh, K. P., Gupta, S., Ojha, P., & Rai, P. (2012). Predicting adsorptive removal of chlorophenol from aqueous solution using artificial intelligence based modeling approaches. *Environmental Science and Pollution Research International*, *20*(4), 2271–2287. doi:10.100711356-012-1102-y PMID:22851225

Sivakumar, B. (2011). Water crisis: From conflict to cooperation—an overview. *Hydrological Sciences Journal*, *56*(4), 531–552. doi:10.1080/02626667.2011.580747

Sonune, A., & Ghate, R. (2004). Developments in wastewater treatment methods. *Desalination*, *167*, 55–63. doi:10.1016/j.desal.2004.06.113

Su, Y., Chen, H., Qi, L., Gao, S., & Shang, J. K. (2013). Strong adsorption of phosphate by amorphous zirconium oxide nanoparticles. *Water Research*, *47*(14), 5018–5026. doi:10.1016/j.watres.2013.05.044 PMID:23850213

Tchobanoglous, G., Burton, F. L., Eddy, M. &., & Stensel, H. D. (2003). *Wastewater engineering: Treatment and Reuse*. Academic Press.

Thirumalini, S., & Joseph, K. G. (2009). Correlation between Electrical Conductivity and Total Dissolved Solids in Natural Waters. *Malaysian Journal of Science. Series B, Physical & Earth Sciences*, *28*(1), 55–61. doi:10.22452/mjs.vol28no1.7

Thiyagarajan, K., Pappu, S., Vudatha, P., & Niharika, A. V. (2017). Intelligent IoT based water quality monitoring system. *ResearchGate*. https://www.researchgate.net/publication/328802276_Intelligent_IoT_Based_Water_Quality_Monitoring_System

Tirkey, P., Bhattacharya, T., Chakraborty, S., & Baraik, S. (2017). Assessment of groundwater quality and associated health risks: A case study of Ranchi city, Jharkhand, India. *Groundwater for Sustainable Development*, *5*, 85–100. doi:10.1016/j.gsd.2017.05.002

Tomasso, J. R. (1997). Environmental requirements and noninfectious diseases. *Developments in Aquaculture and Fisheries Science*, *30*, 253–270. doi:10.1016/S0167-9309(97)80012-9

Tomić, A. Š., Antanasijević, D., Ristić, M., Perić-Grujić, A. A., & Pocajt, V. (2018). A linear and non-linear polynomial neural network modeling of dissolved oxygen content in surface water: Inter- and extrapolation performance with inputs' significance analysis. *The Science of the Total Environment*, *610–611*, 1038–1046. doi:10.1016/j.scitotenv.2017.08.192 PMID:28847097

Topare, N. S., Attar, S. J., & Manfe, M. M. (2011). Sewage/wastewater treatment technologies: A review. *Scientific Reviews and Chemical Communications, 1*(1). https://www.tsijournals.com/abstract/sewagewastewater-treatment-technologies-a-review-11194.html

Tütmez, B., Hatipoglu, Z., & Kaymak, U. (2006). Modeling electrical conductivity of groundwater using an adaptive neuro-fuzzy inference system. *Computers & Geosciences*, *32*(4), 421–433. doi:10.1016/j.cageo.2005.07.003

Von Haefen, R. H., Van Houtven, G., Naumenko, A., Obenour, D. R., Miller, J. A., Kenney, M. A., Gerst, M. D., & Waters, H. (2023). Estimating the benefits of stream water quality improvements in urbanizing watersheds: An ecological production function approach. *Proceedings of the National Academy of Sciences of the United States of America*, *120*(18), e2120252120. Advance online publication. doi:10.1073/pnas.2120252120 PMID:37094134

Walczak, S. (2019). Artificial neural networks. In Advances in computer and electrical engineering book series (pp. 40–53). doi:10.4018/978-1-5225-7368-5.ch004

Wen, X., Gong, B., Zhou, J., He, Q., & Qing, X. (2017). Efficient simultaneous partial nitrification, anammox and denitrification (SNAD) system equipped with a real-time dissolved oxygen (DO) intelligent control system and microbial community shifts of different substrate concentrations. *Water Research*, *119*, 201–211. doi:10.1016/j.watres.2017.04.052 PMID:28460292

Whelton, A. J., Dietrich, A. M., Burlingame, G. A., Schechs, M., & Duncan, S. E. (2007). Minerals in drinking water: Impacts on taste and importance to consumer health. *Water Science and Technology*, *55*(5), 283–291. doi:10.2166/wst.2007.190 PMID:17489421

Wissel, B., Boeing, W. J., & Ramcharan, C. W. (2003). Effects of water color on predation regimes and zooplankton assemblages in freshwater lakes. *Limnology and Oceanography*, *48*(5), 1965–1976. doi:10.4319/lo.2003.48.5.1965

Yahya, A. S. A., Ahmed, A. N., Othman, F., Ibrahim, R. K., Afan, H. A., El-Shafie, A., Fai, C. M., Hossain, S., Ehteram, M., & El-Shafie, A. (2019). Water Quality Prediction Model Based Support Vector Machine Model for Ungauged River Catchment under Dual Scenarios. *Water (Basel)*, *11*(6), 1231. doi:10.3390/w11061231

Yang, F., Moayedi, H., & Mosavi, A. (2021). Predicting the degree of dissolved oxygen using three types of Multi-Layer Perceptron-Based artificial neural networks. *Sustainability (Basel)*, *13*(17), 9898. doi:10.3390u13179898

Zainab, A., Amina, I., Abdulmuhaimin, M., Sadiku, A. S., & Baballe, M. A. (2023). The water monitoring system's disadvantages. Zenodo *(CERN European Organization for Nuclear Research)*. doi:10.5281/zenodo.8161049

Zaresefat, M., & Derakhshani, R. (2023). Revolutionizing Groundwater Management with Hybrid AI Models: A Practical Review. *Water (Basel)*, *15*(9), 1750. doi:10.3390/w15091750

Chapter 8
Revolutionizing Water Quality Monitoring:
The Smart Tech Frontier

Ambati Vanshika
Sri Padmavati Mahila Visvavidyalayam, India

B. Ramya Kuber
Sri Padmavati Mahila Visvavidyalayam, India

Nalluri Poojitha
Sri Padmavati Mahila Visvavidyalayam, India

ABSTRACT

Safe water is becoming a scarce resource, due to the combined effects of increased population, pollution, and climate changes. Due to the vast increase in global industrial output, rural to urban drift and the over-utilization of land, and high use of fertilizers in farms and sea resources, the quality of water available to people has deteriorated greatly. Around 40% of deaths are caused due to contaminated water in the world. Hence, there is a necessity to ensure supply of purified drinking water for the people both in cities and villages. Smart water quality monitoring systems have gained significant attention due to their ability to enhance water management practices and safeguard water resources. These systems integrate advanced technologies such as IoT sensors, data analytics, and machine learning algorithms to continuously monitor and assess water quality parameters in real-time. Smart water quality monitoring harnesses cutting-edge technologies, including internet of things (IoT), sensors, and data analytics, to revolutionize traditional water quality assessment methods.

DOI: 10.4018/979-8-3693-1194-3.ch008

1. INTRODUCTION

Fiji's seas have experienced an increase in pollution over the past few decades, mostly as a result of chemical waste and oil accidents. During a serious event in 2014, broken pipes allowed 200 gallons of untreated sewage to flood the Samabula River every second. Although comprehensive pollution eradication is difficult, its effects can be reduced. This project proposes to use IoT and RS technologies to monitor seawater quality, assisting Fiji in the fight against pollution. Temperature, conductivity, Oxidation Reduction Potential (ORP), and pH will all be measured by the Smart Water Quality Monitoring System. Anomalies could indicate the presence of pollutants, but consistent results are predicted. Even if transient aberrations won't set off alarms, IoT alerts will let people know about such instances. With the deployment of numerous monitoring stations for extensive coverage, this strategy might provide an early warning system for water pollution.

Numerous human activities, including eating, agriculture, and transport, depend heavily on water. However, these activities have a substantial impact on water quality. Monitoring water quality is necessary to address this issue. In this method, the quantities of ammonium, chloride, dissolved oxygen, pH, redox potential, and other chemical parameters are evaluated. Due to the presence of organic and nutritional components, surface water bodies are especially prone to quality problems. Agriculture is a major cause of pollution, according to River Basin Management Plans (RBMP), whether through diffuse or point source inputs of organic matter, fertilizers, pesticides, or hydro-morphological effects. The nitrogen and phosphorus load from diverse sources, such as agricultural wastewater, metropolitan areas, and other sources, is described in the RBMP. However, as laboratory procedures are frequently slow to produce operational replies and might not provide real-time public health protection, there is a compelling need to improve the current monitoring system. This emphasizes how critical it is to enhance and broaden monitoring and evaluation methods in order to get a statistically sound and thorough picture of the state of the aquatic environment for future planning.

1.1 Water Quality Degradation: Causes and Global Impact

Impact of Natural Elements on Water Quality

Natural processes have various effects on the quality of surface water and groundwater, including climate change, natural disasters, geological causes, soil matrix, and hyporheic exchange.

Water systems are significantly impacted by geological reasons, natural disasters, and climate change. Surface and groundwater are affected by changes in temperature, evapotranspiration, and precipitation. Filtration in treatment procedures is hampered by cold temperatures, and river water is diluted by heavy rain. Due to variable rainfall, semi-arid zones experience lower groundwater recharge and greater solute concentration. Water contamination is a result of natural disasters such as earthquakes, floods, and tsunamis. Water was a factor in more than 73% of the world's disasters from 2001 to 2018, which cost $1.7 trillion and resulted in 300,000 fatalities. Over 60% of these disasters were triggered by floods and droughts. Water quality is influenced by topography, soil types, and mineral dissolution. Radioactive compounds provide a concern for contaminating groundwater. Soil composition is shaped by geological and climatic processes, which affect pollutant persistence. In streambeds, hyporheic exchange has an impact on a variety of processes. The sustainability of groundwater is threatened by seawater intrusion in coastal places. The overall effect of these variables on water quality is significant.

Table 1. The loss of water quality caused by the natural forces listed above is thought to be the most pervasive

Water Resources	Processes/Factors	Important Processes
Surface water Ground water All water resources	Hydrological process	Evaporation, suspension, and setting Transpiration, infiltration, and leaching Dilution
All water resources Mainly rivers and lakes	Physical process	Adsorption and desorption, diffusion Heating and cooling, vitalization, gas exchange with the gas exchang
Ground water All water resources	Chemical process	Ionic exchange Acid-base reactions, redox reactions, Precipitation of minerals, photo degradation, Dissolution of particles
Surface water All water resources Mainly rivers	Biological process	Primary production Microbial die-off and growth Bioaccumulation, decomposition of organic matter, biomagnification

Figure 1. Anthropogenic activities on surface water and ground water

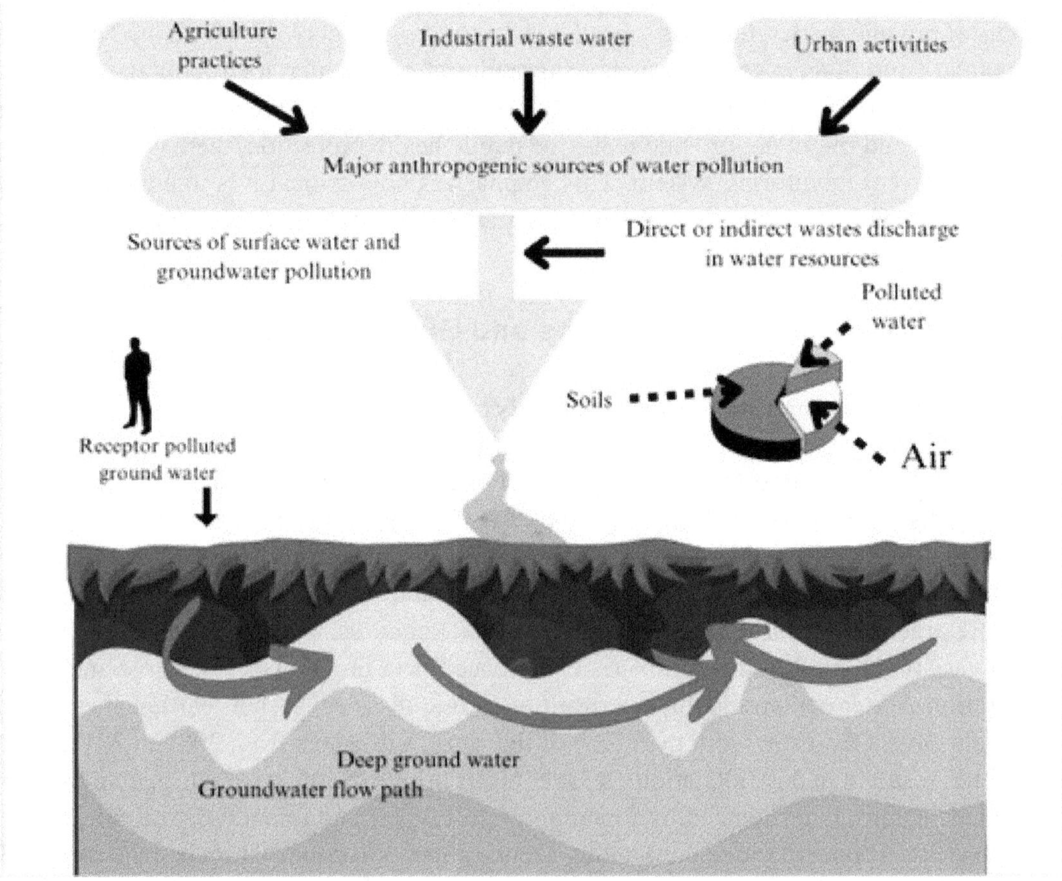

Anthropogenic elements, primarily resulting from land-use patterns, cause contaminants in surface water, which greatly contaminate the environment. Leakages, irrigation, extraction, and wastewater operations are all impacted by these human-caused changes, which have a huge impact on the water cycle. Local social factors connected to land-use techniques and infrastructure must be taken into account. Both surface and groundwater systems are contaminated by a variety of processes, including agriculture, industry, and urban activities. Figure 1 illustrates this and the various routes that contaminants can travel (Akhtar et al., 2021).

Conversion to agriculture in semiarid and arid areas mobilizes subsurface salts, dissolves minerals like gypsum, and raises water tables, causing salinization. In humid regions, impacts stem from fertilizer transport. A 600% rise in global nitrogen fertilizer use from 1961 to 2000 greatly affects nutrient cycles. A 50% reduction in wetlands diminishes denitrification, a key nitrate sink (Scanlon et al., 2007).

1.2 The Need for Continuous Water Quality Monitoring

The dynamic nature of aquatic systems necessitates ongoing water quality monitoring. Real-time testing is necessary because both organic and manmade factors can cause abrupt changes in water quality. Regular monitoring makes early contamination identification certain, enabling prompt action to protect ecosystems and public health. Continuous monitoring offers vital information for efficient use of water resources in light of tighter environmental regulations.

2. CONVENTIONAL TESTING LIMITATIONS AND THE DEMAND FOR REAL-TIME ASSESSMENT

Global shortages of safe drinking water have gotten worse in the twenty-first century as a result of growing issues including pollution and global warming. This has made it extremely difficult to guarantee a steady fresh water supply. Municipal corporations in many large Indian towns treat river water chemically before dispensing it, yet this treated water frequently reaches consumers without being sufficiently tested for safety.

There are many obstacles to real-time water quality monitoring, such as the effects of global warming, a lack of water resources, and a rising population. More efficient techniques for real-time monitoring of water quality parameters are therefore urgently needed.

One important parameter is pH, which measures hydrogen ion concentration and establishes the acidity or alkalinity of water. A pH of 7 indicates that the water is pure, with values lower or higher suggesting acidity or alkalinity. The optimal pH range for drinking is between 6.5 and 8.5 on the pH scale, which runs from 0 to 14.

Another important statistic is turbidity, which measures the number of suspended particles that are frequently imperceptible to the naked eye. Lower turbidity implies healthier water, but higher levels raise the danger of cholera and other waterborne illnesses, including diarrhea.

To assess if the water is hot or cold, temperature sensors are also used. Conventional water quality monitoring methods confront substantial difficulties despite these critical criteria. Uncertainty exists because of the reliance on chemical treatment and the scant post-treatment testing. The safety of the water supply may also be further jeopardized by the fact that conventional procedures might not provide real-time findings. Therefore, it is necessary to review and improve water quality evaluation methods (Gikas & Grant, 2013).

2.1 Importance of Real-Time Feedback in Water Quality Management

A key component of improving water quality management is real-time feedback obtained through high-frequency data gathering and near-real-time networking technology. The following crucial details highlight its significance:

1. More Data at Lower Cost:
    - The long-term declining cost per measurement more than offsets the initial investment in sensor equipment.
    - Automated sensor networks optimize resource allocation by lowering travel, collecting, and analysis costs
    - Allows for frequent data collection even in difficult-to-reach or remote places.
2. Less Knowledge Barriers:
    - Easy access and interpretation are made possible by integrating data into a user-friendly interface (Model Interfacing).
    - Without the requirement for specialist statistical or visualization abilities, immediate visualization promotes informed decision-making.
    - Automation reduces the need for technical competence on-site.
3. Quick Action:
    - Decision-makers can respond quickly to new problems thanks to real-time data.
    - Automated alarm systems improve site safety by promptly notifying users of potential risks or significant changes in the quality of the water.
    - Quick post-event analysis is made possible, which helps to better understand changes in water quality following significant accidents.
4. Adaptive Monitoring:
    - Quick access to high-resolution data enables quick modifications to monitoring plans in response to observed system behavior.
    - Maximizing the significance and impact of the data gathered is made possible by systematic network administration, which assures efficiency and adaptability.

In conclusion, real-time feedback systems provide a practical, adaptive, and affordable method of managing water quality. They give decision-makers the resources they need to act quickly in the face of shifting circumstances, thereby protecting water supplies and guaranteeing their sustainable usage.

3. UNVEILING SMART WATER QUALITY MONITORING (SWQM)

Creating a large-scale, intelligent prediction technique has been made possible by the revolution in water quality monitoring brought about by the fusion of IoT and AI. The focus of this study is on real-time water quality evaluation for sophisticated water monitoring in contemporary cities. The system promises to improve worldwide water safety by using data analytics, cloud computing, and smart water quality sensors.

Figure 2. Smart water system using IoT and cloud

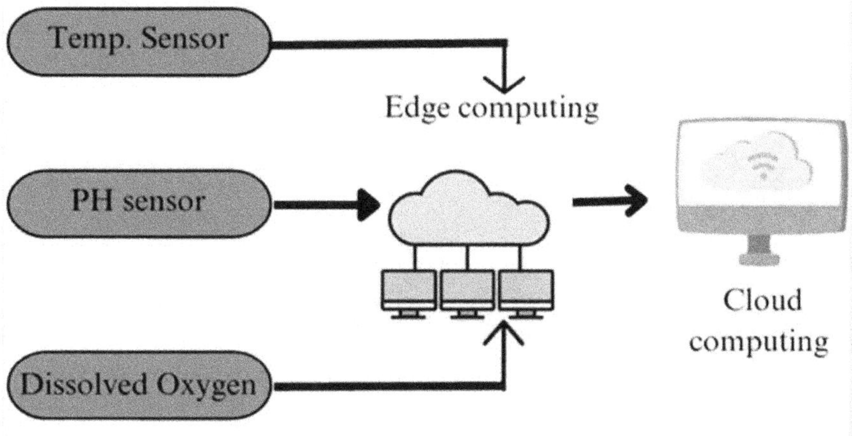

A cost-effective monitoring system, using machine learning for early drought detection, and a thorough examination of water quality forecasting methods are notable contributions. Through enabling effective and pro-active water quality management, this integration provides crucial insights for promoting sustainable urban growth. Communities can protect their water supplies by using these technologies, which will ultimately lead to a healthier, more sustainable future (Hemdan et al., 2023).

3.1 Understanding SWQM: IoT Sensors, Data Analytics, and Machine Learning

IoT, cloud, and machine learning are some of the most advanced technologies being used in the field of smart water quality monitoring (SWQM). It highlights how important clean water is for human survival and describes the problem of worldwide water contamination. Important elements like sensing hardware, a data transmission network, edge computing for real-time analytics, and powerful data processing capabilities are all included in the SWQM system. Real-time analytics and reporting for effective decision-making are made possible by the combination of cloud computing and IoT-based water quality sensors. The goal of the research is to create a low-cost, effective IoT-based water quality system that offers an original approach to water management.

Key Contributions:

1. The development of an IoT-based water quality system for implementation at a reasonable cost.
2. The creation of a low-cost IoT-based water quality system as a drought early warning system.
3. Thorough data preprocessing, which includes feature selection, handling of outliers, and missing data management.
4. The use of forecasting algorithms and their evaluation, with a particular emphasis on Facebook Prophet's capability to analyze water quality.
5. A comparison of the LSTM and Facebook Prophet to help data suppliers and analysts decide on the water quality.

Figure 3. Hardware rendering of the suggested IoT system

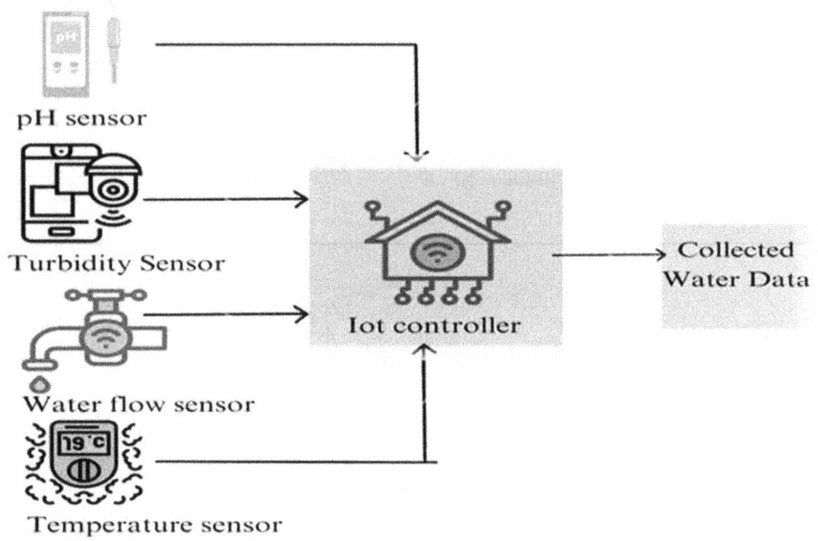

Overall, the study presents a comprehensive approach to revolutionize water management through the amalgamation of IoT sensors, data analytics, and machine learning techniques, aiming to ensure safe and clean water accessibility worldwide.

3.2 Advantages of SWQM for Water Quality Assessment

Continuous real-time monitoring of water quality metrics is made possible by SWQM's use of cloud computing and IoT-based sensors. Due to the fast input, any deviations or anomalies can be addressed right away.

1. Drought Early Warning System: The SWQM's machine learning integration enables the early detection of drought conditions. This vital skill gives communities advance warning, enabling them to implement water conservation measures swiftly.
2. Economicalness: The creation of an affordable IoT-based water quality system makes it possible to implement this technology in a variety of settings. It is a desirable solution for both urban and rural areas because of its cost.
3. Detailed Data Analysis: SWQM applies stringent data pretreatment methods, such as handling outliers, managing missing data, and feature selection. This guarantees the validity of the data utilized for analysis, resulting in more precise evaluations of the water quality.
4. Forecasting Algorithms for Informed Decisions: This study assesses forecasting algorithms with a particular emphasis on Facebook Prophet's performance in analyzing water quality. Data suppliers and analysts are now more equipped to control water quality because of this information.

5. Improvement of Global Water Safety: Communities can actively contribute to improving global water safety by utilizing SWQM. The combination of IoT sensors, data analytics, and machine learning methods is a potent tool for protecting water resources all around the world.

With real-time monitoring, early drought identification, cost-effectiveness, extensive data analysis, and capability for informed decision-making, SWQM represents a substantial development in the assessment of water quality. This all-encompassing strategy could alter current water management methods and lead to a more sustainable future.

4. TRANSFORMING WATER QUALITY ASSESSMENT WITH SWQM

An extensive research project called "Transforming Water Quality Assessment with SWQM" is being conducted to solve the crucial problem of water quality on a global scale. The study emphasizes that just a small portion of the water on Earth is fit for human use, which causes a number of issues, such as pollutant concentrations, scarcity, and waterborne infections. The study highlights the critical need for real-time water body monitoring and discusses the difficulties brought on by elements like growing emissions and global warming. The study suggests using cutting-edge technology, in particular the Internet of Things (IoT), to improve water management in order to address these concerns. It underscores how crucial security is when putting IoT solutions into place to protect against potential online threats. The study highlights its main goals, which include alarms for harmful parameters, pollution identification, and real-time water quality monitoring. To monitor trends and improve device management, machine learning models are used, producing remarkably accurate results.

The study is divided into a number of components, including a summary of smart water management options, a review of related literature, and a thorough explanation of the suggested IoT-based framework. Along with the design of the machine learning model for analyzing water quality, it also discusses the crucial parameters that must be monitored and controlled. A detailed presentation of the findings and their consequences is followed by a summary of the future directions for this important field of research.

4.1 Early Detection and Prevention of Contamination Events

Early Warning Systems (EWS) are increasingly being used by regulatory bodies and suppliers to monitor water quality, underscoring its significance in preserving public health. High false alarm rates and false negatives may result from using conventional approaches based on exceedance-criteria event detection. The use of characteristics like pH, conductivity, chlorine concentrations, and TOC in anomaly-based detection techniques has showed potential. These techniques can be divided into three groups: data mining, artificial intelligence, and statistics. High false alarm rates, processing requirements, and the requirement for a large amount of training data are only a few of the problems that currently exist with existing techniques. This paper suggests an expansion of the Dempster-Shafer (D-S) evidence theory to overcome these flaws. With its explicit estimations of imprecision and conflicts, this theory excels at handling knowledge that is insufficient and uncertain. In order to resolve evidentiary disputes, the revised D-S theory makes use of time series residuals from water-quality predictions as well as weighted averaging and time-dimension information. Variously severe simulated and actual contamination incidents are used to test the methodology. Overall, this novel approach has the potential to improve the early detection and avoidance of water contamination incidents (Hou et al., 2013).

4.2 Swift Responses to Pollution Incidents and Public Safety

the major difficulties in managing water resources that different countries confront, with Taiwan as a case study. The need for sophisticated water quality monitoring systems to balance consumer demands and protect natural ecosystems is brought home by this. Inability to deliver timely, high-resolution data is a criticism leveled about conventional approaches. Real-time monitoring and analysis are made possible by the development of IoT-driven wireless sensor networks (WSNs), which is regarded as a game-changer.

1. Taiwan's Particular Water Resource Issues: Effective water resource development is hampered by heavy rainfall and difficult terrain.
2. Industrial Impact on Water Quality: The rapid industrialization of the world has resulted in serious water contamination occurrences, which have a considerable financial impact on businesses.
3. Restrictions on Conventional Monitoring Techniques: Conventional manual in-situ (TMIS) techniques and costly equipment are not timely, do not cover all relevant areas, and are not cost-effective.
4. The Evolution of Water Quality Monitoring Around the World: In the middle of the 20th century, many industrialized nations started implementing continuous monitoring programs.
5. Parameters Monitored: Water level, conductivity, turbidity, ORP, temperature, and dissolved oxygen are all monitored by mature online systems. Environmental and toxicological concerns are addressed by biological monitoring.
6. The importance of Wireless Sensor Networks (WSNs) and IoT: WSNs are the foundation of IoT technology and are crucial for data collecting, processing, and wireless transmission.
7. IoT in Environmental Applications: IoT technology is widely used in a variety of environmental fields, such as the management of air and water resources.
8. WSN and IoT's Growing Purpose: WSN technology has developed from device-level applications to autonomous continuous monitoring and integration with IoT for anomaly detection.
9. Challenges in Current Water Quality Monitoring: Sensor production limitations, maintenance costs, size constraints, and weak early warning capabilities hinder effective environmental management.
10. Total Solution Approach: The proposed solution advocates for the widespread deployment of domestically-produced, cost-effective, and miniaturized water quality monitoring devices integrated with IoT technology.
11. Value-added Analysis for Pollution Management: Utilizing cross-environmental data, the system can swiftly identify pollution sources, issue early warnings, and offer value-added applications for public safety.

Nations may substantially improve their capacity to identify, address, and manage environmental hazards by strategically deploying monitoring sites and utilizing IoT-driven WSN technology. This will protect the public's safety and preserve natural ecosystems (Chen et al., 2020).

4.3 Ensuring Adherence to Regulatory Standards

Given the global water problem, ensuring compliance with legal requirements is crucial when managing water resources. It is essential to maintain water quality because less than 2.5% of the water on Earth is fit for human use. This study promotes using sensors, IoT devices, and machine learning in water infrastructure. Such developments promise to greatly increase efficiency through real-time monitoring, pollution alarms, and trend analysis. Prioritizing protection against future cyberattacks is necessary, nevertheless. To protect public health and the environment in the face of growing water scarcity and pollution issues, strict adherence to regulatory rules is crucial.

5. OPTIMIZING RESOURCE ALLOCATION AND ENHANCING PUBLIC AWARENESS

The development of smart cities depends on effective resource allocation, notably in the area of Internet of Things (IoT)-enabled smart building water management. This involves a number of essential elements.

First, sensors that continuously track water use in real-time are used in smart water consumption management. This makes it possible to allocate this important resource precisely and reduces waste. The second crucial element is early issue detection. IoT systems quickly locate any pipe problems, preventing water loss and potential damage. In addition to saving resources, this also avoids structural damage.

Additionally, data analysis and machine learning are crucial. Using machine learning algorithms, centralized systems analyze the gathered data to identify potential leaks, identify anomalies, and identify usage trends. This preventive strategy guarantees prompt remedial action, improving overall effectiveness. Another crucial element is resource efficiency. Utilizing technology, smart buildings optimize water consumption, saving resources and thus cutting expenses. This strategy supports objectives for environmental and economic sustainability.

Integration with Wider Water Monitoring is also crucial. The management of urban water resources as a whole must be combined with smart building water management. This provides thorough information on water levels, flow rates, and pressure to ensure that methods are successful. In the end, these projects' main goal is to promote sustainability. Smart water resource utilization in urban settings becomes a reality through lowering consumption and waste. This all-encompassing strategy not only protects this priceless resource but also guarantees a future for urban areas that is more sustainable and resilient (Palermo et al., 2022).

Recognizing the crucial role of women in daily water consumption is fundamental. Tailored campaigns for both urban and rural communities, accounting for climate-induced disparities, are essential. Comprehensive initiatives are required to address the global water crisis. These initiatives include educational efforts, with a focus on schools, and the promotion of conservation practices to boost public awareness of intelligent water quality monitoring. Participating in citizen science initiatives gives people more authority and develops a sense of responsibility. It is essential to communicate concise information on environmental effects. The European Water Framework Directive also emphasizes the importance of cross-sector cooperation in order to effectively protect water quality. These tactics work together to promote responsible water stewardship, thereby addressing the global water crisis (Seelen et al., 2019).

5.1 Resource Allocation Efficiency Through SWQM

In light of rising population demands, climate change, and pollution, efficient water resource management is crucial. It draws attention to how the growth of smart cities—made possible by improvements in ICT and IoT—can improve urban sustainability. The discussion switches to smart buildings, where the idea of "intelligent buildings" with advanced resource management technology is introduced. The paper focuses on the importance of sensor devices for maximizing water consumption, finding leaks, and guaranteeing sustainability in general. It summarizes a number of studies and surveys conducted in the area and offers a distinctive viewpoint on IoT-based solutions for water monitoring at the building scale.

1. As a result of pollution, climate change, and growing population, effective management of water resources is essential.
2. The growth of smart cities, driven by ICT and IoT, is essential for improving urban sustainability.
3. For resource optimization, "intelligent buildings" use cutting-edge systems like Building Management Systems (BMS).
4. In smart buildings, sensor devices allow for accurate water management and early leak detection.
5. Centralized systems analyze water meter data to provide information about potential leaks and to implement machine learning algorithms for improvement.
6. Water level, flow, and pressure data are essential for efficient operational management and affordable solutions.
7. The paper presents an overview of innovative systems for smart water monitoring at the building scale, focusing on IoT-based strategies for measuring water levels, monitoring consumption, and detecting leaks.

5.2 Empowering Public Awareness via Real-Time Data Accessibility

Public participation in decisions involving water resources, with a focus on the need for better engagement and communication tactics. It analyzes a number of variables that affect water awareness, such as education level, age, career, gender, geography, and firsthand knowledge of water-related calamities like drought or flooding. By providing fast, accurate information on water quality and related stressors, real-time data accessibility can play a crucial role in boosting public awareness. This can promote a more knowledgeable and involved populace by improving understanding and encouraging proactive participation in efforts to address water quality challenges.

6. REAL-LIFE APPLICATIONS AND CASE STUDIES

The vital importance of monitoring and preserving water quality for preserving life and preventing pollution-related risks. It also emphasizes the initiatives taken by groups like the Central Pollution Control Board (CPCB) in India to set up permanent monitoring facilities for water bodies.

1. Central Pollution Control Board (CPCB) Monitoring Stations

The continuous monitoring stations operated by the CPCB are essential for determining the quality of the country's water supplies. These stations transmit real-time data to a central site using GPRS/GSM or 3G cellular serviceability technologies. This guarantees a prompt response and intervention in the event of any departure from desired water quality standards.

2. Ganga River Valley Monitoring Network: The CPCB's attempt to establish a water standard monitoring network throughout the Ganga River valley reflects a commitment to solving the unique difficulties this historically significant water body faces. One of India's most significant rivers' water quality is to be protected through this network.
3. Testing for Water Quality in India: The startling number of public water sources in India (50 lakh) and the amount of samples that must be examined (120 lakh/year, according to NRDWP) highlight the enormity of the task. This shows the necessity of effective, automated water quality monitoring systems.
4. Evolution of Water Testing Methods: A historical overview of water testing techniques, from substrate techniques to epidemiological methods, demonstrates the development of analytical philosophies. This evolution is a result of ongoing work to improve the precision and efficacy of water quality measurement.
5. BOD Values and Water Quality: Data showing the proportion of specimens recorded between 1995 and 2007 with BOD values less than 3mg/liter illustrate a good trend in water quality. This could indicate a decrease in organic pollution levels, which is good news for the environment.

Figure 4. Examination of purity

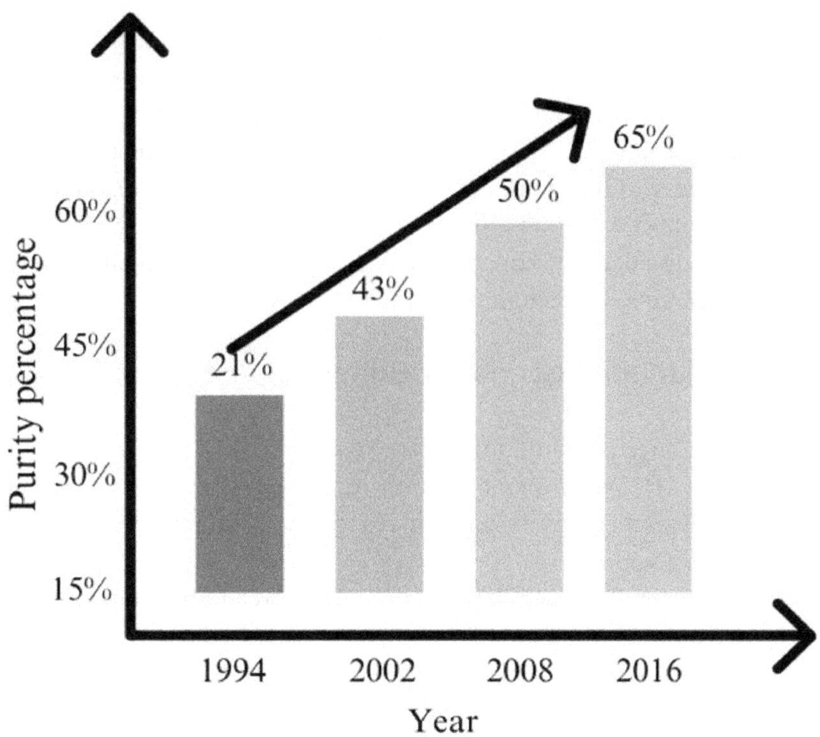

With a focus on technological developments, legislative actions, and the sheer size of the task, the information offers an overall complete assessment of the efforts done in India to monitor and preserve water quality. These case studies and real-world examples show the value of ongoing, automated water quality monitoring systems for preserving environmental and public health.

7. ADVANCEMENTS AND INNOVATIONS IN SWQM

In order to deal with the increasing environmental constraints on water bodies, smart water quality monitoring has made amazing strides. This development involves monitoring crucial variables, including pH, conductivity, dissolved oxygen, turbidity, and nutrients in real-time, enabling quick responses in emergency situations.

Modern sensors reduce the need for manual sampling by providing high precision and continuous operation. Wireless monitoring technologies increase monitoring coverage, especially in rural or coastal areas, while transmitting data to the cloud for thorough analysis. Integrating different indicators offers in-depth insights into trends in water quality, which are essential for reducing particular concerns. Intelligent monitoring systems save costs by lowering vessel and labor expenditures, while cloud computing streamlines data storage and processing. Fundamentally, these technologies transform how aquatic ecosystems are assessed and managed, ensuring quick response to ecological emergencies and protecting priceless water supplies (Park et al., 2020).

IoT sensors are used by smart water quality monitoring systems to deliver real-time data on pH, temperature, and other critical parameters. AI-driven analytics that identify patterns and abnormalities and enable proactive actions underpin these systems. Predictive modeling also foresees changes, assisting in becoming ready for upcoming changes.

Users of mobile applications can keep an eye on the quality of their drinking water, and operational procedures are streamlined by treatment facilities' seamless connection. Ecosystems are protected by widespread coverage provided by distributed networks, with prompt response to early warnings. The block-chain recording ensures strong data security. Affordably priced sensor technology encourages diversity and democratizes access to information on water quality. The UN Sustainable Development Goals for worldwide access to clean water are aligned with sustainability through integration with conservation activities. These cutting-edge solutions promote a better, more sustainable future for everyone while also addressing water scarcity and contaminants.

7.1 Integration of Artificial Intelligence for Enhanced Insights

Water quality monitoring has a huge potential for increasing perceptions and decision-making by integrating Artificial Intelligence (AI). Water quality metrics may be assessed and predicted in real-time thanks to the analysis of large datasets by AI systems. In reaction to changing environmental conditions, this enables stakeholders to move quickly.

One major benefit of artificial intelligence is its capacity to interpret high-frequency data streams from diverse sensors, resulting in a more thorough understanding of water bodies. AI-driven systems provide continuous, fine-grained monitoring, in contrast to traditional methods with low-frequency sampling, and may detect subtle fluctuations in physico-chemical-biological parameters. With the help of this high-resolution data, it is possible to spot minute alterations that may be crucial for spotting new problems.

Furthermore, AI algorithms are particularly adept at building intricate cause-and-effect connections in dynamic systems. AI can find hidden patterns and foresee possible deterioration events by comparing data on water quality with environmental factors such as climate, human activity, and industrial processes. For proactive decision-making and the implementation of focused mitigation measures, this analytical capacity is essential.

Models powered by AI can also change to accommodate changing circumstances. AI can swiftly include these variables in its analysis as new determinants are added to water systems as a result of elements like climate change and industrial behavior. This versatility makes sure that monitoring systems are still applicable and efficient in the face of new difficulties.

AI deals with the problem of data interoperability as well. AI assists in harmonizing data gathered from many sources and areas by using standardized analysis techniques. This encourages a common knowledge of water quality on a global scale, making cooperative conservation and management activities possible.

In conclusion, including AI into water quality monitoring transforms how we approach protecting this vital resource. It supports global data consistency, provides real-time, high-resolution analysis, locates intricate relationships, and adapts to new problems. By utilizing AI, people can make better judgments and put targeted interventions into place to safeguard water quality and the health of communities and ecosystems around the world (Manjakkal et al., 2021).

7.2 Sensor Technology Innovations: Miniaturization and Efficiency

The urgent issues surrounding water quality monitoring have seen substantial progress because to developments in sensor technology. Miniaturization and efficiency developments are principally responsible for this progress. Due to the occasional sample collection and lab analysis used in traditional approaches, it was difficult to establish cause-and-effect correlations. The interval between sampling and analysis also contributed to the results' unpredictability. Autonomous sensors that can detect physical, chemical, and biological factors in real time have appeared to fill these gaps. Satellite photography helps with regional water quality evaluation, and sensorized buoys, boats, and even aerial vehicles have been used for in-situ monitoring.

Although problems still exist, especially in extreme situations like underwater ones where sensing technology is used. Significant obstacles include things like restricted range, communication problems, and corrosion. As a result, it's critical to design multimodal systems and deployment methods that make use of mobile robots. By capturing spatiotemporal fluctuations, these technologies hope to provide a thorough and timely picture of water quality. In the end, the use of clever algorithms, such artificial intelligence (AI), for standardized data processing and forecasting, promises to significantly increase the efficiency of sensor technologies in protecting water resources.

7.3 Collaborative Initiatives and Partnerships for SWQM Advancements

Collaborations and Partnerships for SWQM (Smart Water Quality Monitoring) Developments:

Efforts to improve water resource management through cutting-edge monitoring methods have sparked collaborations and cooperative efforts amongst many sectors. The effectiveness and accessibility of data on water quality are being improved through these initiatives. Important points include:

1. Interdisciplinary Collaboration: To promote SWQM, environmental scientists, engineers, technologists, and policymakers must work together. This multidisciplinary approach encourages creativity and makes sure that monitoring systems are in line with both accepted scientific practices and real-world uses.
2. Public-commercial Partnerships (PPP): Public and commercial organizations are collaborating to create and use SWQM technology. Public sector knowledge of environmental rules and water resource management is complemented by private sector competence in sensor development, data analytics, and ICT.
3. Research Organizations and Universities: Academic establishments are essential to the development of SWQM technology. Research collaborations help to develop cutting-edge sensors, data analysis methods, and machine learning integration for predictive modeling.
4. Governmental Organizations and Regulatory Bodies: Working together with environmental organizations guarantees that SWQM systems adhere to legal requirements and support the development of policies that are supported by factual information. These collaborations assist in connecting monitoring activities to more general environmental objectives.
5. Technology Providers and Start-ups: Cutting edge sensor, communication, and data analytics platforms are created and provided by cutting edge technology providers and start-ups, which contribute to the advancement of SWQM. Partnerships with these organizations foster the development of new technologies.
6. Environmental NGOs: Environmental NGOs frequently play a crucial role in promoting sustainable water management methods. By promoting the adoption of cutting-edge monitoring technologies, aiding pilot projects, and bringing attention to the significance of water quality monitoring, they make a contribution.
7. Global Partnerships: In the area of SWQM, global alliances promote information sharing and technology transfer. Addressing issues with transboundary water quality is made easier by shared research projects and data-sharing agreements.
8. Grants and Funding Organizations: Funding for research and development of SWQM technology is provided by charitable and governmental organizations. These financial alliances are essential for driving innovation and wide-scale adoption.
9. Standardization Bodies: Cooperation with groups in charge of establishing industry standards ensures data consistency and interoperability between various SWQM systems. An integrated framework for data collecting and analysis is developed as a result of standardization initiatives.
10. Community Engagement and Citizen Science: Involving neighborhood groups and amateur researchers in water quality monitoring programs fosters a sense of responsibility and environmental stewardship. Collaboration enables communities to provide useful data and expand the scope of monitoring initiatives.

The development of Smart Water Quality Monitoring depends heavily on collaborative efforts and collaborations. These partnerships foster innovation, guarantee legal compliance, and ultimately support more efficient and sustainable use of water resources by bringing together a variety of resources and skills.

8. ENSURING WATER SAFETY AND SUSTAINABILITY FOR FUTURE GENERATIONS

A crucial objective is to ensure water security and sustainability for future generations, especially in India, where freshwater demand is expected to outpace supply by 2050. India only has a tiny share of the world's freshwater reserves despite being home to a sizeable portion of the world's population. Over 60% of Indians live in rural areas where it is difficult to acquire domestic tap water, thus exacerbating this serious scenario.

Rural women typically have to carry heavy loads to get water from far-off sources, which has a negative impact on their health, livelihoods, and general well-being. Additionally, due to overuse and climate change, groundwater resources in rural areas are fast disappearing, with a stunning 61% decrease between 2007 and 2017.

A significant percentage of families now have functional household tap connections thanks to initiatives like the Jal Jeevan Mission. These efforts have been strengthened through partnerships with groups like Tata Trusts, but more of these kinds of initiatives are essential.

It is crucial to make investments in high-quality infrastructure and set up reliable maintenance procedures in order to ensure a dependable and sustainable water supply. Solutions that are specific to the context are required to ensure the sustainability of water sources, especially in various geographies. Furthermore, it is crucial to manage the excessive water usage in agriculture when compared to other nations. For effective water management, utilizing cutting-edge technology such as satellite data, remote sensing, and AI can be crucial.

For water services to be sustained over the long term, financial management is also essential. Along with federal financing for infrastructure, local governments must raise money for operations and upkeep. Integrating local knowledge and practices into water resource management requires coordinated planning at all levels of governance.

Last but not least, a paradigm shift is necessary to turn citizens, local governments, and water agencies from passive implementers to active stakeholders. This calls for broad-reaching communication campaigns, multi-stakeholder involvement, and capacity-building initiatives similar to those seen during the Swachh Bharat Mission.

It is crucial to take broad, bottom-up action now to protect water resources for future generations. Before it's too late, take action right away.

8.1 Contributions to Sustainable Development Goals (SDGs)

The vital significance of water sustainability for India's future, emphasizes the difficulties brought on by population increase, unequal access to water resources, and groundwater depletion. It also addresses current initiatives by the Indian government to provide access to clean water in rural regions, particularly through the Jal Jeevan Mission (JJM) and partnerships with businesses like Tata Trusts.

The initiatives described are in accordance with a number of important objectives that contribute to the Sustainable Development Goals (SDGs):

Clean Water and Sanitation SDG 6: The aforementioned programs, including the joint efforts of the Tata Trusts and the Jal Jeevan Mission, directly advance this objective by focusing on ensuring that all citizens have stable, long-term access to clean water.

A vital component of good health and well-being is increased access to clean water. It lowers the prevalence of waterborne illnesses, encourages good hygiene, and improves general health outcomes.

Access to dependable water sources can have a significant influence on livelihoods, especially for rural women who frequently shoulder the task of water collection in accordance with SDG 1: No Poverty and SDG 8: Decent Work and Economic Growth. It may result in greater output and economic development in these communities.

SDG 11: Sustainable Cities and Communities: Although the article's primary focus is on rural areas, its emphasis on high-quality infrastructure and environmentally friendly water sources is relevant to urban areas as well, which are also dealing with water-related issues.

SDG 13: Climate Action: The reference to the effects of climate change on water availability highlights the need for adaptable measures, which are consistent with the objectives of climate resilience.

SDG 17: Partnerships for the Goals: The Government of India, state governments, Tata Trusts, and other stakeholders' collaborations provide as an example of the type of multi-stakeholder partnerships required to realize these goals.

9. CHALLENGES AND FUTURE DIRECTIONS

Challenges:

1. Deployment Costs: The initial cost of establishing smart water monitoring systems, including hardware and software components, might be a major roadblock to their wider acceptance.
2. electricity Usage: Making sure that these systems are operating as efficiently as possible is essential, especially in distant or off-grid settings where the availability of electricity may be constrained.
3. Maintenance: To maintain accurate and dependable data collection, it is crucial to perform routine maintenance and service on sensors, actuators, and communication equipment.
4. Privacy and Security: As more data is gathered, the risk of cyberattacks during data transmission to the server increases, underscoring the necessity of strong privacy and security measures.
5. Connectivity Coverage: The accessibility of dependable network connectivity, particularly in remote locations, may be a barrier to the efficient operation of these devices.
6. Complexity and Ease of Operation: For systems to be widely adopted, they must be user-friendly and simple to use. Overly complicated systems may discourage users.
7. Adaptability and Replicability: It is difficult to design smart water systems that can be readily modified and copied in many settings and places.

Future Directions:

1. Low-Cost, Energy-Efficient Devices: Widespread IoT applications in the field of water management will be supported and driven by the development of low-cost devices with improved energy efficiency.
2. Privacy and Security Enhancements: Ongoing research and innovation are required to solve privacy and security issues brought on by the massive amounts of data received and communicated by IoT-based systems.

3. Continued advances in communication technology, including wireless protocols, will be crucial for enhancing connectivity and getting around restrictions on battery usage and communication range.
4. Standardization and compatibility: By establishing industry-wide standards and protocols for smart water monitoring systems, it will be simpler to integrate new technologies into existing systems by facilitating compatibility between various parts and systems.
5. Data Analytics and Artificial Intelligence: By utilizing cutting-edge data analytics and AI technologies, it will be possible to analyze water quality data in a more sophisticated manner, providing more precise and timely information for decision-making.
6. Scalability and Flexibility: Systems of the future should be flexible and easy to scale, allowing for varied degrees of monitoring, from specific buildings to whole urban regions.
7. Community Engagement and Education: The key to effective deployment and long-term sustainability will be to involve communities and educate them about the advantages and operation of smart water systems.
8. Environmental Impact Assessment: Thorough analyses of how smart water systems affect the environment will guarantee that they have a favorable impact on long-term water management strategies.

Finally, in order to fully utilize IoT technologies for smart water quality monitoring, it will be crucial to overcome these issues and accept these new directions. It will result in more effective and sustainable water management techniques, which will be advantageous to both the environment and communities.

9.1 Scaling Up SWQM Implementation: Infrastructure and Affordability

Strong water infrastructure that can handle rising consumption needs is required for the transition to smart cities. Integrating machine learning models that have been fed with a wealth of data is essential for achieving robust, autonomous operations. Smart metering is made possible by the Internet of Things, which is emerging as a key technology. While extensive monitoring has many advantages, household-level monitoring at the individual level gives insightful consumer data. A promising alternative that strikes a balance between effectiveness and affordability for water utility providers is a hybrid strategy that combines central water quality monitoring with home flowmeters. The difficulty of maintaining sensors, however, continues to be an important factor (Syrmos et al., 2023).

9.2 Future Prospects: Innovations, Research, and Policy Integration

Future opportunities for research, innovation, and policy integration in water management seem bright. The use of IoT technology in smart metering systems, in particular, has the potential to fundamentally alter how we track and control water consumption. This development may enable people to choose wisely how much water they use, promoting more sustainable habits on a worldwide level. A wider adoption and accessibility are also made possible by the development of highly modular and affordable solutions, as shown by the provided IoT system.

The use of machine learning for water disaggregation and other advancements in water quality monitoring are essential steps toward guaranteeing safe and dependable water distribution. This research lays the groundwork for developing new markets and business models based on service-oriented methodologies.

As smart cities advance, optimizing water distribution networks will depend on tackling interoperability issues and utilizing technologies like Low-Power Wide-Area Networks (LPWANs).

From a policy standpoint, it is obvious that proactive steps are required to facilitate the integration of smart water metering systems. Recognizing these technologies' potential to lessen water problems, policymakers ought to investigate incentives and rules that promote their implementation. In order to improve and scale up these advances, continuing research efforts like the one presented should be given ongoing funding and support.

In general, the future of water management is moving toward being more data-driven, effective, and sustainable. We can address the world's water concerns comprehensively and effectively by utilizing developments in IoT, machine learning, and smart city technology. The management of water resources will be shaped in the future by this fusion of technology, research, and policy.

10. CONCLUSION

In order to effectively battle water pollution and provide sustainable water management, smart water quality monitoring (SWQM) systems that integrate Internet of Things (IoT) and Artificial Intelligence (AI) technologies have been developed. These devices enable precise, in-the-moment evaluations of water quality, which is a substantial improvement over traditional testing technique.

The collaboration of cutting-edge sensors, cloud computing, and data analytics is the key to SWQM's efficacy. This combination makes it possible to manage water resources pro-actively, which is essential for protecting aquatic ecosystems and people's health. Decision-makers are given the tools they need to act quickly and strategically by SWQM, which quickly detects abnormalities, foresees changes, and provides early warnings.

In addition, SWQM has an impact that goes beyond technology, as it is essential to smart city resource allocation optimization. It supports initiatives to reduce waste and advance sustainability in general. It's not just a technological undertaking, though. Public awareness campaigns are essential for involving regional populations, encouraging a sense of accountability, and encouraging cross-sector cooperation. This all-encompassing strategy makes sure that SWQM's advantages are fully realized.

Even yet, there are obstacles in the way of achieving SWQM's full potential. Cybersecurity issues and deployment costs are pressing issues that necessitate creative solutions. Looking ahead, the emphasis switches to developing affordable gadgets that are protected by strong security measures, utilizing renewable energy sources, and utilizing cutting-edge data analytics. In order to make these technologies accessible and adaptive across various situations, expanding network coverage is also crucial.

To achieve this flexibility, efforts in standardization and localization are essential. Regardless of the unique requirements or limitations of a particular place, SWQM aims to be a flexible solution that is accessible everywhere. This versatility is essential in managing the wide range of water management issues that many communities worldwide deal with.

SWQM stands out as a ray of hope in a world where there is a shortage of fresh water and it is contaminated. It serves as a blueprint for a sustainable future with ample water for future generations by exemplifying the peaceful coexistence of environmental management and cutting-edge technology. Adopting and developing these disruptive solutions must be done as soon as possible. In our shared effort to ensure a thriving, water-rich planet for everybody, every second counts.

REFERENCES

Akhtar, N., Syakir Ishak, M. I., Bhawani, S. A., & Umar, K. (2021). Various Natural and Anthropogenic Factors Responsible for Water Quality Degradation: A Review. *Water (Basel), 13*(19), 19. Advance online publication. doi:10.3390/w13192660

Chen, F.-L., Yang, B.-C., Peng, S.-Y., & Lin, T.-C. (2020). Applying a deployment strategy and data analysis model for water quality continuous monitoring and management. *International Journal of Distributed Sensor Networks, 16*(6), 1550147720929825. doi:10.1177/1550147720929825

Gikas, J., & Grant, M. M. (2013). Mobile computing devices in higher education: Student perspectives on learning with cellphones, smartphones & social media. *The Internet and Higher Education, 19*, 18–26. doi:10.1016/j.iheduc.2013.06.002

Hemdan, E. E.-D., Essa, Y. M., Shouman, M., El-Sayed, A., & Moustafa, A. N. (2023). An efficient IoT based smart water quality monitoring system. *Multimedia Tools and Applications, 82*(19), 28827–28851. doi:10.100711042-023-14504-z

Hou, D., He, H., Huang, P., Zhang, G., & Loaiciga, H. (2013). Detection of water-quality contamination events based on multi-sensor fusion using an extented Dempster–Shafer method. *Measurement Science & Technology, 24*(5), 055801. doi:10.1088/0957-0233/24/5/055801

Manjakkal, L., Mitra, S., Petillot, Y. R., Shutler, J., Scott, E. M., Willander, M., & Dahiya, R. (2021). Connected Sensors, Innovative Sensor Deployment, and Intelligent Data Analysis for Online Water Quality Monitoring. *IEEE Internet of Things Journal, 8*(18), 13805–13824. doi:10.1109/JIOT.2021.3081772

Palermo, S. A., Maiolo, M., Brusco, A. C., Turco, M., Pirouz, B., Greco, E., Spezzano, G., & Piro, P. (2022). Smart Technologies for Water Resource Management: An Overview. *Sensors (Basel), 22*(16), 16. Advance online publication. doi:10.339022166225 PMID:36015982

Park, J., Kim, K. T., & Lee, W. H. (2020). Recent Advances in Information and Communications Technology (ICT) and Sensor Technology for Monitoring Water Quality. *Water (Basel), 12*(2), 2. Advance online publication. doi:10.3390/w12020510

Scanlon, B. R., Jolly, I., Sophocleous, M., & Zhang, L. (2007). Global impacts of conversions from natural to agricultural ecosystems on water resources: Quantity versus quality. *Water Resources Research, 43*(3), 2006WR005486. Advance online publication. doi:10.1029/2006WR005486

Seelen, L. M. S., Flaim, G., Jennings, E., & De Senerpont Domis, L. N. (2019). Saving water for the future: Public awareness of water usage and water quality. *Journal of Environmental Management, 242*, 246–257. doi:10.1016/j.jenvman.2019.04.047 PMID:31048230

Syrmos, E., Sidiropoulos, V., Bechtsis, D., Stergiopoulos, F., Aivazidou, E., Vrakas, D., Vezinias, P., & Vlahavas, I. (2023). An Intelligent Modular Water Monitoring IoT System for Real-Time Quantitative and Qualitative Measurements. *Sustainability (Basel), 15*(3), 3. Advance online publication. doi:10.3390u15032127

Chapter 9
Data-Driven Aquatics: The Future of Water Management With IoT and Machine Learning

V. Dankan Gowda
https://orcid.org/0000-0003-0724-0333
BMS Institute of Technology and Management, India

Anil Sharma
https://orcid.org/0000-0002-7115-6278
Amity School of Engineering and Technology, Amity University, Noida, India

Rama Chaithanya Tanguturi
https://orcid.org/0000-0002-9923-7360
PACE Institute of Technology and Sciences, India

K. D. V. Prasad
https://orcid.org/0000-0001-9921-476X
Symbiosis Institute of Business Management, Symbiosis International University, India

Vasifa Sameer Kotwal
Dr. D.Y. Patil Polytechnic, Kolhapur, India

ABSTRACT

The water management industry has undergone a sea change since the advent of machine learning (ML) and internet of things (IoT) technology. In this chapter, the utilization of ML and IoT applications for assisting with the fundamentals of water management data gathering and preprocessing will be explored. In order to make educated decisions toward water sustainability, sensors and gadgets connected to the IoT have improved monitoring and evaluation of water resources. In the initial paragraphs, the primary focus of the chapter is introduced: the importance of data collection in water management and the challenges of using traditional data collection techniques. However, before the data acquired from these sensors can be used for analysis and modeling, it must frequently undergo some form of preprocessing. Important data preparation tasks including data cleansing, outlier identification, and data fusion are discussed in this chapter. The reliability of future ML algorithms is enhanced by preprocessing the data to verify its correctness and consistency.

DOI: 10.4018/979-8-3693-1194-3.ch009

INTRODUCTION

Water is a finite resource that is crucial to human survival and important to a broad variety of human activities, from agriculture and industry to domestic duties and hygiene. However, complicated challenges in water management have arisen as a result of environmental alterations, population increase, and rising water demands. Examples of advanced technologies that can aid in this regard include the IoT and ML.

The benefits of applying ML to the subject of Water Resources Management (WRM) have recently seen an upswing in the study. The emergence of big data has been a major factor in this shift, as it has provided hydrologists with novel approaches to old issues and encouraged innovative applications of ML. By 2025, the world's data is expected to have ballooned to a staggering 175 zettabytes. This massive data deluge marks the beginning of a new era in WRM. The next challenge for hydrological researchers is to integrate established hydrology principles with these cutting-edge, data-driven methods. Various ML approaches are currently used to guide decisions ranging from the most fundamental to the most complex in the scientific realm. Machines are best suited to process and leverage big data because of their size, velocity, veracity, and variety Koditala and Pandey (2018). Hydrologists have not been immune to ML's appeal in recent years, with many adopting the technology across multiple domains to take advantage of its superiority in handling complex scenarios.

In the coming years, hydrologists around the world will need to innovate and strategize for WRM security in response to the challenges posed by climate change, increasing water resource constraints, expanding populations, and natural threats. The ninth phase of the Intergovernmental Hydrological Programme (IHP) (IHP-IX, 2022-2029) has recently begun. To ensure water security in a time of climate change, this plan puts hydrologists, academics, and policymakers in the driver's seat, with the goal of fostering a resilient and sustainable water ethos. Furthermore, new opportunities for refined hydrological assessments by simplifying complex issues are opening up as a result of the proliferation of available hydrological datasets and the development of sophisticated ML algorithms Sugumar, R, Phadke, Prasad, and R (2021).

There has been a shift from single-step to multi-step prediction, from short-term to long-term forecasting, from deterministic to probabilistic models, from univariate to multivariate systems, from structured to unstructured data utilization, and from simple spatial analyses to more complex spatiotemporal and geo-spatio-temporal environments Perumal et al. (2022). The unpredictability and complexity of natural hydrological processes are well captured by ML models, which have made great strides in supporting optimal decisions in WRM. Since ML is so effective at computational tasks, it can drastically reduce processing needs, making it possible to switch from physically driven models to ML-based ones for more complex problems. Therefore, new hydrological problems, such as droughts and floods, can be better understood and handled with the help of ML developments.

Water Management Challenges

Environment, society, and infrastructure all play a role in creating complex problems for water management to solve. The complexity and scope of these problems call for creative solutions, such as the IoT and ML. Some of the most pressing problems in water management are depicted in Figure 1.

- **Water Scarcity:** With a growing population comes a greater need for water in many forms, including agriculture, industry, and domestic consumption. Inadequate water supply, over-extraction of groundwater, and competing demands have created water scarcity in many regions.
- **Water Pollution:** Agriculture, industry, and sewage that hasn't been treated all contribute to pollution in water sources. The quality of water, aquatic ecosystems, and human health are all put in jeopardy by this pollution.
- **Aging Infrastructure:** The water treatment and distribution systems in many communities are in disrepair and need to be upgraded or replaced. Water is lost and wasted due to leaking pipes.
- **Climate Change and Extreme Events:** Extreme weather, such as floods and droughts, are becoming more common as a result of climate change, which has also altered precipitation patterns. Water management problems are made worse as a result of these shifts in the natural water cycle.
- **Inefficient Water Use:** Excessive irrigation and inadequate water conservation are two examples of inefficient water use that contribute to water waste. Sustainable water management requires the elimination of inefficiencies.
- **Lack of Real-time Monitoring:** Oftentimes, real-time monitoring is not available in conventional water management systems Kolli, Ranjani, Kavitha, Daniel, and Chandramauli (2023). This slows down our ability to react to changes in the water supply, distribution, or quality.
- **Complex Regulatory Environment:** Effective water management can be complicated by the need to navigate complex regulatory frameworks pertaining to water rights, pollution control, and water allocation.
- **Data Fragmentation:** The data collected and stored regarding water resources often show incompatibility with one another. It isn't easy to bring together and make sense of all this disparate information.
- **Limited Predictive Capabilities:** Predictive abilities are often lacking in conventional approaches to water management. Without empirical evidence, forecasting shifts in water supply or quality can be difficult.
- **Stakeholder Engagement:** It can be difficult to strike a balance between the needs of governments, industries, farmers, and communities when formulating long-term plans for water management.
- **Resource Inequities:** Inequalities in access to water and resources can have a disproportionate impact on already vulnerable populations.
- **Financial Constraints:** Many regions face economic difficulties because of the high cost of investing in the advanced water management technologies and infrastructure upgrades necessary to implement them Kirankumar, Keertana, Sivarao, Vijaykumar, and Shah (2021).

The IoT and ML are two of the cutting-edge technologies that can help us overcome these obstacles. More efficient and adaptable water management strategies that can respond to shifting conditions and lessen the impact of these threats may be possible with the help of these technologies, which promise to collect, analyze, and use data in real-time.

The Role of Technology

Emerging as a potent answer to the difficult problems of water management is technology, and in particular, the combination of IoT and ML. Better water management can be achieved through the use of these cutting-edge tools, which provide novel approaches to data collection, analysis, and use. Figure 2 describes technology's importance in a well-organised water management system.

- **Data Collection and Monitoring:** By making it easier to install sensors and other devices across water-related infrastructure, IoT plays a crucial role in water management. In addition to measuring water quality, quantity, temperature, pressure, and flow rates, these sensors can track a plethora of additional metrics Prasad et al. (2023). Collecting data in real-time allows for a more complete picture of water systems, which improves the quality of decisions and the speed with which they can be implemented.
- **Real-Time Insights:** As a result of IoT's real-time data collection and transmission capabilities, water resource stakeholders now have access to the most current and reliable data available. With this data-driven strategy, water managers can allocate and distribute resources more efficiently and effectively.
- **Predictive Analytics:** ML is particularly effective at sifting through large datasets in search of trends and patterns Palanikkumar, Mary, and Begum (2023). Using ML algorithms applied to both historical and real-time data, water managers can foresee shifts in water availability, pinpoint causes of water quality problems, and predict the occurrence of extreme weather. Planning and allocating resources in advance is made possible by this.
- **Anomaly Detection:** Algorithms based on ML can spot unusual patterns in a water system's operation. In the event of a leak, contamination, or other abnormal event, this real-time detection is invaluable for taking corrective action as soon as possible.
- **Optimization of Resource Use:** Optimizing water distribution networks is possible with the help of ML algorithms by analyzing data on consumption patterns, distribution efficiency, and demand. As a result, less is wasted, more water is conserved, and more resources are put to good use.
- **Decision Support Systems:** Together, IoT and ML help create state-of-the-art decision-making tools. These systems provide water managers with actionable insights and recommendations, allowing them to balance the needs of the present with the needs of the future in terms of sustainability.
- **Improved Environmental Monitoring:** Sensors enabled by the IoT can track pollution levels, dissolved oxygen concentrations, and pH balances in water systems. This extensive monitoring helps find pollution early and guarantees that rules are followed.
- **Adaptive Management Strategies:** Adaptive methods of water management are now possible thanks to technological advancements N. Kumar, Reddy, and Ashreetha (2023). Using real-time data and predictive models, businesses can quickly adapt to changing conditions and lessen the severity of disruptions.
- **Public Awareness and Engagement:** IoT technologies can be used to update the public instantly on water availability, quality, and conservation initiatives. This raises consciousness and inspires conservation among the general populace.

- **Research and Policy Development:** Science and policy making can benefit from the data gathered and analyzed by IoT and ML. Better water management and conservation policies can be implemented with the help of technology and an emphasis on data-driven decision-making.

In the field of water management, technology serves as an intermediary between data and policy making. When IoT and ML are combined, data is transformed into useful information that can help water managers, policymakers, and communities solve water problems more effectively, efficiently, and sustainable.

IoT in Water Management

By allowing for the interconnection of devices, sensors, and systems to collect, transmit, and analyze data related to water resources, the IoT has revolutionized water management. The IoT has the potential to improve water management in a number of ways, including the provision of real-time insights, the optimization of resource utilization, and the quality of decision-making.

- **Sensor Networks:** Rivers, reservoirs, treatment plants, and distribution networks are just some of the places where sensors for the IoT can be installed Begum, Tanguturi, and Ahmed (2022). Water quality, flow rates, pressure, temperature, and water levels are just some of the things that can be measured by these sensors.
- **Real-Time Data Collection:** Connected sensors in the water distribution network are constantly gathering information. The dynamic and precise understanding of water conditions made possible by this real-time data collection allows for prompt reactions to changes or anomalies.
- **Remote Monitoring:** IoT technology enables water system remote monitoring. Managers of water systems can improve efficiency by accessing data from sensors located in inconvenient or far-flung areas.
- **Data Transmission and Communication:** The information is sent through various channels to various hubs. The wireless nature of this data transfer makes it possible to maintain a constant connection even in outlying areas.
- **Data Integration:** By compiling information from a wide variety of sensors, IoT systems can provide a bird's-eye view of our water supply. Integrating data from numerous sensors yields a holistic picture of water availability, quality, and infrastructure conditions.
- **Early Warning Systems:** Early warning systems for disasters like floods, contamination, or infrastructure failures can be activated by data from IoT sensors that are updated in real-time. This paves the way for speedy action and the implementation of preventative measures.
- **Water Quality Monitoring:** Sensors connected to the internet keep an eye on things like acidity, turbidity, oxygen levels, and pollution Reddy, P, and S (2022). This constant checking helps locate the origins of pollution and guarantees that water quality standards are met.
- **Demand Management:** IoT data can be used to examine water use and demand patterns. This data is helpful for planning and managing water supply systems during times of high demand.
- **Smart Irrigation:** Smart irrigation systems, which rely on the IoT, monitor soil moisture and weather forecasts to make watering decisions. This reduces unnecessary water use while protecting crop quality.

- **Sustainability and Conservation:** The IoT helps ensure water supplies for the future by encouraging more responsible consumption. Stakeholders can better manage water use and allocate resources with real-time data.

Water managers, policymakers, and stakeholders are given a leg up in developing proactive strategies for water management thanks to the IoT capacity to provide real-time, accurate, and comprehensive data Sridhara et al. (2023). The IoT allows for a shift from reactive to predictive water management practices, improving the sustainability of water stewardship.

ML In Water Management

ML has become a game-changing innovation in the water management industry, providing cutting-edge resources for statistical analysis, forecasting, and policy making. To improve many facets of water resource management, ML algorithms process massive amounts of data to find patterns, make predictions, and generate insights.

- **Data Analysis and Pattern Recognition:** Algorithms based on ML examine both historical and real-time data in search of trends and patterns that would otherwise be difficult to spot. Water managers can benefit from recognizing these patterns because it allows them to see connections between previously unseen variables.
- **Predictive Modeling:** Future water-related events, such as changes in water quality, flooding, or droughts, can be predicted using ML models. These models learn from the past to make predictions about the future, allowing for better preparation and faster responses.
- **Anomaly Detection and Event Recognition:** Algorithms based on ML can detect irregularities and outliers in water networks Sharma and Arun (2022). With this ability, leaks, contamination events, and infrastructure malfunctions can be spotted before they cause extensive damage and waste.
- **Water Quality Monitoring and Forecasting:** Changes in water quality can be predicted using ML models that take into account factors like weather, pollution sources, and human activities. This allows for prompt measures to be taken to keep water quality at acceptable levels.
- **Leak Detection and Infrastructure Maintenance:** AI-driven leak detection systems examine sensor data for telltale signs of water or gas leaks in pipes J. Ramesha, Kudari, and Samal (2021). Equipment failures can be anticipated with the help of predictive maintenance models, which allow for prompt servicing and reduce downtime.
- **Water Demand Forecasting:** Based on past consumption patterns, demographic trends, and environmental factors, ML can forecast both immediate and future water needs. This helps with the efficient administration of water supply and distribution.
- **Decision Support Systems:** Insights generated by ML give decision-makers access to useful data. ML is used by decision support systems to advise on the best ways to approach water management.
- **Resource Allocation and Conservation:** With the help of ML, water resources can be distributed fairly while still being conserved. It aids in preserving resources and maintaining a steady supply of goods.

- **Climate Change Adaptation:** The effect of climate change on water supplies can be evaluated using ML models that analyze climate data. Having this knowledge is useful for making adjustments to suit shifting circumstances.

When ML is applied to water management, it enables stakeholders to improve resource efficiency, respond to water-related challenges, and make data-informed decisions. ML's ability to glean useful information from large data sets makes it a powerful resource for building water management systems that can withstand the test of time.

Benefits and Impacts

There are many positive outcomes that can be achieved through the combination of IoT and ML in water management. Technology advancements in data collection, analysis, and decision-making have led to greater water resource management effectiveness, sustainability, and resilience.

Water quality, quantity, and infrastructure conditions can all be monitored and responded to in real-time thanks to IoT-enabled sensors. This paves the way for prompt reactions to shifts, anomalies, and emergencies, reducing the likelihood of harm and making the best use of available resources.

Decisions that are informed by data are made with the help of IoT and ML. If water managers have access to reliable data, they can craft policies and programs that actually work.

Analyzing the future, ML algorithms foresee things like water quality shifts, infrastructure breakdowns, and extreme weather S. Pai, Shruthi, Naveen, and K (2020). These forecasts allow for preventative planning, which lessens risks and guarantees punctual interventions.

IoT and ML Improve Water Distribution Network Efficiency, Decrease Waste, and Distribute Resources More Efficiently. As a result, there is potential to save money by reducing energy and water consumption. IoT sensors monitor water quality parameters in real-time, allowing for rapid identification of pollution sources or contamination events, leading to more effective water quality management. This improves safeguards for public health and adherence to rules and regulations.

Evidence-based decision-making is aided by data generated by the IoT G. N. Ramesha and Pai (2020). Water management policies can be made more specific and efficient with the help of ML analysis.

Smart irrigation systems, enabled by the IoT, help farmers save water by monitoring soil moisture and predicting precipitation. Increased crop yields and more environmentally friendly farming methods are the results.

The environmental toll of water resource extraction, treatment, and distribution is lightened thanks to IoT and ML-enhanced efficiency in water management.

Distribution of Water Resources: The IoT and ML Help Make Water Available to More People in More Places Swetha et al. (2021). Transformative changes are achieved through the integration of IoT and ML technologies in water management to address water scarcity, pollution, and climate change. Water resources and the communities that rely on them will have a more secure and stable future as a result of these positive outcomes for the environment and the economy.

DATA SOURCES AND SENSORS IN WATER MANAGEMENT

Research into the nation's water resources relies heavily on systematic, long-term observation and documentation of hydrologic systems. Understanding physicochemical and biological dynamics relies heavily on this kind of information, which is also the backbone of many prediction models. Multiple systems and instruments collect information on water and its many uses. Soil hydration, snow cover, precipitation, river flow, groundwater levels, aquifer replenishment, and plant water loss are all examples of hydrological reserves and flow metrics. Physical, chemical, biological, and ecological factors are all taken into account when making assessments that span water, land, and air B. S. Kumar, Ramalingam, Balamurugan, Soumiya, and Yogeswari (2022).

There are a lot of reasons why it's important to collect this information. It is used in the issuance of flood warnings, the support of health and safety monitoring, the facilitation of weather predictions, the direction of engineering projects, and the backing of business, industry, and academic research. It can serve as an indicator of potential risks to health and safety, prompt policy shifts, and support studies of hydrological and related phenomena's temporal and spatial dynamics Omambia, Maake, and Wambua (2022).

Extreme weather events like floods and droughts, for example, require specific data in order to be properly studied. Only by carefully monitoring precipitation and river flow in a wide range of climatic and hydrologic environments over decades or even centuries can such insights be gained Bharadwaj et al. (2022) . Understanding the effects of global climate change on groundwater and surface water supplies requires similar long-term monitoring infrastructures.

However, non point source pollution of waterways due to runoff containing nitrate, pesticides, and sediment may necessitate hourly data collection. This is because polluted runoff is directly related to when and how hard it rained. Localized differences in pollutant concentration often reflect shifts in land utilization and agricultural methods, making it crucial to capture all relevant spatial dimensions.

In conclusion, the United States faces a wide variety of hydrological challenges now and in the future, all of which demand the implementation of monitoring systems that can reliably and accurately operate across large and small scales of time and space.

Satellites carrying remote sensing instruments collect information about bodies of water like lakes, rivers, and reservoirs. Large areas can be observed, and changes in water quality, surface temperature, and level can be tracked with the aid of remote sensing K, Ajith, K, and Ramesh (2022).

The parameters of water temperature, pH, dissolved oxygen, turbidity, and pollutants are monitored by ground-based sensors that are installed in water bodies and infrastructure. These sensors supply real-time readings for in-depth regional analysis and research.

Sensors that measure the depth of water are useful for monitoring how much water has been used or lost due to factors like weather and population growth. These sensors are indispensable for tracking flood levels and controlling water storage facilities. pH, turbidity, electrical conductivity, and nutrient levels are just some of the things that can be measured by water quality monitors Abusukhon and Altamimi (2021). They conduct analyses of water quality and locate the origins of pollution.

Global Positioning System (GPS) technology and Geographic Information System (GIS) platforms aid in mapping and analyzing spatial data relating to water resources, infrastructure, and land use. These resources are useful for pinpointing problem areas and formulating rescue strategies.

Unmanned Aerial Vehicles (UAV) drones can take detailed pictures and gather information in inaccessible places where humans simply can't go. They can be used to keep an eye on water quality, measure pollution, and check on buildings and other structures.

Citizens can help with data collection and problem reporting for water systems by using mobile applications. This method encourages community involvement in water management and increases data coverage.

Connected sensors and other devices that wirelessly collect and transmit data make up what are known as IoT networks. The real-time data from various nodes in these networks enhances monitoring capabilities.

Together, these sensors and data sets paint a more accurate picture of the hydrologic cycle. Water managers, academics, and politicians are better equipped to distribute water resources effectively because of the data they collect.

IOT INFRASTRUCTURE FOR WATER DATA COLLECTION

Technology circles have been using the phrase IoT for some time, but its true potential has not yet been tapped. In particular, its effects may be felt in the field of water quality monitoring.

There are several advantages to using IoT devices to track water quality. It's useful in drought-stricken regions since it gives real-time data on water quality and usage that's essential for making decisions regarding water treatment. Identifying the origins of pollution is an important step in preventing its spread and protecting human and environmental health, and the IoT can help with that.

Improving water quality is only one use case for IoT possibilities. The potential for the IoT to revolutionize our lives cuts across many different fields. Imagine, for instance, the introduction of high-tech sensors to monitor various contaminants and pollutants in our water supplies. These sensors, when strategically placed in water bodies like rivers, lakes, and reservoirs, can continuously monitor conditions like pH, temperature, dissolved oxygen, and the presence of specific chemicals and microbes. This information is transmitted immediately after being collected and analyzed by centralized systems. Such immediate responses allow for the detection of irregularities and the prompt implementation of corrective measures.

IoT's live monitoring and data accumulation capabilities hold the greatest promise for improving the reliability and consistency of water distribution networks. It provides the flexibility needed for water management entities to detect and effectively counteract new threats, guaranteeing a steady supply of clean, sustainable water for the populace.

The IoT allows for the continuous monitoring of several critical water quality parameters, including:

- Water's acidity or basicity can be measured by its pH level.
- As a function of temperature, solubility, toxicity, and the habits of aquatic organisms change.
- Healthy aquatic ecosystems are characterized by high levels of dissolved oxygen.
- Total dissolved solids (TDS) are the dissolved substances that can alter the flavor and clarity of water.
- The turbidity of water reflects the presence or absence of contaminants.
- Evidence of ion presence affecting water quality is indicated by conductivity.
- Eh, the oxidation-reduction potential serves as an indicator of ion presence.
- Chlorophyll is a marker of algal growth.

- Biological Oxygen Demand (BOD) emphasizes the importance of water's organic material.
- The presence of ammonia nitrogen suggests a possible hazard to marine life.

Figure 3 describes the IoT Water Monitoring System Sensors Flow Diagram Deployment. Lakes and rivers are prime locations for installing water quality sensors. The functional blueprint for an IoT-centric water monitoring apparatus could be depicted as follows:

- These sensors will record critical information.
- The data is transmitted to a central server or the cloud using various communication protocols (such as WiFi or cellular).
- All sensor-generated data is archived in a safe location, such as a central server or the cloud.
- Data analytics entails the use of algorithms or specialized software to examine the collected data for meaningful patterns.
- Those in charge of water purification, for example, are notified when abnormalities are detected, such as when the pH level is too high or too low.
- After receiving a notification, the proper authorities will act to rectify the situation.
- Dashboard tools provide stakeholders with access to real-time data and longitudinal trends visualization interface.
- In dissemination of data, both processed and raw data collected by IoT devices can be made available to the public, government agencies, NGOs, and other interested parties.
- This seamless application of technology to water quality management exemplifies the disruptive potential of IoT and portends a future in which precise and preventative measures are taken to protect our limited water supplies.

CHALLENGES FOR IOT AND ML IN WATER QUALITY MONITORING

There are a number of significant hurdles that must be overcome when designing an IoT system for water quality monitoring. The first step in developing an IoT water quality monitoring system is to determine which parameters are crucial for measuring. The specifics of the application and its conformity to rules and regulations will be taken into account. The design of such a system must take into account potential emergency situations.

It is also crucial to set up a reliable network and communication system for sending sensor data to a centralized server or cloud service. Wireless technologies like WiFi, Bluetooth, and cellular connectivity could be incorporated into this infrastructure. The processing and analysis software and algorithms for the collected sensor data are just as crucial. In order to find patterns and trends in the data, ML algorithms may be used at this stage.

IoT devices have the ability to continuously monitor water quality by collecting data on variables like temperature, pH, and oxygen levels. This method outperforms more conventional approaches, such as manual sampling and analysis that is performed on a more sporadic basis.

Water quality data collected by IoT devices can be accessed and analyzed remotely thanks to their wireless data transmission capabilities. This is an especially helpful function for assessing water quality in inaccessible areas.

IoT-enabled, always-on water quality monitoring contributes to rapid problem identification and resolution. By taking this preventative measure, problems won't grow into costly catastrophes.

More efficient and cost-effective monitoring practices are possible thanks to the widespread use of IoT technologies in the water quality assessment process.

However, there are a number of obstacles that prevent full IoT integration in water quality monitoring. The lack of established norms presents a major challenge. Many different types of sensors and other devices flood the market, making it difficult to find items that work well together. Problems with interoperability further muddy the waters, preventing the free flow of information between different platforms.

The expense of setup and upkeep is also a problem. The high price of sensors and the frequent need for replacements or upgrades drive up the cost of acquiring and maintaining IoT systems. The ongoing cost of maintaining and supporting these systems is a major factor.

The vulnerability to cyber attacks and security breaches is a further possible drawback. Due to their reliance on distributed networks of devices and sensors, IoT systems can be an easy target for cyber criminals. Accidental disclosure of sensitive information related to water supply and distribution systems could be at risk if such incidents occurred.

The Mirai botnet attack in 2016 is a prominent example of a cyberattack related to the IoT. Distributed Denial of Service (DDoS) attacks were carried out on several well-known websites and internet services by a botnet formed from a network of compromised devices like security cameras and routers. Attackers used these compromised endpoints to flood their intended targets with traffic and render them inaccessible. This was accomplished by exploiting software flaws. As an example of malware designed to infiltrate industrial control systems that oversee critical infrastructure like power plants, the Stuxnet worm was discovered in 2010. By infecting computers, this worm gave attackers the ability to interfere with their operation and possibly cause harm.

Although there are many benefits to using IoT for water quality assessment, it is essential to assess any risks that may be involved carefully.

Within an IoT framework, data security is of utmost importance. Maintaining data privacy, security, and accessibility is essential. Data security can be improved by a number of methods:

To protect data privacy, only authorized users should have access to the information. Only those who should have access can gain it, thanks to encryption and other authentication measures.

Unauthorized changes to data must be detected and prevented in order to maintain data integrity. The use of cryptographic hashing and other similar techniques can aid in preserving data integrity by revealing any tampering.

Having data readily available for authorized users at the right time is crucial. Downtime risks can be reduced through the use of redundancy and solid architecture.

The precision and dependability of IoT-based water quality monitoring systems is another area of concern. During the system design process, several considerations should be made to address these worries:

Accurate and repeatable water quality readings rely on regular calibration of the IoT system's sensors and equipment. This method involves comparing data to established norms or reference materials.

Implementing quality control measures aids in locating and fixing any inconsistencies in the data collection process, which in turn increases the data's credibility.

Training Employees Proper training of employees working with the IoT system ensures smooth operations, reliable data collection, and prompt resolution of any issues that may arise.

By using multiple sensors or backup sensors, reliability can be increased through redundancy. This strategy makes use of overlapping measurements to pinpoint where in the system problems may be originating. For instance, if one sensor reports a change in water quality, additional measurements can verify the shift and confirm the alarm. This method greatly reduces false alarms and improves the system's dependability.

The collected data from the IoT system can be analyzed to reveal patterns that may indicate faulty machinery or inaccurate readings. Because shifts in trends may indicate contamination, they are especially useful in the context of water quality monitoring. The timely detection of these variations is made possible by continuous data analysis over time, which allows for swift corrective actions.

APPLICATION OF IOT AND ML IN WATER RESOURCE MANAGEMENT

The most common classifications for ML algorithms are supervised, unsupervised, and Reinforcement Learning (RL). For classification and prediction tasks, supervised learning algorithms make use of labeled data sets for training. The input and output values for this technique are known in advance. In unsupervised learning, algorithms are trained to perform clustering using data that has not been labeled. Without any additional help from a human analyst, these algorithms are able to unearth previously hidden data patterns and clusters. The field of ML, known as RL, investigates how a smart agent might choose actions in a given setting so as to maximize the rewards available to it. Both supervised learning and RL involve connecting inputs and outputs in order for the agent to learn which actions will lead to the desired results. In RL, rewards and punishments are used to indicate positive and negative behaviors. Therefore, RL agents explore environments without prior training, supervised learning allows machines to learn from labeled data, and unsupervised learning detects patterns in data without labels. As a result, engineers need to make sure the type of ML they use is appropriate for their project. The term "prediction" refers to any technique that utilizes data processing to make an informed estimate about the future.

The words "prediction" and "forecasting" mean the same thing in the context of this review chapter. Predictions made by ML models can be very precise because they are grounded in a large body of prior knowledge across many different fields. The algorithm determines the likelihood of each possible outcome as new data is entered, allowing the modeler to settle on the most likely outcome.

Time series data, big data, univariate datasets, and multivariate datasets are all valid inputs for ML algorithms developed for predictive purposes. Observations of a variable are collected in a time-ordered series over a period of time. Structured, unstructured, and semi-structured data all fall under the umbrella term "Big Data," which is defined by its massive size and complexity. These algorithms are useful for a wide range of prediction tasks because of ML's flexibility in accommodating different data formats.

In 2005, the phrase "Big Data" was invented to characterize massive datasets that were unmanageable using traditional database administration software. Significant difficulties arise when processing large data since its complexity and volume outstrip those of current data processing methods. Structured, unstructured, and semi-structured categories can be used to describe this data. Hydrologists are used to dealing with easily available, organized data that is often stored in spreadsheets. Images, videos, and audio are examples of unstructured data that can't be analyzed mechanically. Machines can process semi-structured data, such as user-defined XML files, to a certain extent. The "five Vs," or volume, variety, velocity, veracity, and value, are the defining characteristics of big data. In order to fully exploit the potential of big data and conquer the complexities of high-dimensional environments, ML and its subsets can now take advantage of modern Graphics Processing Unit (GPU) advancements.

Raw data is processed by being shaped, combined, restructured, and rearranged so that it meets the needs of the model. The metrics used by researchers to assess and compare the efficacy of ML algorithms vary. The vast field of water resource management is chronicled in a wide variety of articles each year. Data like streamflow, precipitation, and temperature are recorded and processed for use in hydrology studies, along with data like large-scale atmospheric data derived from gauge and satellite information.

Due to their superior performance in complex environments, data-driven models perform better than statistical models when making predictions in the field of WRM. This demonstrates the significance of employing data-driven methods in water resource management for producing precise forecasts and insights.

SMART SOLUTIONS FOR WATER MANAGEMENT

By enabling real-time data collection, instant alerts, and preemptive actions to prevent potential problems, smart sensor devices play a pivotal role in improving efficient and secure water management for consumers and workers. Numerous processes, depending on the underlying infrastructure, can be automated, leading to the generation of alerts and insights in real-time. Utilities and their teams benefit greatly from the incorporation of these devices and sensors into water industrial infrastructure, particularly in the areas of infrastructure maintenance and worker safety. Due to the large volume of data being collected, service teams can efficiently manage remote infrastructure with less frequent on-site visits required for maintenance. When any part of the infrastructure is compromised, the service teams are immediately alerted so that they can take preventative measures and make the required adjustments automatically. By using this method, service teams are not dispatched to the site unless absolutely necessary; instead, automated processes handle the necessary adjustments in real-time. Therefore, smart sensor devices help with better water management, easier maintenance, and more stringent safety protocols for utility workers and the infrastructure as a whole.

Monitoring Water Quality

The incorporation of IoT sensors and embedded technologies has revolutionized water management. Water distribution systems face energy conservation challenges that could be mitigated through the combination of energy harvesting techniques and the IoT. Smart water systems can be improved with the help of visual application layers and technologies like solar cells, piezoelectric, electromagnetic, and thermoelectric harvesting. Studies of advanced water quality management can be propelled by the development of micro-digital-electro-mechanical systems like lead zirconate-titanate films, piezoelectric nanowires, multi-parameter sensors, and iridium oxide films, as well as analytical tools like X-ray photoelectron spectroscopy.

The safety of our connected devices and infrastructure is crucial. Creating a comprehensive security plan begins with mapping and identifying assets. The IT department will be able to react quickly to potential security breaches by using real-time alerts and threat detection mechanisms. This all-encompassing strategy protects IoT-enabled water management systems from potential threats.

Data Processing, Analysis, and Computation

New data needs to be collected and processed as a result of the widespread deployment of IoT devices across the infrastructure landscape. The time and effort required to move this data to cloud servers can be substantial, delaying the processing and alerting of crucial data that could otherwise prompt timely responses to potential hazards. Embedding intelligence at the point of data collection is a new idea that promises to reduce network costs and speed up data processing. This method, also known as edge computing, paves the way for a wide variety of automated processes to run across the system. Remotely monitoring device performance, keeping up with updates, and applying patches are all simple tasks that can become difficult and time-consuming. Local decision-making at the control center is another cause for concern because it could slow down responses to local issues. In addition, issues with data reduction can arise from improper deployment of edge computing. It's possible that sending too much data over the network will tax the resources available or cost more than necessary. Careful planning and management of edge computing are required to reap its full potential while minimizing any drawbacks.

Water Processing and Storage

Every country's economic growth is directly tied to the efficiency with which its water supply is distributed. However, problems like water leakage and contamination are major causes for alarm. These problems have serious effects on economic expansion as well as public health and quality of life. Integrating IoT technology becomes essential to ensure water safety, minimize waste, and address these challenges. Common hydrological parameters used in the collection of water-related datasets include precipitation, streamflow, recharge, ocean-land-atmosphere interactions, water quality indicators, and energy requirements. These methods try to quantify accuracy, but they have trouble keeping up with and managing data in real-time. This is because of the wide variety of data, the length of time needed to collect it, the number of resources required to transmit it, and the breadth and depth of available networks. Using state-of-the-art technology, water quality can now be monitored in real-time, facilitating immediate notifications and the implementation of preventative measures. Using IoT solutions improves the efficiency and effectiveness of water quality monitoring, allowing for early detection of potential problems and prompt remediation. This innovation in technology has the potential to drastically improve water management by making it more secure, less wasteful, and more protective of public health and welfare.

CASE STUDIES AND APPLICATIONS

Integrating IoT technology, Cloud computing, and ML for holistic water management forms the basis of the proposed framework. In this setup, ML is used for in-depth analysis, and the Cloud is used as a monitoring and alerting platform. Plant pumps and pressure points are just two examples of the places where the framework's sensors can be put to use. Key parameters are measured by these sensors, including pH, dissolved oxygen, TDS, metals, salinity, and turbidity, as well as data on pipe leakage, emission flows, and chemical contamination indicators like head, static, and stagnant pressure, flow (pressure and depth), flow velocity, and more.

In order to assess water quality and manage the devices in real time, the data generated is tracked and analyzed in real-time. The data collected by the sensors and probes can then be processed by the IoT devices thanks to their connection to the central controllers. The processed data is then transmitted via wireless networks from these devices to Cloud portal systems. Dashboard analysis, trend identification, and insight extraction are all performed by means of artificial intelligence and ML algorithms within these Cloud-based systems. These realizations allow for deliberate decision-making and immediate action.

Hardware and software are both included in the framework. Hardware-wise, sensors and probes are positioned strategically at three stages: water harvesting (pressure sensors, motor pumps), water storage (ultrasonic water level sensors), and water distribution.

Connectivity between these IoT devices and the plant's infrastructure ensures that data can be gathered in real-time from a wide variety of hardware sensors. With the help of a sensor, a Raspberry Pi can perform edge computing and store data locally. It takes analog measurements, digitizes them, and sends the partially processed data to remote servers in the cloud.

Multiple processing modules make up the system's software. The IoT relies on a number of components to function properly, including edge devices for data processing, monitoring agents for supervision, Cloud portal applications for management, and ML algorithms and datasets for producing actionable results. Through streamlined data collection, analysis, and actionable insights, this all-encompassing method guarantees efficient water management. Figure 4 shows the overall structure for water management integration with IoT framework for prediction of turbidity in water.

The five-tiered framework of the proposed model is described below:

- ***Obtaining the Data:*** Data collection and processing in real-time rely on this layer. Data is gathered from the plant's systems as well as IoT devices. This layer consists of the sensors, embedded devices, and energy-harvesting devices used in the IoT. The characteristics of the project, the budget, and the parameters to be measured all play a role in deciding which commercially available sensors to use. These sensors are indispensable for gathering information about water quality.
- ***The Energy Harvester Layer:*** It is concerned with collecting power from the environment, such as the wind, the sun, and the earth's vibrations. By transforming the collected energy into electricity, the devices can be powered, or their battery life extended. At this level, components harness power from mechanical, thermal, biochemical, and electromagnetic processes. The focus of this effort is on controlling the operations of embedded systems, network nodes, and IoT sensors. The goal is to deliver low-cost solutions that minimize energy consumption.
- ***The Network Layer:***It is in charge of managing the transmission of data from the lower layers and performing any necessary value processing. It takes care of the wired and wireless connections required for data transfer. When traveling greater distances, people often resort to cellular communication (2G, 3G, 4G) despite the fact that this requires more energy. On the other hand, short-range protocols are used, such as Zigbee, 6lowpan, and Radiofrequency identification, with the help of pre-programmed tags. Microcontrollers can connect to WiFi with the help of devices like the ESP8266 WiFi module.
- ***The Cloud Layer:*** Its primary goal is to protect users' personal information while it is being processed, stored, and transmitted between various electronic gadgets. This layer makes use of a Raspberry Pi microcontroller board (Pico RP2040) and its complement of 14 digital I/O pins. This guarantees that data is encrypted and handled securely as it travels across the network.
- ***The Analytics Layer:*** This is the last layer, and it's responsible for processing data, making decisions, and solving problems. Neural networks trained with neuro-fuzzy algorithms predict the

quality of wastewater at this level. Water quality problems are also identified using an adaptive network fuzzy system.

This five-tiered model presents an all-encompassing strategy for water management data collection, energy management, communication, cloud integration, and advanced analytics. To improve the system as a whole, it takes into account vital factors like energy efficiency, data security, and predictive abilities.

Water management presents significant challenges in the context of agriculture and industries, including issues of access, efficiency, and sustainability. Industries consume a lot of water and contribute to pollution, especially in India. Availability, cost, and demand are all considerations when deciding between groundwater and surface water. Water pollution is a result of increasing industrial needs and insufficient wastewater treatment.

For industries to treat and use water ethically and for public authorities to deal with water scarcity and quality, effective water management strategies are required. These difficulties are amplified by prolonged droughts, highlighting the need for strategic responses. In order to overcome these obstacles, intelligent methods are indispensable.

Intelligent algorithms can model and optimize water distribution infrastructure, ensuring a steady supply of clean water to cities without compromising on safety or sustainability. The model can suggest different pricing and metering schemes, as well as the most water-efficient home appliances.

Analyzing physical, biological, and chemical characteristics is essential in determining water quality. Chlorophyll, pH, dissolved oxygen, metals, chloride, and lead are just some of the indicators used to evaluate water quality. Water quality can be monitored, pollution levels predicted, and regulations followed with the help of AI technologies like the IoT, deep learning, and ML.

Water distribution system leak detection and flow rate monitoring are two functions that can be managed by intelligent systems. As a result, water waste is reduced, and efficiency is maximized.

Industry can detect and avoid water contamination with the help of smart techniques.

Monitoring systems may detect pollutants and send alerts in real-time so that corrective measures can be implemented promptly.

The creation of methods for economical water usage is aided by intelligent systems.

Implementing tools for real-time consumption monitoring, limiting excessive use, and demanding responsible conduct are all examples of such strategies.

Data-driven decision-making in water management, consumption patterns, and distribution methods might benefit from the use of intelligent algorithms that analyze past data.

Figure 5 illustrates the need for smart solutions in addressing the complicated issues of water management in agriculture, industry, and urban areas. Together, deep learning, ML, and IoT can make water management more sustainable and efficient.

FUTURE TRENDS AND CHALLENGES

Several new developments in technology are influencing the direction of data collecting and preprocessing in water management. These patterns may be traced back to the need for more precise and long-term water resource management. The proliferation of IoT devices and networks will lead to a rise in the number of sensors used to keep tabs on our water supply. This enlargement will increase data coverage and allow for a deeper dive into water system dynamics.

In order to reduce the amount of time spent sending raw data to centralized servers, businesses are increasingly turning to edge computing and edge analytics. With the help of edge analytics, data can be analyzed in real-time at the sensor level, drastically cutting down on latency and speeding up responses.

To improve the quality of data analysis and prediction, AI methods like deep learning and neural networks will be combined with IoT information. Models powered by artificial intelligence can decipher intricate water-related patterns and relationships.

Water management could benefit from Blockchain Technology's increased data security, transparency, and traceability. It can be used to safely store and transmit information about water use, ownership, and transactions.

Data from multiple sources, such as satellites, ground stations, and citizen scientists, will be fused into a single cohesive whole in the systems of the future. A more complete picture of our water supply will be possible with the help of cutting-edge data fusion methods.

Augmented and Virtual Reality (AR/VR) technologies can help with decision-making, training, and public engagement through immersive visualization of water systems. Increased security measures to protect private data will be built into water management systems in response to rising privacy concerns.

Autonomous drones and underwater vehicles with sensors will collect information in hazardous conditions. Robots can help with data collection, maintenance, and inspections of infrastructure. As more and more water data is collected, stored, and analyzed, cloud-based solutions will become increasingly important. Scalability, accessibility, and the ability to work together are all benefits of cloud-based platforms.

Difficulties in Gathering and Cleaning Data

- Accurate, trustworthy, and actionable information cannot be achieved without the help of data acquisition and preprocessing, which play a pivotal role in water management. However, several obstacles must be overcome to guarantee the accuracy and utility of the data processed. Some of the major obstacles include the following:
- It's a huge challenge to keep data accurate, consistent, and reliable. Erroneous readings can be produced due to sensor failures, calibration problems, or data drift, all of which impact decision-making.
- Calibration and maintenance of sensors are critical steps in ensuring reliable readings. However, in less-frequently-visited or inaccessible areas, these activities can be time-consuming and labor-intensive.
- Missing values and incomplete data can occur when a sensor fails, or there is a communication breakdown. It is difficult to handle missing values and impute them without introducing bias.
- Noise and outliers in the data can skew sensor readings, making conclusions about the world around them incorrect. It can be difficult to find and remove these outliers without losing valuable information.
- Managing incompatible data formats, units, and protocols is a common challenge when attempting to combine information from a wide variety of sensors and sources. It can be not easy to guarantee data compatibility and interoperability for efficient analysis.
- As the number of sensors and data sources proliferates, the resulting data could be extremely large in size. Scalable infrastructure is required for efficient processing and management of this data, especially in real-time scenarios.

- Problems with latency can arise when sending a lot of data from dispersed sensors to centralized servers. Finding a happy medium between the demands of real-time data and limited bandwidth is difficult.
- It is critical to prevent unauthorized access to and cyber threats to private data pertaining to water. It can be difficult to implement strong security measures without affecting data accessibility.

Data Fusion and Integration

To avoid biased or skewed results, it is necessary to integrate data from a wide variety of sources and sensors, each with its own characteristics and uncertainties. Sensor placement and demographic representation are two examples of potential sources of bias in data collection. Accurate analysis relies on collecting data that is both representative and objective.

Issues of data ownership, sharing, and access rights can arise when dealing with multiple parties interested in the same data. Collaboration relies on the establishment of clear protocols.

Regulatory Compliance

When dealing with potentially sensitive information like water quality or usage data, it can be difficult to ensure compliance with data privacy and regulatory requirements.

Especially in outlying areas, it is important to ensure that the power sources for IoT devices and sensors are efficient and reliable.

It will take the combined efforts of engineers, data scientists, domain experts, policymakers, and stakeholders to overcome these obstacles. Water resource management would suffer greatly without the creation of stringent data-gathering methods, sensor maintenance plans, and cutting-edge preprocessing approaches.

CONCLUSION

This chapter discussed the revolutionary possibilities of merging Data Acquisition and Preprocessing with cutting-edge technology like ML and the IoT in the field of water management. This comprehensive investigation of synergy was undertaken to gain insights into its potential and the novel solutions it offers to the complex challenges of water resource management. To begin, an examination was conducted on how the inherent challenges in water management have led to the introduction of innovative solutions. The conversation that followed was preceded by a brief summary of the current state of water management. The potential of the IoT and ML to revolutionize water management practices and play a role in alleviating these issues has been highlighted. Good evidence of synergy between IoT's capacity to connect sensors and equipment for real-time data collection and ML's competency in data analysis and prediction suggests improved decision-making, more effective resource allocation, and preventative measures. IoT's use in water management for real-time monitoring, data transfer, and distant sensing became clear after a thorough investigation.

With a more thorough knowledge of water quality and quantity provided by this dynamic mode of data collecting, water managers are better equipped to react quickly to anomalies. The following step included investigating the possible uses of ML algorithms in fields including prediction, process optimization, and data analysis, as well as their usage in water management. This information might help water management improve the distribution system, find leaks, and plan for future water quality.

The potential of data acquisition and preprocessing for use in water management has been further illuminated by the application of ML and the IoT. These technologies provide a roadmap for a sustainable and resilient future at a time when the world is dealing with unheard-of water management issues. This book is a tribute to the scientists, engineers, policymakers, and stakeholders who have worked together to protect and utilize one of our most valuable resources: water.

REFERENCES

Abusukhon, A., & Altamimi, F. (2021). Water Preservation Using IoT: A proposed IoT System for Detecting Water Pipeline Leakage. In *2021 International Conference on Information Technology (ICIT)* (pp. 115–119). 10.1109/ICIT52682.2021.9491667

Begum, Y., Tanguturi, R. C., & Ahmed, B. (2022). Implementation of an Early Warning System using Internet of Things and Rainfall Threshold. In *2022 6th International Conference on Electronics, Communication and Aerospace Technology* (pp. 491–496). Academic Press.

Bharadwaj, K. V., Dusarlapudi, K., Sudhakar, K., Polasi, V. S. K., Tiruvuri, C., & Raju, K. N. (2022). Novel IoT & Machine Learning Base Water Flow Monitoring System. In *2022 IEEE 2nd Mysore Sub Section International Conference (MysuruCon)* (pp. 1–5). IEEE.

K, A., Ajith, V., K, N., & Ramesh, M. V. (2022). Design of IoT and ML enabled Framework for Water Quality Monitoring. In *2022 3rd International Conference on Electronics and Sustainable Communication Systems (ICESC)* (pp. 452–460). Academic Press.

Kirankumar, P., Keertana, G., Sivarao, S. U., Vijaykumar, B., & Shah, S. C. (2021). Smart Monitoring and Water Quality Management in Aquaculture using IOT and ML. In *2021 IEEE International Conference on Intelligent Systems, Smart and Green Technologies (ICISSGT)* (pp. 32–36). 10.1109/ICISSGT52025.2021.00018

Koditala, N. K., & Pandey, P. S. (2018). Water Quality Monitoring System Using IoT and Machine Learning. In *2018 International Conference on Research in Intelligent and Computing in Engineering (RICE)* (pp. 1–5). 10.1109/RICE.2018.8509050

Kolli, S., Ranjani, M., Kavitha, P., Daniel, D. A. P., & Chandramauli, A. (2023). Prediction of water quality parameters by IoT and machine learning. In *2023 International Conference on Computer Communication and Informatics (ICCCI)* (pp. 1–5). 10.1109/ICCCI56745.2023.10128475

Kumar, B. S., Ramalingam, S., Balamurugan, S., Soumiya, S., & Yogeswari, S. (2022). Water Management and Control Systems for Smart City using IoT and Artificial Intelligence. In *2022 International Conference on Edge Computing and Applications (ICECAA)* (pp. 653–657). 10.1109/ICECAA55415.2022.9936166

Kumar, N., Reddy, N. S., & Ashreetha, B. (2023). Technologies for Comprehensive Information Security in the IoT. In *2023 International Conference for Advancement in Technology (ICONAT)* (pp. 1–5). Academic Press.

Omambia, B., Maake, A., & Wambua. (2022). Water Quality Monitoring Using IoT & Machine Learning. In *2022 IST-Africa Conference* (pp. 1–8). Academic Press.

Pai, S., Shruthi, M., Naveen, B., & K. (2020). Internet of Things: A Survey on Devices, Ecosystem, Components and Communication Protocols. *2020 4th International Conference on Electronics, Communication and Aerospace Technology (ICECA)*, 611–616.

Palanikkumar, P. A., Mary, A. Y., & Begum, D. G. (2023). A Novel IoT Framework and Device Architecture for Efficient Smart city Implementation. In *2023 7th International Conference on Trends in Electronics and Informatics (ICOEI)* (pp. 420–426). Academic Press.

Perumal, B., Nagaraj, P., Raja, S. E., Sunthari, S., Keerthana, S., & Muthukumar, M. V. (2022). Municipality Water Management System using IoT. In *2022 International Conference on Automation, Computing and Renewable Systems (ICACRS)* (pp. 317–320). 10.1109/ICACRS55517.2022.10029181

Prasad, K. D. V. (2023). A novel RF-SMOTE model to enhance the definite apprehensions for IoT security attacks. *Journal of Discrete Mathematical Sciences and Cryptography*, *26*(3), 861–873. doi:10.47974/JDMSC-1766

Ramesha, G. N. (2020). Internet of things: Internet revolution, impact, technology road map and features. *Adv. Math. Sci. J*, *9*(7), 4405–4414. doi:10.37418/amsj.9.7.11

Ramesha, J., Kudari, M., & Samal, A. (2021). Smart Agriculture and Smart Farming using IoT Technology. *Journal of Physics: Conference Series*, *2089*(1), 12038–12038. doi:10.1088/1742-6596/2089/1/012038

Reddy, S., P, & S, P. (2022). Data Analytics and Cloud-Based Platform for Internet of Things Applications in Smart Cities. *2022 International Conference on Industry 4.0 Technology (I4Tech)*, 1–6.

Sharma, K., & Arun, M. R. (2022). Priority Queueing Model-Based IoT Middleware for Load Balancing. *2022 6th International Conference on Intelligent Computing and Control Systems (ICICCS)*, 425–430.

Sridhara, P., Yadav, S. B., Chaudhary, S., Gahlot, K., Arya, A., Dahiya, Y., . . . N. (2023). Internet of Things and Cognitive Radio Networks: Applications, Challenges and Future. In Recent Advances in Metrology (Vol. 906). Springer.

Sugumar, S. J., R, S., Phadke, S., Prasad, S., & R, S. G. (2021). Real Time Water Treatment Plant Monitoring System using IOT and Machine Learning Approach. In *2021 International Conference on Design Innovations for 3Cs Compute Communicate Control (ICDI3C)* (pp. 286–289). 10.1109/ICDI3C53598.2021.00064

Swetha, T. M., Yogitha, T., Hitha, M. K. S., Syamanthika, P., Poorna, S. S., & Anuraj, K. (2021). IOT Based Water Management System For Crops Using Conventional Machine Learning Techniques. In *2021 12th International Conference on Computing Communication and Networking Technologies (ICCCNT)* (pp. 1–4). 10.1109/ICCCNT51525.2021.9579651

APPENDIX

Figure 1. Key water management challenges

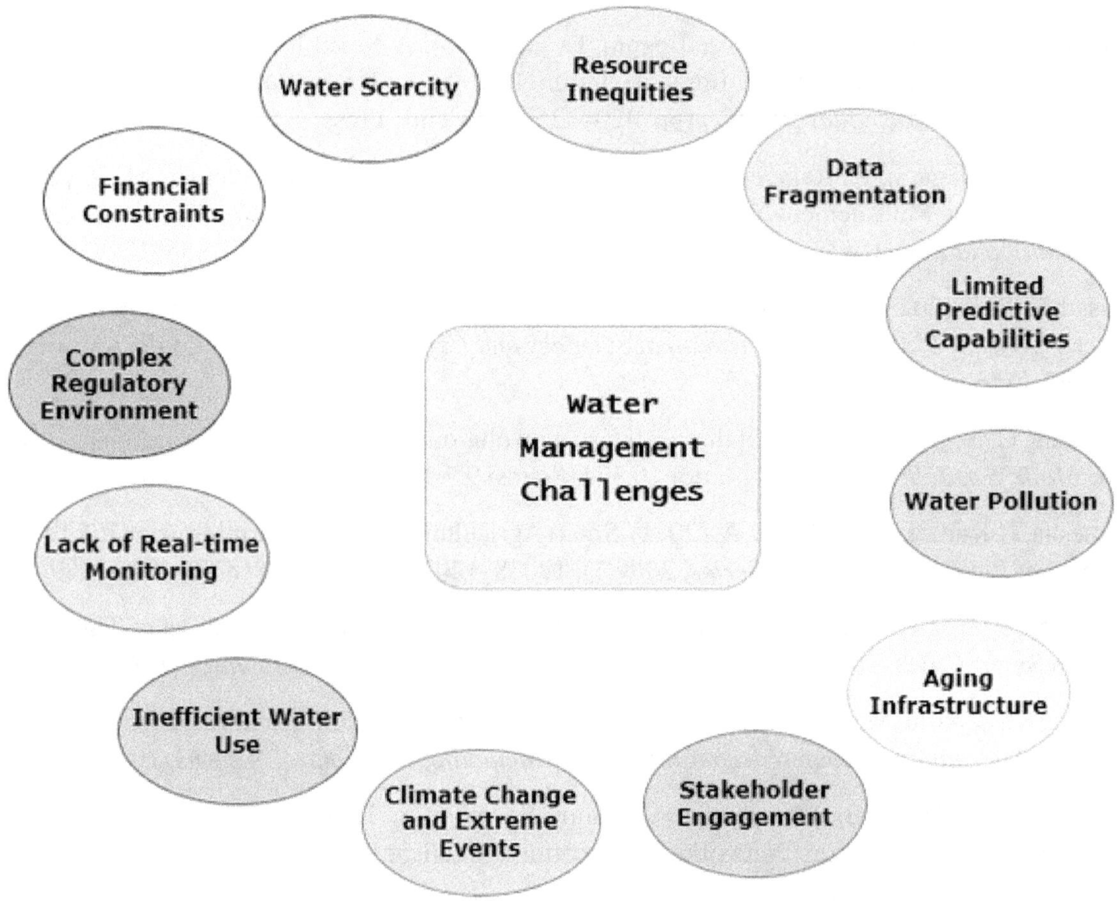

Figure 2. Role of technology in water management system

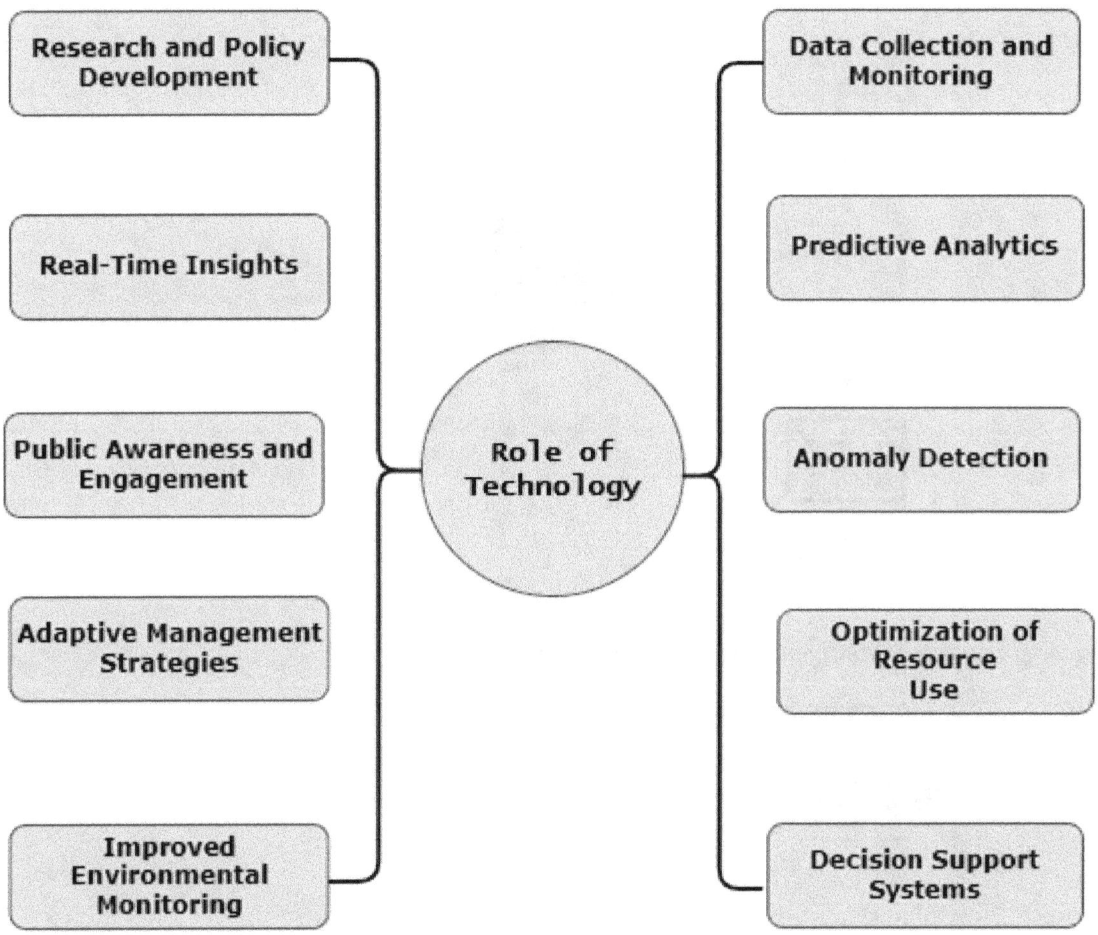

Figure 3. Water monitoring system flow diagram deployment

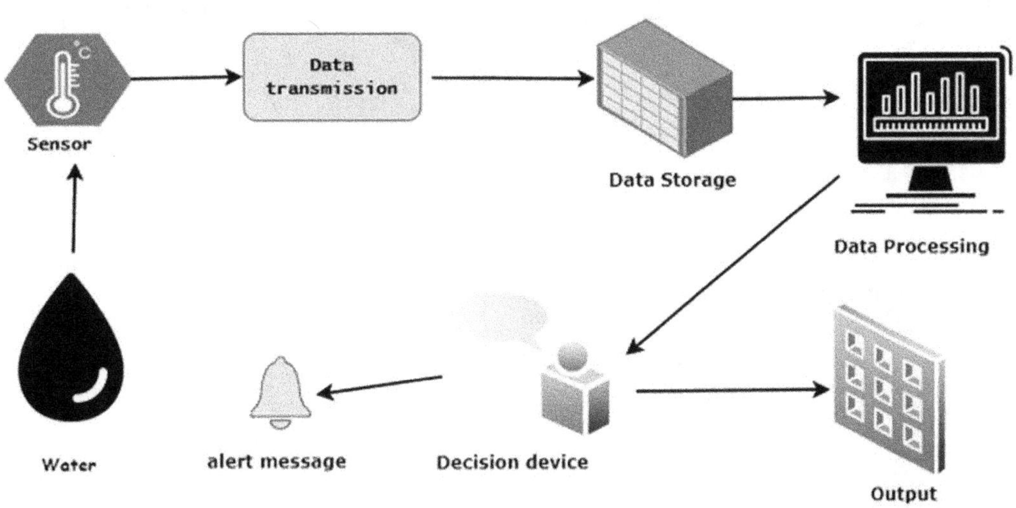

Figure 4. Water management integration with IoT framework for prediction of turbidity in water

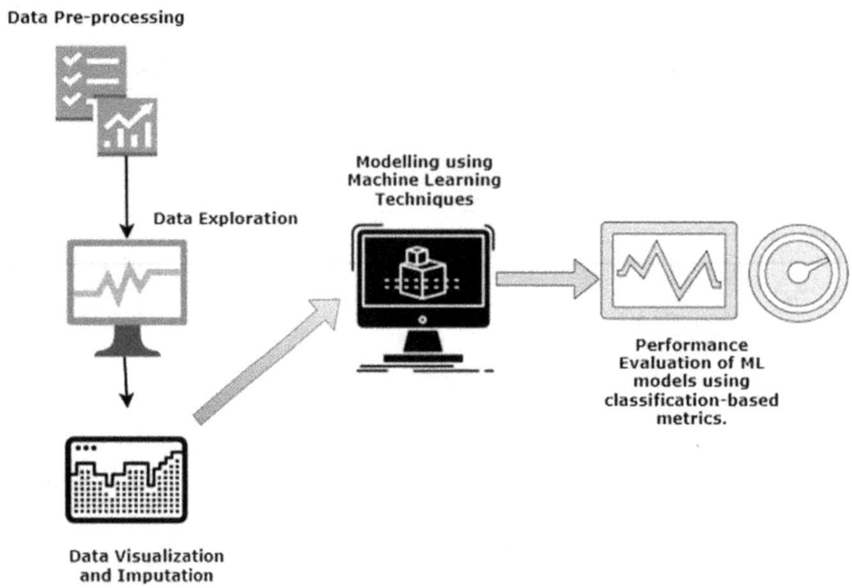

Figure 5. Harnessing intelligent systems for water management

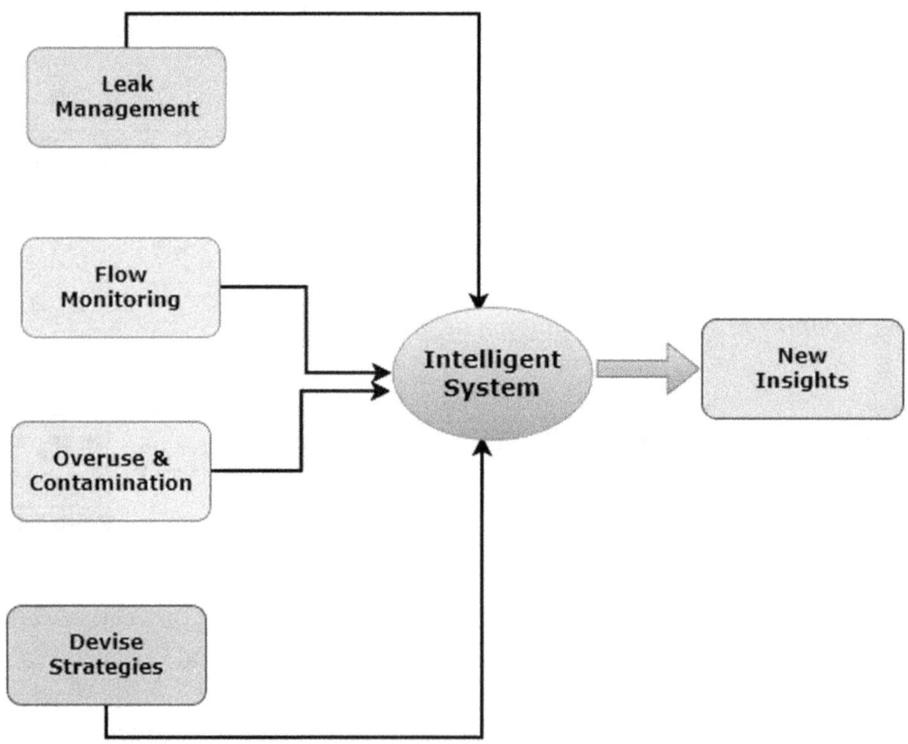

Chapter 10
Detection of Ephemeral Sand River Flow Using Hybrid Sandpiper Optimization-Based CNN Model

Arunadevi Thirumalraj
https://orcid.org/0009-0003-5396-6810
K. Ramakrishnan College of Technology, India

V. S. Anusuya
New Horizon College of Engineering, India

B. Manjunatha
New Horizon College of Engineering, India

ABSTRACT

Ephemeral sand rivers are a major supply of water in Southern Africa that flow continuously all year. The fact is a sizeable fraction of this water permeates the silt in the riverbed, protecting it from evaporation and keeping it available to farmers throughout the dry season. This study set out to investigate the usefulness of satellite optical data in order to assess the possibility for discovering unexpected surface flows. The spatio-temporal resolution required to identify irregular flows in the comparatively small sand rivers typical of dry regions. A hybrid pre-trained convolutional neural network is used to execute data categorization using the hybrid sandpiper optimization technique. Sentinel-2's higher spatial and temporal resolution allowed for accurate surface water identification even in conditions where river flow had drastically decreased and the riverbeds were heavily hidden by cloud cover. The model suggested in this study fared better than rival models in this field, obtaining a remarkable accuracy rate of 99.77%.

DOI: 10.4018/979-8-3693-1194-3.ch010

1. INTRODUCTION

Ephemeral rivers, often referred to as non-perennial rivers, have a special seasonal pattern in which they have a hidden stream in the dry season and an exposed riverbed in the rainy season. Due to causes including climate change, changes in land use, and water abstraction, this trait is becoming more and more important on a worldwide scale (He et al., 2019). Due to the variety of sediment particle sizes, these rivers often have sandy riverbeds with high permeability and varied porosity along their courses (Liang et al., 2021; Baswaraju et al., 2023; Aprisal et al., 2019). These dry transitory alluvial streams, especially in arid and semi-arid areas, are crucial for replenishing local and regional aquifers despite their patchy flow (Su et al., 2019). One method for addressing water shortage concerns is changing the design of the river channel to improve axial connection and increase the distance for flood route (Chen et al., 2021). This approach improves the whole river ecosystem as well as the riparian and urban surroundings. In this situation, it is essential to accurately anticipate flood movement under various circumstances, particularly route lengths.

Flood routing models frequently employ equations like the Saint-Venant equation or its simplified variants (Ngoma et al., 2021; Thirumalraj et al., 2023) and the Muskingum equation (Mpala et al., 2020), although commonly used software tools like MODFLOW (Huang et al., **2020**) and HYDRUS (Sacre et al., 2020) are mostly focused on groundwater movement. To account for flood routing with leakage (Hughes, 2019), these models contain infiltration models such as the Green-Ampt model, the Philip model, Kostiakov's empirical formula, and the Horton solution (Gong et al., 2020). Accurately estimating transmission loss is a major difficulty in the development of groundwater flow models for characterising liquid equilibrium inside alluvial deposits (ISRIC World Soil Information (2023) WebsiteS. https://www.isric.org/. Accessed 1 May 2019). The phrase "transmission loss" refers to streamflow reductions brought on by evaporation from streams, infiltration into riverbeds, and reductions in streamflow to upstream embankments or floodplains. Through gearbox loss studies, flood-induced vertical patterns of water content in riverbed sediments have been investigated (Issoufou et al., 2023). Infiltrometers have also been used to determine if particles from the riverbed may re-enter the river channel during dry spells (Kendon et al., 2019). These techniques provide insightful analyses of geographical and temporal data at small sizes. However, there is still need for further study and improvement in terms of enhancing these indicators for catchment water balance calculations.

The main contributions of this paper are:

- This study proposes satellite optical images, in particular Sentinel-2's high spatial resolution, as a practical approach for sand river surface flow detection.
- The study uses a hybrid pre-trained CNN for classification, showcasing how well deep learning methods can identify these sporadic and intermittent surface flows.
- In order to acquire accurate findings in surface flow detection, it is crucial to effectively fine-tune the model. This study introduces the Hybrid Sandpiper Optimization Technique (HSPOA) optimisation for hyperparameter optimisation.
- For their examination along the Shingwidzi River in Limpopo, researchers used Sentinel-2, a satellite with a stunning 10-meter spatial resolution, to get around this restriction.
- Results are evaluated using five parameter metrics such as accuracy, precision, recall, specificity and F-measure.

The study's remaining sections are structured using shadows: Section 2 summarises the pertinent works, Section 3 provides a brief explanation of the proposed model, Section 4 displays the results and validation analysis, and Section 5 provides a summary and conclusion.

2. RELATED WORKS

In their work, Lindle et al. (2023) examined the dynamics of recharge mechanisms and their relative implications. Ten sites in the South African province of Limpopo were used to collect hydrograph data, with sites chosen for their closeness to river systems and rain gauges. The many topography, vegetative, hydrogeological, and climatic features of the area were represented by the hydrographs that were chosen. The hydrographs' research demonstrated the groundwater levels' sensitivity to changes in rainfall throughout the rainy season. It is noteworthy that bigger rises in groundwater levels were seen during years with significant intermittent rainfall events, which are crucial for replenishing groundwater stocks. The study used both the water-table fluctuation method and the HYDRUS-1D computer model to calculate yearly recharge quantities. As it was shown to increase recharge rates by up to five times when compared to scattered recharge approaches, the findings highlighted the importance of focused recharging near ephemeral rivers. A clear connection and linear association between yearly river flow and recharging along the river were also found by the study. This confirmed the value of giving targeted recharging activities close to river networks top priority. The study's conclusion emphasises the need of taking concentrated and episodic recharge into account when managing groundwater supplies in dry locations. Recharging in deserts is unpredictable in terms of timing and location, which emphasises the significance of taking a holistic strategy at different scales.

A perennial river in Romania and an intermittent torrent channel in Greece were assessed using the (Diaconu et al., 2023) technique to determine the typical types of hydrographic networks in the Mediterranean and Black Sea area. The technique used was effective in determining the exact location, scope, and duration of the bed materials extraction. These encouraging results offer a monitoring strategy for illegal human behaviour that the relevant authorities may simply apply. Adopting the method would improve river protection monitoring by allowing authorities to quickly and accurately detect places where unlawful river bed extraction is occurring.

The paper by Huang et al. (2023) presents a thorough framework that covers the complete procedure from the collection of radar data to the computation of flow. The study carefully examines, applies, and discusses the limits of state-of-the-art signal sampling methods. It also explores important topics including signal processing, the Doppler effect, spectrum estimation, and flow inversion. The article, which is significant, also discusses possible difficulties that these technologies could run into in the future. In order to do this, it suggests workable methods targeted at improving the accuracy and real-time capabilities of hydrometric measurements. This field is expected to continue to advance significantly as a result of these developments.

The research by Woźnica et al. (2023), involving data from four different sources, provided an efficient method for determining the river's susceptibility. Patrol surveillance and state environmental monitoring were used to gather the data. Statistical methods for grouping and correlation were employed in the study. Of the 20 physical and chemical components that were considered to be crucial for salinity, magnesium ions, calcium sulphate, and chloride are three. Despite the need for flow studies to calculate the load, data made available by multi-parameter sensors allowed for the identification of areas with high salt pressure. Conclusions: the levels of the examined ions were highest in the Vistula River in the Silesia Region. Patrol monitoring might be a highly helpful strategy for identifying the root causes for clean water quality issues and helping state surveillance of the environment.

The goal of the work carried out by Ha et al. (2023) was to propose a thorough set of 17 Hydro-Ecological System Services (HESS) indicators that could be successfully obtained utilising satellite imaging technology in conjunction with Geographic Information Systems (GIS) and hydrological models. These metrics support CGIAR's strategic objectives and guiding ideals. The HESS17 framework was thoughtfully created to take into account variables that are directly related to water flow, fluctuations, and storage, while highlighting their natural qualities and minimising human effect. These indicators are found by fusing GIS, hydrological modelling methods, and earth observation models. In the context of water management, where water is used for both consumptive and non-consumptive uses, the need of building the HESS framework is made clear. The generation of dry matter and evapotranspiration are examples of consumptive uses, whereas stream flow, recharge, and water storage are examples of non-consumptive uses. HESS17 should be included into spatial land-use planning, water footprint assessments, water accounting, transboundary water resource management, and food security programmes, according to governments and organisations in charge of integrated water management. The HESS17 framework also offers a useful tool for assessing conservation initiatives for land, soil, and water. The study framework may lack some particular, but it is nonetheless flexible and usable across sizes, proving its usefulness in a variety of scenarios.

The quadratic function was changed to the dual exponential function sum, which was created by Hanand and Basu (2023). The energy that most closely matched the data was then generated using this function. Second, two separate grayscale characteristics that are more precise and stable were used to compute the adaptive area-fitting centres (AFCs). Thirdly, an edge indicator function was utilised in place of the Dirac function, which was previously employed to assist gradient descent flow in stopping at the appropriate edges. Several regularisation parameters were added to the goal function to improve the model's stability. The results of the tests for river channel detection showed that the recommended model performed better in terms of detection accuracy and efficiency than current state-of-the-art techniques.

The elasticity coefficient technique was used in the work by Wu et al. (2023). To evaluate the influence of both human activity and climate change on runoff changes. They examined temporal and spatial runoff patterns in addition to a thorough investigation of the basin's meteorological conditions. Their data showed that yearly runoff has been steadily declining over the past 60 years. The functioning of the reservoir in conjunction with the Middle Route Programme of South-to-North Water Diversion (MRP-SNWD) conducted throughout the nation significantly altered the temporal runoff patterns upstream and above the Danjiangkou Reservoir, according to the research. The decrease in runoff and ensuing changes in rainfall patterns were mostly caused by human activities. It's interesting to see that evapotranspiration had a relatively less impact. The MRP-SNWD in particular showed a significant human-induced influence, accounting for around 20.3% of the overall runoff drop. Overall, the study showed how human activity negatively affects the hydrology of the Hanjiang River and offered helpful advice for turning vulnerability into resilience.

3. PROPOSED METHODOLOGY

In this section, the brief explanation of the study is discussed.

3.1. Dataset Description

Sand rivers presented a difficulty for the study because of their short flow times—often only a few days—and tiny channel widths, which are generally just a few tens of metres. Low-resolution satellite images from sources like Landsat and MODIS became unfeasible as a result (Walker et al., 201923). Sentinel-2 satellite data, which was just released, however, provided a possible remedy by providing greater temporal and geographical resolution. This study looked at how high-resolution optical satellite imaging may be used to follow sand river flow patterns and, in turn, provide accurate predictions for replenishing the alluvial aquifer. The study concentrated on two unique catchment regions, the Molototsi and the Shingwidzi, which are separated by around 35 km in the eastern state of Limpopo. A single rainy season, lasting from November to March throughout the summer months, is all that this area's hot, dry environment has to offer. However, this wet season demonstrates substantial year-to-year fluctuation. Notably, there is a distinct gradient in rainfall, with the eastern lowveld being drier than the two rivers' steep headwaters in the southwest of the nation. Past severe droughts in Limpopo have been sporadic and unpredictable, leading to agricultural failures and significant financial losses. The sporadic occurrence of these drought occurrences highlights the urgent need for better water resource management in the area. Geological rocks from the Archaean epoch, including gneisses, greenstones, and weathered regolith, underlie the catchment regions under study.

The Shingwidzi watershed is mostly contained within the borders of Kruger National Park and has its origins in the Vhembe District. It moves eastward until it ultimately merges with Mozambique's Olifants River. This catchment area inside the Kruger National Park is managed by the South African Department for Water and Sanitation (DWS), particularly the portion that is located above the Silwervis gauge. This basin part has an area of 810 square kilometres, a river that is about 90 km long, and a channel that is between 20 and 40 metres wide close to the gauge. The Mopane-veld vegetation type dominates this low-altitude watershed, despite the existence of a few settlements and small farms in the upper parts of this rather confined basin.

The Olifants River joins the Molototsi Basin in Mopani District, and then runs through Kruger National Park before draining into the Groot Letaba River. Over its 120-kilometer length, the river's channel width fluctuates from Fifty to ninety metres at its shortest point. 1170 km2 make up the catchment basin. The highest heights of the Molototsi basin contain rocky terrain, populated lowlands, and forested slopes. The Modjadji Dam, which was constructed in the neighbourhood in 1997 to provide the Groot Letaba municipal area with potable water, is beyond these higher reaches, which have an area of around 70 km2. The topography grows flatter, drier, and sparserly populated as one travels down from the the mountains, with mostly Mopane-veld vegetation.

3.1.1. Shingwidzi Ground Observations

The Shingwidzi River, which has its source in the Vhembe District, flows mostly through the Kruger National Park before joining the Olifants River in eastern Mozambique. The South African DWS is in charge of managing the catchment area inside Kruger National Park, in particular the portion located

Table 1. Summary of data used in the study

<table>
<tr><th>Type</th><th>Source</th><th>Period</th><th>Location</th></tr>
<tr><td>Sentinel-2 1C products</td><td>Copernicus Open Access Hub</td><td>Oct 16 to Jan 18</td><td>Shingwidzi watershed</td></tr>
<tr><td>Rainfall</td><td>SANParks</td><td>Oct 16 to Jan 18</td><td>Woodlands and Shangoni</td></tr>
<tr><td>Sentinel-2 1C products</td><td>Copernicus Open Access Hub</td><td>Oct 16 to Sep 17</td><td>MS</td></tr>
<tr><td>River flow</td><td>DWS</td><td>Oct 16 to Jan 18</td><td>Shingwidzi River @ Silwervis</td></tr>
<tr><td rowspan="3">Rainfall</td><td rowspan="3">DWS, ARC, local observers</td><td>Oct 16 to Sep 17</td><td>Modjadji Dam, Giyani, A hi</td></tr>
<tr><td></td><td>tirheni Mqekwa Farm and</td></tr>
<tr><td></td><td>Duvadzi Farm</td></tr>
<tr><td>Flow occurance</td><td>Local observer</td><td>Oct 16 to Sep 17</td><td>Molototsi River, Duvadzi Farm</td></tr>
</table>

above the Silwervis gauge. With a river that is around 90 km long and an opening that is between 20 and 40 m wide close to the gauge, this drainage zone is rather small, comprising an area of only 810 km2. This low elevation watershed is dominated by mopane-veld vegetation, despite the presence of a few communities and small-scale agriculture in the higher reaches.

The Molototsi watershed (MW) is located in the Mopani District and travels through the Kruger National Park before meeting the Olifants River and draining into the Groot Letaba River. The river's width of channel fluctuates between 50 and 90 metres at its narrowest point over its 120-kilometer length. The basin's area measures 1170 km2. The highest altitudes of the Molototsi basin are characterised by rocky terrain, populated lowlands, and forested slopes. The Modjadji Dam, a neighbouring structure completed in 1997 to supply Grand Letaba Municipality with potable water, is beyond these higher reaches, which occupy an area of around 70 km2. The environment flattens, gets drier, and is sparserly populated as one moves downwards across the dam away into the the mountain regions, with mostly Mopane-veld vegetation.

3.1.2. Ground Observations From Molototsi

The Molototsi was chosen as a test example to demonstrate how the notion is employed since it is ungauged, has a huge river, and is believed to run hardly by the locals. Controlling the rapidly diminishing water resources used for irrigation by the Molototsi's local farmers requires an understanding of the movement rate and recharging regularity of the sand river. Because the Molototsi is not gauged, a resident who frequently uses the sandy river aquifers for farming documented and took pictures of every incidence of surface flow throughout the hydrological year 2016-2017. These flow frequency measurements were used to evaluate the reliability of the satellites flow detecting method.

This study sought to determine whether it was possible to forecast flows in ungauged catchments just using rainfall data. This was accomplished by gathering rainfall data and contrasting it with flow records from other sources. A single rain gauge at Modjadji Dam, managed by the Department for Water and Sanitation (DWS), captured pertinent rainfall data despite the area's rough topography. Two community-monitored rain gauges were installed at Ahi Tirheni Mqekwa Farm and Duvadzi Farm in the lowveld region of the Molototsi basin Additional rainfall data was also provided by the Giyani rain gauge station operated by the South African Agricultural Research Council (ARC), which is located around 11 km northeast of the watershed.

3.2. Satellite Data and Data Pre-Processing

When compared to other publicly available satellites a multispectral picture (such MODIS - 250 m and LandSAT - 30 m), which were used for this inquiry, Sentinel-2 imagery provides a higher level of spatial detail. The Sentinel-2 A and -2B satellites, which were launched into space on the 23rd of June the year 2015, and on March 7, the year 2017, accordingly, were created by the European Space Agency (ESA). With three distinct spatial resolutions (10, 20 and 60 m) and a 5-day return period, the multispectral sensor on board the satellites monitor the electromagnetic radiances of the sun as they are reflected in 13 different spectrum bands, spanning the visible to the SWIR range. In order to collect validated top-of-atmosphere (TOA) transparency data, referred to as Level 1 C products, for the study areas for the hydrological years 2016 to 2018 (Shingwidzi) and 2016-2017 (Molototsi), Level 1C products were developed. These reflectance products, referred to as Level 2A, were atmospherically modified to BOA reflectance data using Sen2Cor (version 2.4.0) and the predetermined parameter values. Sen2Cor, an alternative Sentinel-29 toolbox plugin, is accepted by ESA. On Level 1C input data, the DDV (Dark Dense Vegetation) technique is employed to perform basic scene categorization, atmospheric modification, and cloud screening.

(1) A land-cover categorization map and Level 2A data are produced after Level 1C data have undergone atmospheric correction, (2) Normalised Difference Water Index (NDWI) computation, (3) Calculation of the NDWI change from the dry season's NDWI value (ΔNDWI), (4) implementation of a threshold, and (5) masking of cloud- or cloud-shadow-obstructed regions.

3.3. Hybrid Pre-Trained CNN Classification

Layered Architecture

DenseNet-169, any among the 169-layer DenseNet series layouts, is a popular design for DL classification problems. When compared to other DenseNet designs with fewer layers, it features much less trainable parameters. The vanishing gradient problem can be solved with DenseNet-169 alongside the other DenseNet architectures. They also support feature reuse and have a limited number of trainable parameters and a reliable feature propagation technique. Both PyTorandes and Tensorflow (Keras) support DenseNet models (Vulli et al., 2022).

CLs, transitional layers, dense layers (totally linked layers), and dense layers (maxpool layers) are all included in the architecture. The model employs activation of ReLU for the other levels of the structure and SoftMax stimulation for the highest layer. Convolution layers are in charge of removing significant characteristics from the image whereas maxpool layers are used to minimise the three-dimensional nature of their input data. The flatten layer, which comes after these layers and resembles an artificial neural network by effectively transforming the data into a single array input. In the network architecture, fully linked layers are then used. Table 2 contains the particular details of this tiered construction.

Table 2. The layered design of DenseNet-169

Layer	Parameters	Kernal Size	Tensor Size
Convolution	Stride=2,ReLu	7 × 7 (Conv)	112 × 112
Pooling	Stride = 2	3 × 3(MaxPool)	56 × 56
Transition-1 Layer	Stride = 2	1 × 1 (Conv) 2 × 2(AvgPool)	56 × 56 28 × 28
Transition-2 Layer	Stride = 2	1 × 1 (Conv) 2 × 2(AvgPool)	28 × 28 14 ×14
Dense-2 block	Dropout = 0.2	1 × 1 × 12 (Conv) 3 × 3 × 12 (Conv)	28 × 28
Dense-3 Layer	Dropout = 0.2	1 × 1 × 12 (Conv) 3 × 3 × 12 (Conv)	14 × 14
Transition-3 Layer	Stride = 2	1 × 1 (Conv) 2 × 2(AvgPool)	14 × 14 7 × 7
Dense-4 Layer	Dropout = 0.2	1 × 1 × 12 (Conv) 3 × 3 × 12 (Conv)	7 × 7
Dense-1 Layer	Dropout =0.2	1 × 1 × 6 (Conv) 3 × 3 × 6 (Conv)	56 × 56
Classification Layer		1 × 1(Global AvgPool) 1000D (fully-connected softmax)	1 × 1

Convolution Layer (CL)

A Convolutional Layer (CL) essentially moves a filter over an input to activate it. A feature map that shows the existence and intensity of particular characteristics at various locations within the input is produced by this procedure. Following the creation of the feature map through a series of filter operations, activation techniques like Rectified Linear Unit (ReLU) are frequently used to improve it. Since the filter is typically lower in size than the input, these operations in a CL typically entail taking a dot product from the filter and some of the input data. Think of a P-P square neuronal element, a CL, and an m-m filter. The output of the CL would be (p - m + 1) (p - m + 1) in this scenario. To find out the non-linear input to the unit x_{ij}^{l}, as illustrated in Equation (1), the inputs from the cells in the preceding layer must be added up.

$$x_{ij}^{l} = \sum_{a=0}^{m-1}\sum_{b=0}^{m-1}\mu_{ab}\, y_{(i+a)(j+b)}^{l-1} \tag{1}$$

Equation (2) illustrates how the CL implements the determined non-linearity.

$$y_{ij}^{l} = \lambda\left(x_{ij}^{l}\right) \tag{2}$$

Using a maxpool layer in a CNN is primarily intended to lessen the feature map's degree of detail. The maxpool layer summaries the features in the region where the pooling filter has reduced the number of features, similar to how a CL might apply a filter to a features map. Think of a feature map as having dimensions $n_h \times n_w \times n_c$ These measurements correspond to the feature map's height, breadth, and number of channels, respectively. Using a filter size of "f" and a stride of "s," it can calculate the dimensions of the feature map after executing the maximum pooling operation, represented as max_p in Equation (3).

$$max_p = \frac{\left(n_h - f + 1\right)}{s} \times \frac{\left(n_w - f + 1\right)}{s} \times n_c \tag{3}$$

Within a neural network, a dense layer displays a strong link with the layer that came before it. It's crucial to understand that every neuron in the Dense Layer forms connections with every neuron in the layer directly above it in this situation. Each neuron in the Dense Layer transmits information to its counterpart neuron in the layer below after performing a matrix-vector multiplication process. The formula for this matrix-vector multiplication operation may be found in equation (4).

$$M \cdot \lambda = \begin{matrix} m_{11} & m_{12} & \cdots\cdots\cdots & m_{1y} & p_1 \\ m_{21} & m_{22} & \cdots\cdots\cdots & m_{2n} & p_2 \\ \vdots & \vdots & \cdots & \vdots & \vdots \\ \vdots & \vdots & \cdots & \vdots & \vdots \\ m_{x1} & m_{x2} & \cdots\cdots\cdots & m_{xy} & p_y \end{matrix} \tag{4}$$

In the equation above, the variable M stands for a matrix with dimensions x y, while the other matrix is p whose dimensions are $1 \times y$. The trained parameters of the layer before it make up the variable matrix, which may be updated via backpropagation during training. The weights for the layer specified were determined via backpropagation by ωl^y and bias identified by the variable Bl^y of equations (5) and (6) are used to modify the neural network over the rate of learning indicated by α.

$$\omega l^y = \omega ly - \alpha \times d\omega ly \tag{5}$$

$$Bly = Bly - \alpha \times dBly \tag{6}$$

A chain rule procedure that starts at the output layer and travels via the hidden levels back to the input layer determines the values of dω and db. The partial derivatives of the loss function with respect to ω and b are represented by these values for dω and db. Equations (7) through (10), respectively, are used to calculate dω and db.

$$d\omega^{ly} = \frac{\partial L}{\partial \omega^{ly}} = \frac{1}{n} \, d\omega^{ly} A^{[ly-1]T} \tag{7}$$

$$dB^{ly} = \frac{\partial L}{\partial \omega^{ly}} = \frac{1}{n} \sum_{i=1}^{n} dZ^{ly(i)} \tag{8}$$

$$dA^{ly-1} = \frac{\partial L}{\partial A^{ly-1}} = W^{ly^T} dZ^{ly} \tag{9}$$

$$dZ^{ly} = dA^{ly} \times g'\left(Z^{ly}\right) \tag{10}$$

Z^{ly} in the above equations stands for the linear activation at layer ly and $g'(Z^{ly}$ denotes the second derivative that represents the irregular function with respect to Z^{ly}. The non-linear function of activation used at the same layer is also indicated by the letter A^{ly}.

CNNs have a transition layer that helps to streamline the model's design. This common transition layer commonly uses a filter to halve the input's width and height with a stride of 2. It also has a 1×1 convolutional layer to efficiently cut down on the amount of data channels.

Due to its prominence as a non-linear activation function in the field of deep learning, the softmax activation function is often used, especially in the context of classification problems. This non-linear activation function is set up as shown in equation (11) by applying weight and bias parameters, denoted by "w" and "b," respectively, to an input vector "x."

$$y = f(w \times x + b) \tag{11}$$

The output layer of a CNN is where the softmax function is used to forecast the likelihood related to each output class. The softmax function essentially gives each neuron in the output layer a different probability value. As a result, each neuron in this layer has an impact on the probability that a certain node will be chosen as the output. According to Equation (12), the exponential function is applied to the input vector designated as e^{v_i} in order for the softmax function, denoted as, to be applied to the input v_i. This calculation applies throughout a set of m cases and additionally considers the exponentiated values from the output vector, indicated as e^{v_o} .

$$\Theta\left(z\right)_x = \frac{e^{v_i}}{\sum_{y=1}^{m} e^{v_o}} \tag{12}$$

The activation function utilised for this study is softmax, while the loss function is binary cross-entropy. For binary classification issues, binary cross-entropy is typically utilised. The binary cross- Entropy loss operation, that takes into consideration systems with n layers, is correctly described in formulae (13) and (14).

$$K\left(W,\ b\right) = \frac{1}{n}\sum_{i=1}^{n}\left(a^{(i)},\ a^{(i)}\right) \tag{13}$$

$$L\left(\hat{a}, a\right) = -\left(a \times \log\hat{a} + \left(1 - a\right) \times log\left(1 - \hat{a}\right)\right) \tag{14}$$

In this instance, the output for class 0 is represented by (1 - a), whereas the outcome for class 1 is denoted by the variable "a." 1- (â) stands for the probability connected to the result for class 1, and (1 - â) represents the probability connected to the result for class 0.

3.3.1. Weight Optimization

When weights are optimised, variables are modified depending on the current instance and its neighbouring successes and failures at each step to simplify the equation. This method may have an effect on a specific instance's close neighbours by taking into consideration the most recent successes and failures related to the chosen instance at any given time. Equation (15) displays the outcomes of m applications of the k-nearest neighbour approach.

$$\omega_d = \omega_d - \frac{1}{k \times m}\left(\sum_{a=1}^{k} \rho^R\left(e_d^b,\ e_{d_{hit}}^{b_a}\right) + \rho^R\left(e_q^b,\ e_{q_{miss}}^{b_a}\right)\right) \tag{15}$$

where $e_{d_{hit}}^{b_a}$ represents the closest hit of the instance b's dth feature variable and $e_{q_{miss}}^{b_a}$ represents the ath closest miss attribute value for the instance b. The variable ρR denotes the similarity among the instances. Equation (16) may be used to optimise weight instead of using a gradient descent strategy with a constant learning rate of $\frac{1}{k \times m}$.

$$j_\omega^R = \sum_{a=1}^{k}\sum_{d=1}^{Q} \omega_d \rho^R\left(e_d^b,\ e_{d_{hit}}^{b_a}\right) - \sum_{a=1}^{k}\sum_{d=1}^{Q} \omega_d \rho^R\left(e_q^b,\ e_{q_{miss}}^{b_a}\right) \tag{16}$$

3.3.3. Hyperparameters

In this part, this paper explores the learning rate and the loss resulting from batch processing, two crucial hyperparameters connected to the optimised DenseNet-169 model. Choosing the right parameters is essential for improving training and testing results and avoiding under- or overfitting problems. In the present study, we thoroughly investigate loss metrics, training and testing accuracy. Where the model initially diverges is where we identify the ideal learning rate range. After establishing this learning rate, the objective is to track a persistent decline in loss. Notably, the author (Another Data Science Student's Blog—The 1cycle Policy. Available online: sgugger.github.io (accessed on 16 March 2022) advises using the greatest learning rate that enables training to progress successfully, even exceeding the HSPOA with different weight decays (0.01, 0.0001, and 0.000001), mostly because to the L2 penalty provided by the optimizer (weight decay).

3.3.4. Hyper Parameter Tuning Using HSPOA

The Hybrid Sandpiper Optimisation Approach (HSPOA), a unique technique known as Migration updated with Supervisor guidance (MUSG), is used to optimise the weight functions of the Hybrid pretrained CNN with the goal of improving detection accuracy. The well-known Teamwork Optimisation Algorithm (TOA) (Zolanvari et al., 2021) and the Sandpiper Optimisation Algorithm (SOA) (Kohnhäuser et al.,

2021) are both used as inspiration for this suggested hybrid optimisation methodology. The TOA is an example of successful collaboration towards shared goals, whereas the SOA is inspired by the flying and aggressive behaviour of sandpipers. To make the use of TOA and SOA in resolving optimisation problems easier, mathematical representations have been provided. These models are designed to enable quick convergence to global solutions through optimisation, reducing the possibility of becoming stuck in local optima. The output is produced by the CNN's HSPOA 8, which has been adjusted to achieve the goal stated in Eq. (16) based on the input (Behera et al., 2022).

The next section provides the stages used in the HSPOA model.

Step 1: Initialise the P population of the N search agents.

Step 2: Set the settings for *itr, Max^{itr}*. Here, *itr, Max^{itr}* points to the most recent iteration and the number of iterations possible.

Step 3: While $itr<Max^{itr}$ do

Step 4: Determine the target agent's fitness using Eq. (16).

Step 5: In its actual use, the SOA model's migration behaviour stands out in particular. Three major activities take place during this migration or exploration phase: avoiding crashes, travelling in the direction of the ideal neighbour, and remaining near the ideal search agent.

- In order to prevent search agent collisions in equation (17), a new variable A, known as the movement behaviour of the search agent, is introduced.

$$C = A*P(X) \tag{17}$$

In this instance, P denotes the search agent's present location and C denotes its non-colliding location. A can also be calculated using Eq. (35), wherein f_c is a parameter that has been added to regulate the frequency of hiring A eq. 18.

$$A = f_c - \left\{ X, \left[\frac{f_c}{Max^{itr}} \right] \right\} \tag{18}$$

This paper established a strategy in our company's engagement in development that is comparable to going the route of the best neighbour. Our search teams will use this strategy to prevent collisions and then go along the path of the best neighbour. As part of our study, we created a unique mathematical model, designated as $G_{best}(X)$, to improve the accuracy of this movement and avoid solutions from getting stuck in local optima. Eq. (19) presents this freshly developed mathematical equation.

$$M = B*[G_{best}(X) - P(X)] \tag{19}$$

The Supervisor guiding stage found in the TOA model is used to determine $G_{best}(X)$. Participants in the initial phase are given updates depending on the supervisor's instructions. The manager's job is to lead the team towards their goal by providing reports and information. Equations (20) and (21) are utilised to duplicate this degree of updating within the TOA.

$$G_{best}(X) = P(X)r^{*}Z - I^{*}P(X) \tag{20}$$

Indicator guidance I is calculated according to Eq. (21).

$$I = round(1+r) \tag{21}$$

Additionally, Z, P(X), and M point to the supervisor, the search agent's current position, and the search agent's position, respectively.

Last but not least, the SOA model's Eq. (22) states that the search agent's position is updated in reference to the optimal search agent.

$$\vec{D} = \left|\vec{C} + \vec{M}\right| \tag{22}$$

The distance that exists between the most effective search agency and the present search agent is also represented by $\vec{D}$.

Step 6: Let's now turn our attention to the Attacking (exploitation) phase, in which the search agents begin their assault by flying in a spiralling pattern through the air. To roughly simulate this spiral motion in the X, Y, and Z axes, we will use equations (23) and (24).

$$x' = r^{*}Cos\left(K\right) \tag{23}$$

$$y' = r^{*}Sin\left(K\right) \tag{24}$$

$$z' = r^{*}K \tag{25}$$

K is the random variable here.

Step 7: The search agent's updated position is represented using Eq. (26).

$$P\left(X\right) = \left(D^{*}\ x'^{*}y'^{*}z'\right)\ P_{best}\left(x\right) \tag{27}$$

Step 8: Return *P(X)*

Step 9: Terminate

4. RESULTS AND DISCUSSION

4.1. Experimental Setup

This section contrasts the efficacy of the suggested system with the existing approaches. The suggested technique (tflearnCPU) was put into practise using OpenCV, Python, and the Neural Network Model module. On an Intel Core i5 3.2 GHz CPU running Ubuntu 16.04, all trials were conducted.

4.2. Performance Metrics

Accuracy: The accuracy metric, in general, calculates the proportion of correct predictions to all instances assessed using eq (28).

$$Accuracy = \frac{No.\ of\ correctly\ classified\ expressions}{Total\ no.\ of\ images} \times 100 \tag{28}$$

Furthermore, Eqs. (29) – (32) are used to compute the precision, f-measure, and sensitivity.

$$precision = \frac{TP}{TP + FP} \times 100 \tag{29}$$

$$F - measure = 2 \times \frac{Precision \times Recall}{Precision + Recall} \times 100 \tag{30}$$

$$Sensitivity = \frac{TP}{TP + FN} \times 100 \tag{31}$$

$$Specificity = \frac{TN}{TN + TP} \times 100\% \tag{32}$$

TP stands for true positives, TN for true negatives, FP for false positives, and FN for false negatives in this context. Improvements in terms of accuracy, precision, recall, F-measure, and specificity should be present in the predicted result.

From Table 3 and Fig 1, Fig 2, Fig 3, Fig 4, Fig 5, in the analysis of classification without HSPOA, AlexNet had an 90.94% accuracy rate, 91.32% specificity, 90.45% precision, 90.23% recall, and 91.54% F-measure. ShuffleNet had an 92.45% accuracy rate, 92.68% specificity, 91.49% precision, 91.64 recall, and 92.22% F-measure. ResNet50 had an 93.79% accuracy rate, 93.46% specificity, 92.35% precision, 93.46% recall, and 93.43% F-measure. SqueezeNet had an 94.95% accuracy rate, 94.69% specificity,

Table 3. Classification without HSPOA

Methods	Accuracy	Precision	Recall	Specificity	F-Measure
AlexNet	90.94	90.45	90.23	91.32	91.54
ShuffleNet	92.45	91.49	91.64	92.68	92.22
ResNet50	93.79	92.35	93.26	93.46	93.43
SqueezeNet	94.95	94.64	94.33	94.69	94.72
Proposed model	96.77	95.37	95.42	95.52	96.21

Table 4. Classification with HSPOA

Methods	Accuracy	Precision	Recall	Specificity	F-Measure
AlexNet	97.91	97.94	97.67	97.53	97.61
ShuffleNet	98.47	97.69	98.67	98.22	98.34
ResNet50	98.70	98.40	98.68	98.66	98.38
SqueezeNet	97.92	98.68	98.21	98.79	97.67
Proposed model	99.77	99.31	99.24	99.43	99.62

94.64% precision, 94.33% recall, and 94.72% F-measure. The Proposed model had an 96.77% accuracy rate, 95.52% specificity, 95.37% precision, 95.42% recall, and 96.21% F-measure.

From Table 4 and Fig 1 through Fig 5, in the analysis of classification with HSPOA, AlexNet had an 97.91% accuracy rate, 97.53% specificity, 97.54% precision, 97.67% recall, and 97.61% F-measure. ShuffleNet had an 98.47% accuracy rate, 98.22% specificity, 97.69% precision, 98.67% recall, and 98.34% F-measure. ResNet50 had an 98.70% accuracy rate, 98.66% specificity, 98.40% precision, 98.68% recall, and 98.38% F-measure. SqueezeNet had an 97.92% accuracy rate, 98.79% specificity, 98.68% precision,

Figure 1. Accuracy analysis

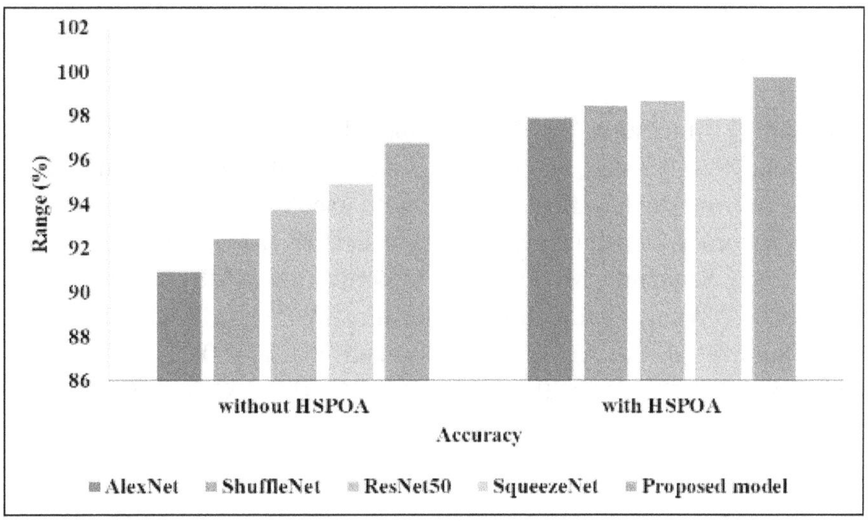

Figure 2. Precision analysis

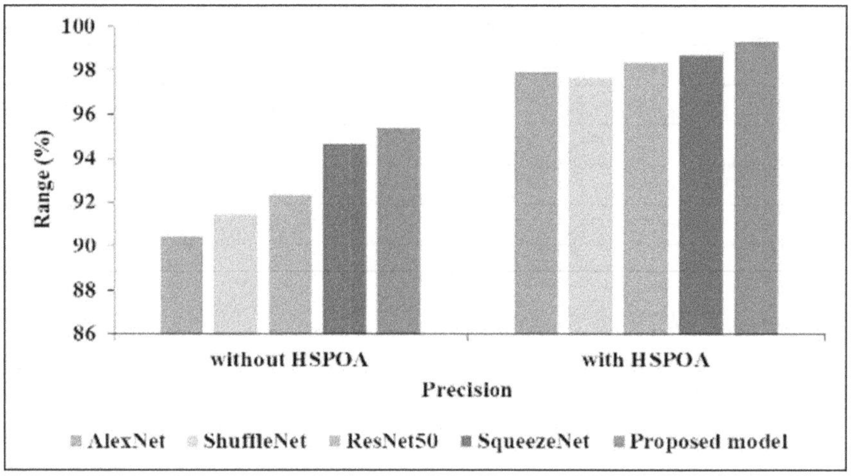

Figure 3. Recall analysis

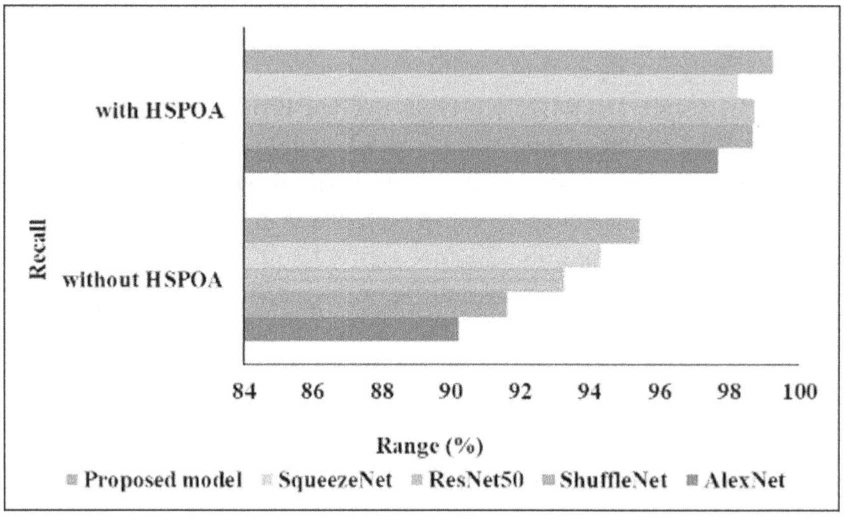

98.21% recall, and 97.67% F-measure. The Proposed model had an 99.77% accuracy rate, 99.43% specificity, 99.31% precision, 99.24% recall, and 99.62% F-measure. It is well known that hybrid pre-trained CNNs perform a range of image classification tasks with excellent accuracy. You may take advantage of their high baseline accuracy by using such models to sand river flow detection.

From Table 5 and Fig 6, Sequential Model-Based Optimization (SMBO) had 93.9% of 0.01, 92.3% of 0.0001, 96.6% of 0.000001. Whale Optimization Algorithm (WOA) had 94.3% of 0.01, 93.4% of 0.0001, 96.5% of 0.000001. Squirrel Search Algorithm (SSA) had 96.7% of 0.01, 95.1% of 0.0001, 97.6% of 0.000001. Proposed model had 98.5% of 0.01, 97.4% of 0.0001, 99.4% of 0.000001.

Figure 4. Specificity validation

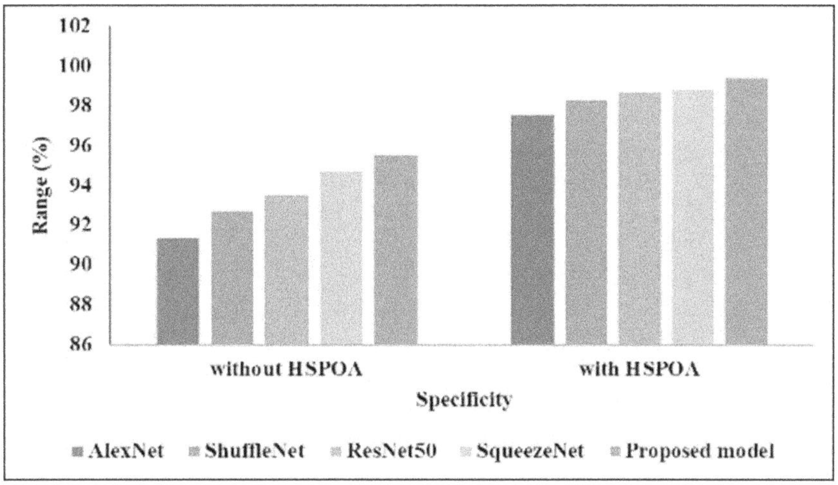

Figure 5. F-measure validation

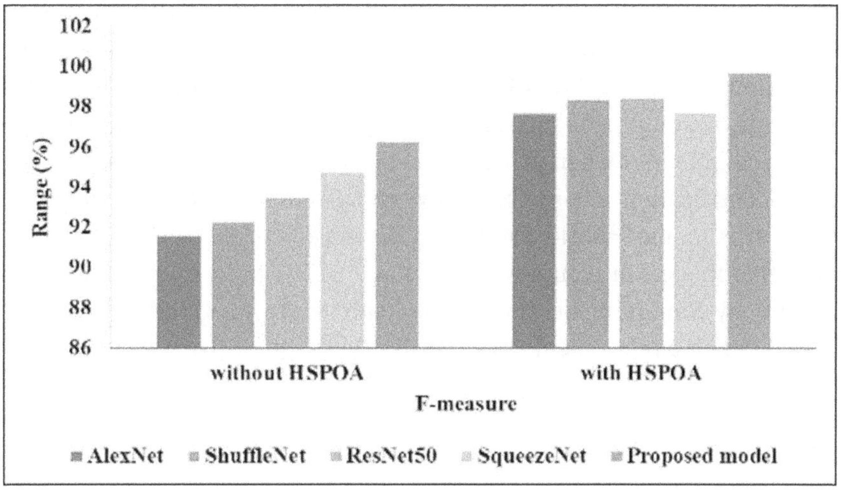

Table 5. Learning rate

Models	0.01	0.0001	0.000001
SMBO	93.9	92.3	96.6
WOA	94.3	93.4	96.5
SSA	96.7	95.1	97.6
Proposed model	98.5	97.4	99.4

Figure 6. Learning rate

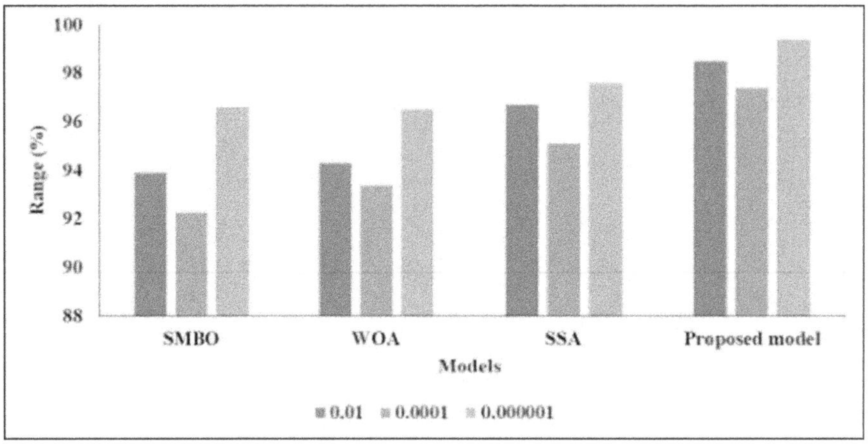

5. CONCLUSION

With a focus on the Shingwidzi River in Limpopo, South Africa, this study has shown the usefulness of satellite optical images, notably Sentinel-2 data, in identifying uncommon surface flows in ephemeral sand rivers. The creation and application of a hybrid pre-trained CNN for image categorization is one of the study's noteworthy accomplishments. Automating the detection of surface water bodies in remote sensing data has showed potential when using deep learning techniques, particularly CNNs. The accuracy of surface flow identification was improved by the hybrid nature of the CNN and the optimisation of hyper-parameters utilising the Hybrid Sandpiper Optimisation Algorithm (HSPOA). Our suggested model outperforms other models in accuracy, achieving 99.77%. Future research may increase detection accuracy and reliability by combining Sentinel-2 data with that from other satellite sensors, such as radar data, particularly in difficult environments like dense vegetation or overcast skies.

REFERENCES

Another Data Science Student's Blog—The 1cycle Policy. (n.d.). Available online: sgugger.github.io

Aprisal, Istijono, B., Ophiyandri, T., & Nurhamidah. (2019). A study of the quality of soil infiltration at the downstream of Kuranji River, Padang City. *Int. J. Geomate*, *16*, 16–20.

Baswaraju, S., Maheswari, V. U., Chennam, K. K., Thirumalraj, A., Kantipudi, M. P., & Aluvalu, R. (2023). Future Food Production Prediction Using AROA Based Hybrid Deep Learning Model in Agri-Sector. *Human-Centric Intelligent Systems*, 1-16.

Behera, B. B., Mohanty, R. K., & Pattanayak, B. K. (2022). Attack Detection and Mitigation in Industrial IoT: An Optimized Ensemble Approach. *Specialusis Ugdymas*, *1*(43), 879–905.

Chen, W. L., Wang, J. J., Liu, Y. F., Jin, M. G., Liang, X., Wang, Z. M., & Ferre, T. P. A. (2021). Using bromide data tracer and HYDRUS-1D to estimate groundwater recharge and evapotranspiration under film-mulched drip irrigation in an arid Inland Basin, Northwest China. *Hydrological Processes*, *35*(7), 14290. doi:10.1002/hyp.14290

Diaconu, D. C., Koutalakis, P. D., Gkiatas, G. T., Dascalu, G. V., & Zaimes, G. N. (2023). River Sand and Gravel Mining Monitoring Using Remote Sensing and UAVs. *Sustainability (Basel)*, *15*(3), 1944. doi:10.3390u15031944

Gong, C., Wang, W., Zhang, Z., Wang, H., Luo, J., & Brunner, P. (2020). Comparison of field methods for estimating evaporation from bare soil using lysimeters in a semi-arid area. *Journal of Hydrology (Amsterdam)*, *590*, 125–334. doi:10.1016/j.jhydrol.2020.125334

Ha, L. T., Bastiaanssen, W. G., Simons, G. W., & Poortinga, A. (2023). A New Framework of 17 Hydrological Ecosystem Services (HESS17) for Supporting River Basin Planning and Environmental Monitoring. *Sustainability (Basel)*, *15*(7), 6182. doi:10.3390u15076182

Han, B., & Basu, A. (2023). Level Sets Guided by SoDEF-Fitting Energy for River Channel Detection in SAR Images. *Remote Sensing (Basel)*, *15*(13), 3251. doi:10.3390/rs15133251

He, Y., Qiu, H. J., Song, J. X., Zhao, Y., Zhang, L. M., Hu, S., & Hu, Y. Y. (2019). Quantitative contribution of climate change and human activities to runoff changes in the Bahe River watershed of the Qinling Mountains, China. *Sustainable Cities and Society*, *51*, 101729. doi:10.1016/j.scs.2019.101729

Huang, P. C., & Lee, K. T. (2020). Refinement of the channel response system by considering time-varying parameters for flood prediction. *Hydrological Processes*, *34*(21), 4097–4111. doi:10.1002/hyp.13868

Huang, Y., Chen, H., Liu, B., Huang, K., Wu, Z., & Yan, K. (2023). Radar Technology for River Flow Monitoring: Assessment of the Current Status and Future Challenges. *Water (Basel)*, *15*(10), 1904. doi:10.3390/w15101904

Hughes, D. A. (2019). A simple approach to estimating channel transmission losses in large South African river basins. *Journal of Hydrology. Regional Studies*, *25*, 100619. doi:10.1016/j.ejrh.2019.100619

ISRIC World Soil Information. (2023) https://www.isric.org/

Issoufou Ousmane, B., Nazoumou, Y., Favreau, G., Abdou Babaye, M. S., Abdou Mahaman, R., Boucher, M., Issoufa, I., Lawson, F. M. A., Vouillamoz, J.-M., Legchenko, A., & Taylor, R. G. (2023). Changes in aquifer properties along a seasonal river channel of the Niger River basin: Identifying potential recharge pathways in a dryland environment. *Journal of African Earth Sciences*, *197*, 104742. doi:10.1016/j.jafrearsci.2022.104742

Kendon, E. J., Stratton, R. A., Tucker, S., Marsham, J. H., Berthou, S., Rowell, D. P., & Senior, C. A. (2019). Enhanced future changes in wet and dry extremes over Africa at convection-permitting scale. *Nature Communications*, *10*(1), 1794. doi:10.103841467-019-09776-9 PMID:31015416

Kohnhäuser, F., Meier, D., Patzer, F., & Finster, S. (2021). On the Security of IIoT Deployments: An Investigation of Secure Provisioning Solutions for OPC UA. *IEEE Access : Practical Innovations, Open Solutions*, *9*, 99299–99311. doi:10.1109/ACCESS.2021.3096062

Liang, J., Yi, Y., Li, X., Yuan, Y., Yang, S., Li, X., Zhu, Z., Lei, M., Meng, Q., & Zhai, Y. (2021). Detecting changes in water level caused by climate, land cover and dam construction in interconnected river-lake systems. *The Science of the Total Environment*, *788*, 147692. doi:10.1016/j.scitotenv.2021.147692 PMID:34022570

Lindle, J., Villholth, K. G., Ebrahim, G. Y., Sorensen, J. P. R., Taylor, R. G., & Jensen, K. H. (2023). Groundwater recharge influenced by ephemeral river flow and land use in the semiarid Limpopo Province of South Africa. *Hydrogeology Journal*, 1–16. doi:10.100710040-023-02682-x

Mpala, S. C., Gagnon, A. S., Mansell, M. G., & Hussey, S. W. (2020). Modelling the water level of the alluvial aquifer of an ephemeral river in south-western Zimbabwe. *Hydrological Sciences Journal*, *65*(8), 1399–1415. doi:10.1080/02626667.2020.1750615

Ngoma, H., Wen, W., Ojara, M., & Ayugi, B. (2021). Assessing current and future spatiotemporal precipitation vari-ability and trends over Uganda, East Africa, based on CHIRPS and regional climate model datasets. *Meteorology and Atmospheric Physics*, 1–21.

Sacre Regis, M. D., Mouhamed, L., Kouakou, K., Adeline, B., Arona, D., Koffi Claude A, K., ... Issiaka, S. (2020). Using the CHIRPS Dataset to Investigate Historical Changes in Precipitation Extremes in West Africa. *Climate (Basel)*, *8*(7), 84. doi:10.3390/cli8070084

Su, P., Wang, X. X., Lin, Q. D., Peng, J. L., Song, J. X., Fu, J. X., Wang, S. Q., Cheng, D. D., Bai, H. F., & Li, Q. (2019). Variability in macroinvertebrate community structure and its response to ecological factors of the Weihe River Basin, China. *Ecological Engineering*, *140*, 140. doi:10.1016/j.ecoleng.2019.105595

Thirumalraj, A., & Rajesh, T. (2023). *An Improved ARO Model for Task Offloading in Vehicular Cloud Computing in VANET*. Academic Press.

Vulli, A., Srinivasu, P. N., Sashank, M. S. K., Shafi, J., Choi, J., & Ijaz, M. F. (2022). Fine-tuned DenseNet-169 for breast cancer metastasis prediction using FastAI and 1-cycle policy. *Sensors (Basel)*, *22*(8), 2988. doi:10.339022082988 PMID:35458972

Walker, D., Smigaj, M., & Jovanovic, N. (2019). Ephemeral sand river flow detection using satellite optical remote sensing. *Journal of Arid Environments*, *168*, 17–25. doi:10.1016/j.jaridenv.2019.05.006

Woźnica, A., Absalon, D., Matysik, M., Bąk, M., Cieplok, A., Halabowski, D., Koczorowska, A., Krodkiewska, M., Libera, M., Sierka, E., Spyra, A., Czerniawski, R., Sługocki, Ł., & Łozowski, B. (2023). Analysis of the Salinity of the Vistula River Based on Patrol Monitoring and State Environmental Monitoring. *Water (Basel)*, *15*(5), 838. doi:10.3390/w15050838

Wu, G., Liu, Y., Liu, B., Ren, H., Wang, W., Zhang, X., Yuan, Z., & Yang, M. (2023). Hanjiang River Runoff Change and Its Attribution Analysis Integrating the Inter-Basin Water Transfer. *Water (Basel)*, *15*(16), 2974. doi:10.3390/w15162974

Zolanvari, M., Yang, Z., Khan, K., Jain, R., & Meskin, N. (2023, February 15). TRUST XAI: Model-Agnostic Explanations for AI With a Case Study on IIoT Security. *IEEE Internet of Things Journal*, *10*(4), 2967–2978. Advance online publication. doi:10.1109/JIOT.2021.3122019

Chapter 11
Design of IoT-Based Automatic Rain-Gauge Radar System for Rainfall Intensity Monitoring

Ahmad Budi Setiawan
The Institute of Public Governance, Economy, and Community Welfare, Indonesia

Danny Ismarianto Ruhiyat
South Tangerang Institute of Technology, Indonesia

Aries Syamsuddin
Distric Government of Blitar, Indonesia

Djoko Waluyo
The Institute of Public Governance, Economy, and Community Welfare, Indonesia

Ardison Ardison
The Institute of Public Governance, Economy, and Community Welfare, Indonesia

ABSTRACT

The occurrence of climate change has become a global problem. These environmental problems then cause many problems for human life such as crop failure in the agricultural sector, the loss of many animal species that are beneficial to human life either directly or indirectly, seasonal changes, and even the occurrence of droughts and irregular season, so this will be difficult activity of human life. Therefore, a rain gauge is needed, which is a tool used to measure rainfall in a location at a certain period of time. This information can be used for various purposes in the community. However, the information generated by these devices also affects the quality of wireless signal transmissions such as free space optics (FSO), GSM, satellite, and outdoor WiFi. This research project aims to create prototypes of IoT-based devices and systems to detect and record rainfall that occurs using a tool created in this research project. The resulting data can be utilized by the community through the website and mobile application.

DOI: 10.4018/979-8-3693-1194-3.ch011

INTRODUCTION

High rainfall is one of the most common hazards of adverse weather and can produce a variety of disasters such as floods and landslides. This can result in a disaster and significant damage to the affected area. To avoid the damage caused by heavy rain, timely response and prevention actions are more crucial. Frick, J., & Hegg, C. (2011) argue that additional consequences of heavy rain include the development of disease as a result of stagnant flood waters for an extended period of time, damage to community housing, major traffic congestion, and even more economic consequences.

Rainfall is the amount of water that falls to the earth during a rainstorm and can be measured in millimeters. Another definition of rainfall is the amount of rainwater that accumulates in a flat area without evaporating, seeping, or overflowing. Indonesia is a country with a humid tropical climate. In Indonesia, the average rainfall is not uniform but still abundant, averaging 2500-3000 mm/year (Narulita & Ningrum, 2017). Even yet, there are many variations in the average rainfall that happens in various regions, which will be varied.

According to several studies, floods kill more people than any other type of weather threat because the water level rises so quickly that victims are caught off guard and have little time to leave. Emam et.al. (2016) argue that floods have a detrimental influence on human health, the environment, cultural heritage, and economic activity. While Kusumastuti et.al. (2017) said that heavy rainfall is less likely to infiltrate and run into rivers, lakes, or other bodies of water. The sooner the water reaches a river or body of water, the greater the likelihood of flooding. As Aryastana and Tasuku (2010) said that water is diverted straight into rivers through drainage systems and gutters, increasing the risk of flooding. There is a lot of land that runs into one big river or, in this example, several smaller rivers or streams. The stream will overflow, resulting in extensive flooding.

Flooding is a long-term phenomenon that can last up to a week. Even a minor delay in taking precautions can hinder rescue operations since treading on flood water without a protective covering can expose you to infection and other ailments. Water is an important resource. However, ordinary people today think that there is no longer a rainy and dry season. Rain can fall anywhere, whether you are in a "rainy month" or not. Many people also notice that the rain is not evenly distributed; rain can fall just in one portion of the territory; and there are certain areas that are genuinely dry and have not been visited by rain. Many research have been undertaken to better understand rain. Rainwater's travel through a specific area is a dynamic process that changes the intensity and shape of rain.

The measuring and sensing of the amount and type of rain allows us to construct detailed physical and dynamic descriptions of rain. As a result, it is preferable to research our environment and plan for potential environmental calamities such as droughts or floods. The primary goal of rainwater management is to reduce economic consequences and life dangers. The majority of underdeveloped countries rely on rainfall to meet their water needs. However, if rainwater management is inadequate, it is difficult to estimate water loss to the earth, water recharge, and how much is still accessible, among other things. As a result, rainfall measurement and monitoring are critical. The rainfall monitoring system will continuously measure rainfall and communicate the data to a platform where emergency services or weather specialists will be able to inform local citizens who are at risk of a rain-related disaster/tragedy.

However, due to the relatively low quality of wireless signal transmission, this information is often limited and only offers aggregate data in the form of rainfall per day/month/year without essential and comprehensive rainfall intensity information. As a result, an IoT-based rain gauge (Internet of Things) is required in which precise data may be disseminated using an internet network or the like, allowing the data to be watched remotely and connected with other objectives.

Based on this, IoT-based rainfall monitoring can be done using several equipment including the AT-Mega128 AVR Microcontroller, SDCOM-5 Module, Wavecom GSM Modem, and Telemetry with SMS system. The first part of the rainfall telemetry system is (agent) which has the function of channeling measurement data for a certain time to the data receiver (server) via the SMS system. Tipping Bucket that has been filled with water sways with the provision that everyone shake from the seesaw or commonly called a tick represents 0.5 mm, so that it can be detected by an interrupt on the microcontroller, then displayed to devices, such as smart phones, PCs, notebooks and various other gadgets.

Knowing these conditions, in this study, a prototype of a variable rainfall measuring device was made in order to make good use of the data obtained by the tool. Annual rainfall data is needed to obtain precise and useful calculations, the more data obtained the more accurate the calculations. Prediction of rainfall variations provides useful information in planning community activities when the rainy season occurs.

LITERATURE REVIEW

Internet of Things (IoT) Automation System

Since the term "Internet of Things" was coined in 1999, there has been a lot of discussion about its implications for our personal and professional life. All of this is due to development drivers such as the advent of ubiquitous computing, widespread usage of the internet protocol (IP), continuing advancement of data analytics, and more. Gartner anticipated that by 2020, around 20.4 billion devices would be connected to the Internet of Things. Despite the benefits, it remains a somewhat unclear idea to some extent (Gilis, 2021).

As mentioned by Shafiq et. al (2022), The Internet of Things (IoT) refers to devices equipped with sensors, computing power, software, and other technologies that communicate and share data with other devices and systems via the Internet or other communication networks. Hendricks (2015) also argue that Electronics, communication, and computer science engineering are all part of the Internet of Things. The term "Internet of Things" has been deemed misleading because devices do not need to be connected to the public internet; instead, they must be connected to a network and be individually addressable (Nilanjan, 2018).

The Internet of Things infrastructure can be used to monitor any events or changes in structural conditions that could jeopardize safety or raise danger. The Internet of Things can help the construction sector save money, save time, have a better workday, eliminate paper, and boost production. In Real-Time Data Analytics, it can aid in making faster decisions and saving money. It can also be utilized to efficiently schedule repair and maintenance activities by coordinating duties between various service providers and users of these facilities (Gubbi et.al., 2013). The use of IoT devices for infrastructure monitoring and operation is likely to improve incident management and emergency response coordination, as well as quality of service, uptime, and operational expenses in all infrastructure-related domains.Even garbage management can benefit from the automation and optimization that the IoT can bring (Chui et. al., 2015).

In one case of using IoT, Li et al (2011) expressed the opinion that Environmental monitoring IoT applications often use sensors to aid in environmental protection by monitoring air or water quality, atmospheric or soil conditions, and can even encompass topics such as tracking wildlife movements and habitats. Hart and Martinez (2015) also argue that the development of resource-constrained devices linked to the Internet also means that additional applications, such as earthquake or tsunami early-warning systems, can be employed by emergency services to provide more effective assistance. In this application, IoT devices often cover a vast geographic region and can even be mobile. It has been proposed that the uniformity brought about by IoT will transform wireless sensing (Ersue et. al., 2015).

Overview Rain Gauge

Rain gauge or udometer is a tool used to measure rainfall in a location for a certain period of time. This data can be utilized for a variety of purposes, including hydroelectric power generation, residential infrastructure development, and outdoor event planning, and it also influences the quality of wireless signal transmission, including free space optics (FSO), GSM, satellite, and outdoor WiFi. However, this data is often limited and only gives aggregate data in the form of rainfall per day/month/year without rainfall intensity information, which is critical for wireless signal transmission quality.

As a result, an IoT (Internet of Things)-based rain gauge is required, with detailed data distributed over an internet network or the like, so that data can be watched remotely and combined with other objectives. In this study, an IoT (Internet of Things) rain gauge system was built. The system consists of a tip-bucket rain gauge, a microcontroller in the Arduino Uno module as a processor through which data is sent to an IoT server via the ESP8266 module, and a WiFi network as a gateway through which data can be accessed by electronic devices such as smartphones and computers connected to the Internet Network. Because it is fitted with a solar panel, the system may run independently outside. The system designed can detect not only rainfall with a detection limit of up to 0.788 mm, but also rainfall intensity with a detection limit of up to 0.788 mm/min and a maximum limit of up to 50 mm/min. The Rain-Gauge system consists of four major components, they are:

1. Hardware; we are currently developing rain detectors (RD-XX Rain Radar) and rainfall gauges (ARG-XX Rain Radar)
2. Software, in the form of: web applications (WebApp), mobile applications (MobileApp) and web-services or APIs (Application Programming Interface)
3. Big Data in the form of information or data that has been collected
4. Brainware, that is: providers and users of the three things above

Rain-Gauge Prototype Design

Radar Rain prototype which will be discussed in this proposal is Rain Radar tool that uses the method of measuring the volume of rainfall by weighing the rainwater that is accommodated and entering the temperature components as well as the air pressure around the rainwater storage container into the calculation of certain formula. The basic formula for calculating the volume of rainwater that is accommodated refers to the following formula of the density of water:

$$\rho = \frac{m}{V}$$

Where, ρ (rho) is the density (gram per mililiter), "m" is the mass (gram) and "V" is the volume (mililiter). The density of water is obtained based on the calculation of temperature and air pressure.

Radar Rain is a system that carries Open Source Software (OSS), Open Source Hardware (OSHW) and Open Data so that it can be ascertained open-source, open-hardware and open-data. The data that has been processed and residing on servers can be accessed by the community.

In the ARG Rain Radar tool, calculations are made referring to the following formula:

1. If the temperature is in the range of 0 °C to 40 °C, assuming the air pressure is close to or equal to 101 325 Pa; then the formula of the *Comité International des Poids et Mesures* (CIPM) or International Committee of Weights and Measures will be used
2. 2. For outside temperature conditions in the range of 0 °C to 40 °C, the revised IAPWS formula will be used in 2016 or also known as ***IAPWS95-2016***

Observing the air temperature (and rainwater) generally in Indonesia—except in certain areas that have extreme temperatures in Indonesia, such as in Puncak Jayawijaya - Papua, Indonesia (can be below 0 ° C) and the highest air temperature ever recorded occurred in Kupang, East Nusa Tenggara, Indonesia (average reaches 39 ° C)—then the formula used is the formula from CIPM.

It can be seen that the range of use of the IAPWS95 formula (Daucik & Dooley, 2011) is much wider, but it has been widely agreed in the international world to use the CIPM formula when the conditions or limitations that have been described are met:

Figure 1. Range of validity of the use of the IAPWS95 formula
Source: Daucik and Dooley (2011)

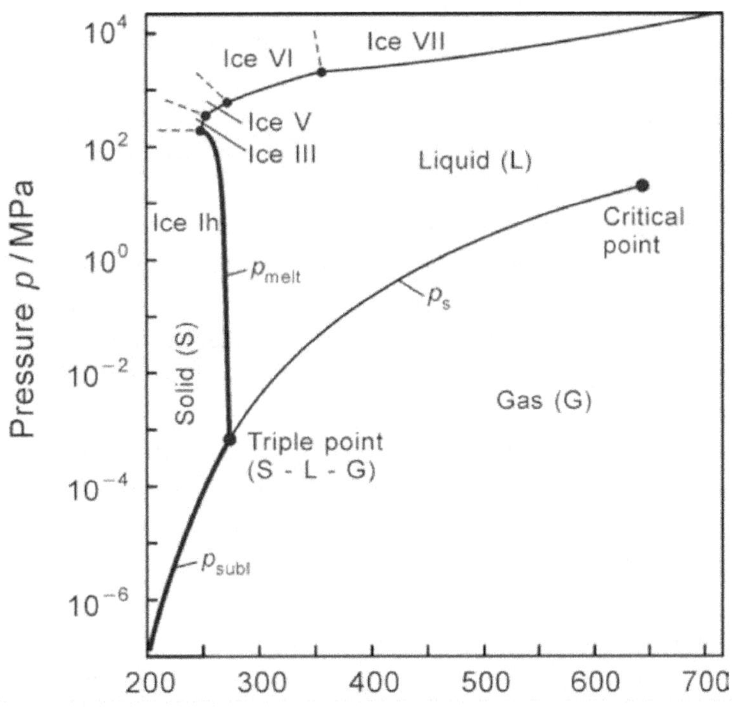

Figure 2. Load-cell and HX711 load-cell driver

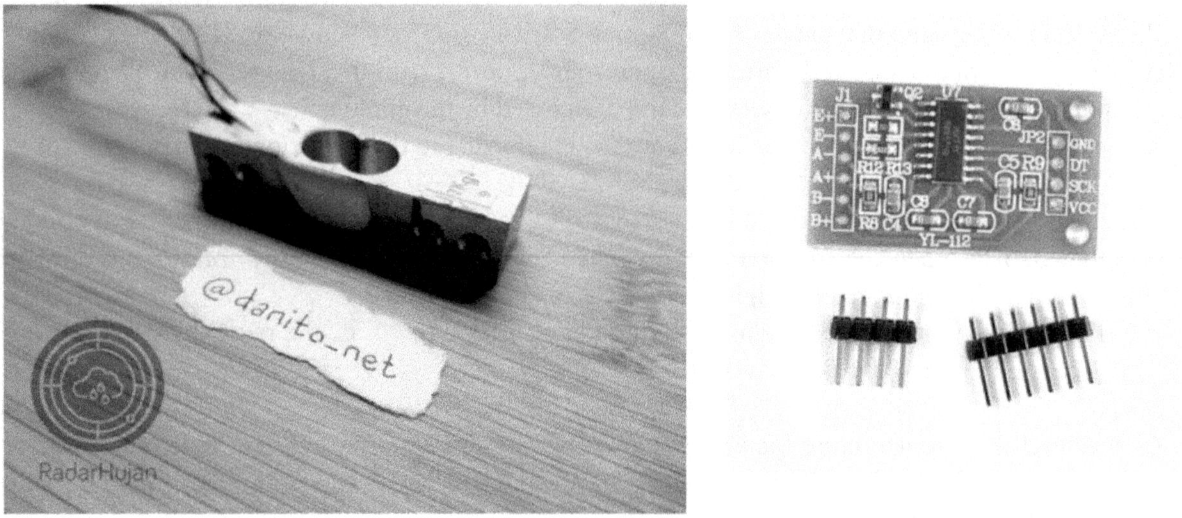

In general among Indonesians, the term "mass" (units: Kg, Kilogram) is often associated with "weight" (units: N, Newton). To facilitate the discussion, it was mutually agreed that the term "weight" in this study is actually "mass" whose unit is Kilogram (Kg) or Gram (g). Measuring the weight of rainwater collected using a weight sensor also known as a "load-cell" which is widely used in digital scales that are widely sold in the community. In order for the sensor reading value to be 'read', the HX711 load-cell driver module must be used (this load-cell cannot function independently). The following is the appearance of the Load-cell and HX711 load-cell driver.

The Load-cell and HX711 load-cell driver used according to needs, which is in accordance with the weight of rainwater collected by rainwater storage containers. In this ARG Rain Radar tool uses a load-cell with a maximum capacity of 7 Kg (or at least a maximum of 5Kg). In addition, a buffer holder for rainwater storage containers is also needed, for example those made of acrylic plastic that is laser-cutting such as the following example (Figure 3).

The ambient air temperature and pressure (around the rainwater storage container) are read using the BMP280 sensor module and if from the test later the volume calculation results are less precise due to the difference in air temperature with the temperature of the rainwater accommodated, it will also be read the temperature of the water accommodated using the waterproof version of the DS18B20 sensor. The following is a look at the BMP280 sensor module and the waterproof version of the DS18B20 sensor (Figure 4).

The following is a design drawing of the prototype of the ARG Rain Radar tool (Figure 5).

The technique used to dispose of or empty rainwater that has been collected is by pumping out rainwater in rainwater storage containers using a peristaltic motor pump. Peristaltic motor pumps have a different way of flowing fluid or water, namely by "sucking" the fluid peristaltis on a special rubber hose in the peristaltic motor pump. Figure 6 is an illustration.

This peristaltic motor pump uses a wireless power source (wireless) or wireless power supply to avoid interference with cables that cross into rainwater storage containers. Wireless power receiving modules are installed on the left, right and rear sides of the container; The front is used to open/install rainwater

Figure 3. The use of Load-cell and HX711 load-cell driver with a buffer holder for rainwater storage containers

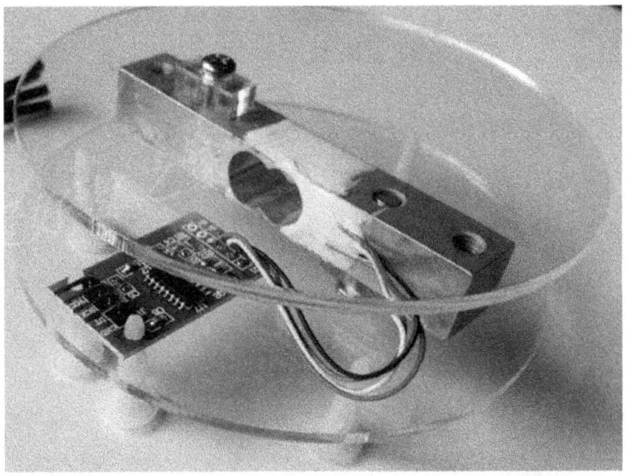

Figure 4. BMP280 sensor module and waterproof version of the DS18B20 sensor

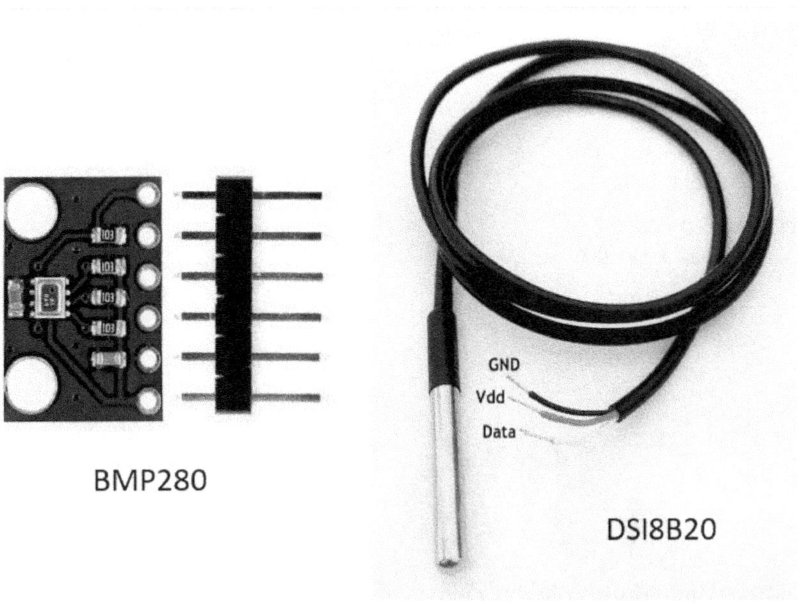

storage containers. While the wireless electricity transmitter or sender module is installed opposite directly in front of each wireless electricity receiver module (with a distance of approximately 5 mm). The following is a photo of a series of wireless electricity transmitters and receivers, where; The photo on the left is the transmitter / transmitter / sender and the photo on the right is the receiver / receiver (Figure 7).

Currently on the market there are many wireless charger devices sold that are used to charge mobile phone batteries that have a wireless-charging feature; Or it could also be for ordinary mobile phones that are installed additional wireless charger receivers. The price of a pair of these devices (wireless charger)

Figure 5. Sketch drawing of the prototype of automatic rain-gauge

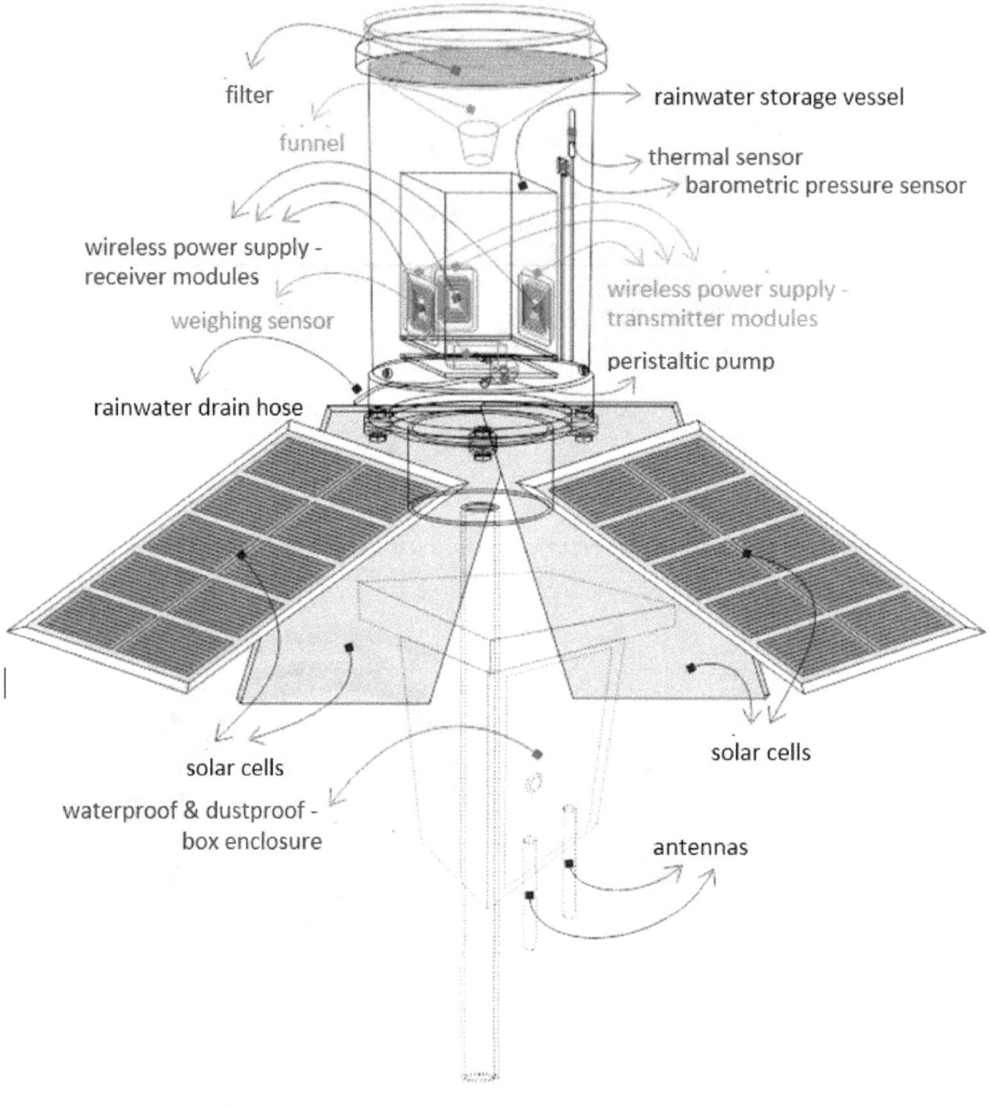

is cheaper than making the circuit yourself (photo above), so we decided to take the circuit inside this wireless charger tool to use.

The following is a detailed design of the Automatic Rain-Gauge tool, to clarify the position of each component (especially sensors and wireless electrical modules):

Features and Specifications of Automatic Rain Gauge

In the implementation of the Automatic Rain-Gauge prototype, the code "ARG" is used on the Rain Radar tool which stands for Automatic Rain Gauge which is a quantitative type of Rain Radar tool that calculates the amount of rainfall; while to simply detect rain (qualitative) used code "RD" or Rain Detector. Automatic Rain Gauge measures rainfall volume by weighing the weight of rainwater collected

Figure 6. Peristaltic motor pump
Source: Zehetbauer et al. (2021)

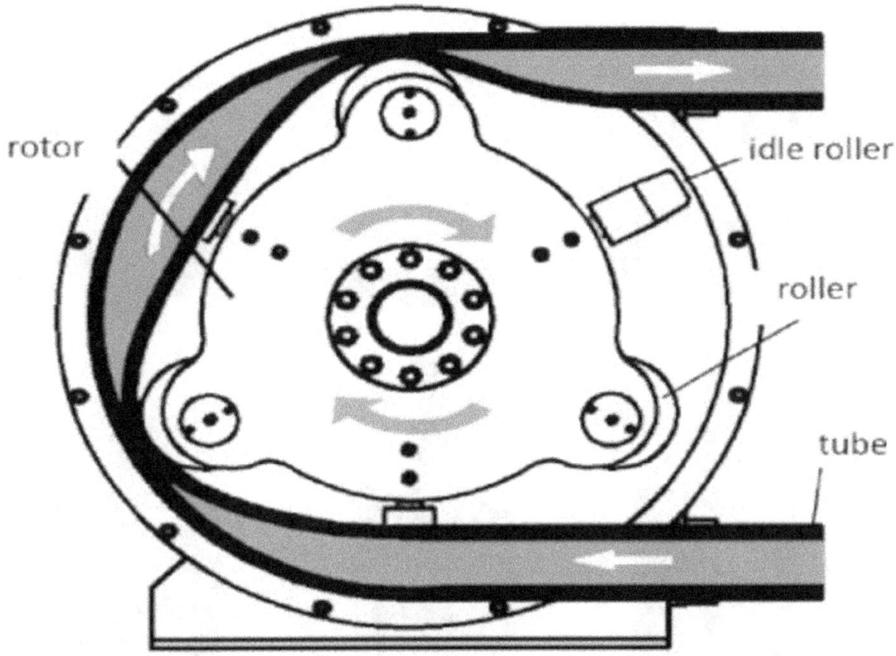

Figure 7. Wireless electricity transmitter

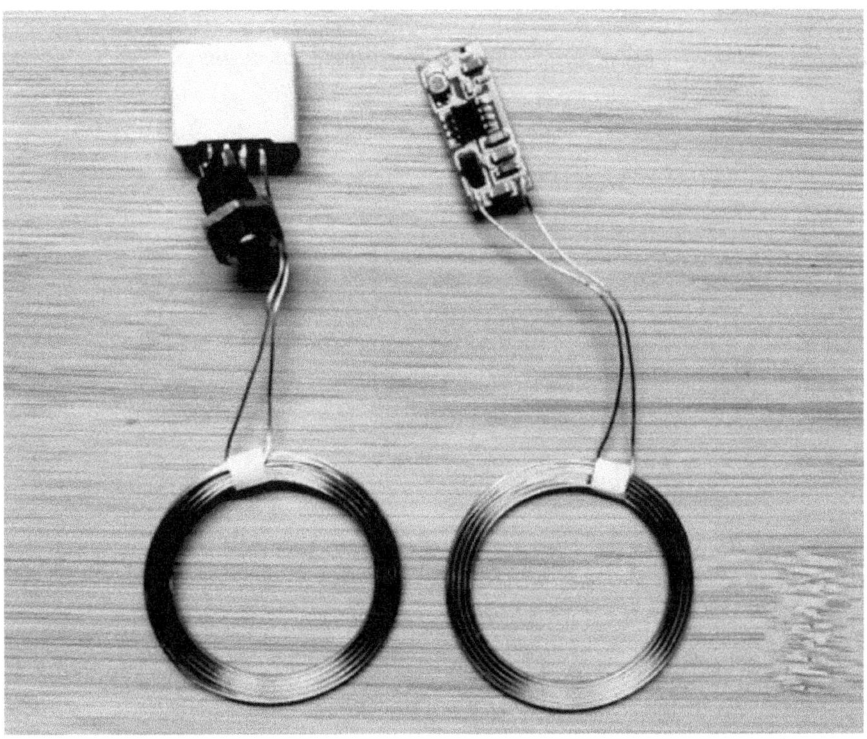

Figure 8. Wireless charger with transmitter and receiver
above: transmitter/sender; below: receiver

Figure 9. Detailed sketch drawings of automatic rain-gauge prototypes

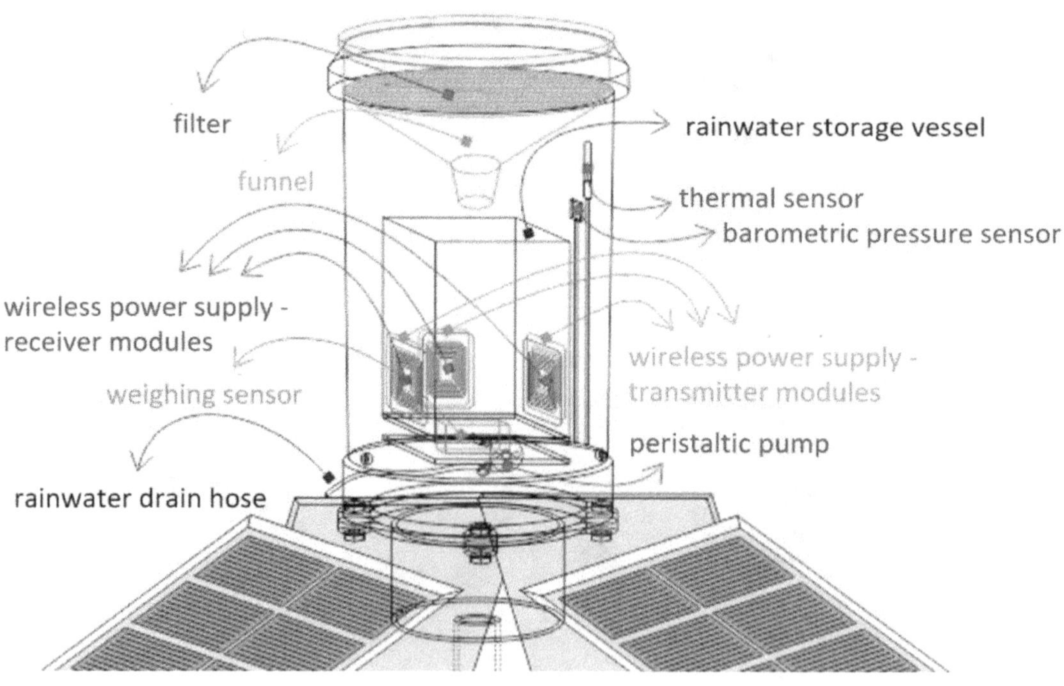

and incorporating temperature components and air pressure around rainwater storage containers into the calculation.

The Automatic Rain Gauge tool has a Control and Computing Unit that uses an Arduino Mega 2560 R3 microcontroller integrated into the single board computer Raspberry Pi 3 Model B. With these technical specifications, it will also have an impact on the amount of electrical energy used, the type and amount of battery power used and the amount of energy that must be obtained to recharge it. The control and computing unit on the Automatic Rain-Gauge based on the Arduino Mega 2560 R3 microcontroller has the following specifications:

1. Two units of 3.6 volt 19,000 mAH Lithium Battery
2. Solar Panel 6 volt 3 watt as many as 2 units
3. DC-to-DC converter modules, solar panel connectors, solar panel cables and other complements are the number is customized for two solar panel units

Meanwhile, the single board computer Raspberry Pi 3 Model B uses the following qualifications:

1. Two units of 5 volt 20,000 mAH Lithium Gel Battery
2. Solar Panel 6 volt 3 watt as many as 4 (four) units
3. DC-to-DC converter modules, solar panel connectors, solar panel cables and other complements are the number is adjusted for 4 (four) solar panel units
4. Mobile Power for Raspberry Pi (MoPi) module as much as 1 unit. This module is used to manage and manage the power source between electricity from solar panels and electricity that has been collected in the battery

As a single board computer used in Automatic Rain-Gauge, of course the Raspberry Pi 3 Model B has an operating system to be used. For the Automatic Rain-Gauge tool, the Raspbian PIXEL Linux operating system is used which has been customized in such a way as to run optimally and more securely (security hardened). The following is the appearance of the Raspberry Pi 3 Model B

Figure 10. Raspberry Pi 3 Model B

Here are the technical specifications of the Raspberry Pi 3 Model B:

- Processor: Quad Core 1.2GHz Broadcom BCM2837 64bit CPU
- Memory: 1GB RAM
- Wireless Modul: BCM43438 *Wireless* LAN and *Bluetooth Low Energy* (BLE)
- Pin Digital I/O: 40-pin GPIO
- Output Video: Full size HDMI
- Port USB: 4 x USB2
- Port Audio/Video: 4 pole stereo output & composite video port
- Other Port: CSI / camera port, DSI / display port
- Data Storage (Storage Device): Slot MicroSD
- Electricity: MicroUSB *power source* 5V 2.5A

Automatic Rain-Gauge with the project name Rain.ID is a community-based project that carries Open Source Software (OSS), Open Source Hardware (OSHW) and Open Data so that this tool can be ascertained as open-source, open-hardware and open-data (data which have been processed and are on the server accessible to the public). The languages used in programming this tool are C/C++, Bash, Python and PHP (do not use low-level programming languages such as assembly). The overall design of the tool is made as simple/simple as possible, using materials (components, sensors, modules, cases/enclosures, storage containers, supports and others) on the Indonesian market; so it is easy to be made by individuals or mass produced.

Data communication on the RadarHujan system (including logs and data recorded / recorded) can be done via (all of them can work if the location allows, some can also be activated or enabled):

a. Data communication via the internet using a modem (using internet quota or credit). The GSM data communication technology used will be taken from the best available in the location starting from: 4G, HUSPA / HSDPA / 3.5G, 3G, EDGE / 2.5G, 2G or GPRS
b. Limited data communication via SMS using a modem (using credit)
c. Data communication via radio waves using a LoRa gateway (does not use internet quota or pulses)
d. Data communication via WiFi on site; so there is no need to disassemble the tool (can wirelessly from a distance some distance from the tool)
e. Take a physical flashdisk on the device during maintenance

The following illustrates how the Automatic Rain-Gauge works:

The process of sending commands and changing configurations via SMS with special protocols and one-time access codes. Automatic Rain-Gauge can send notifications or special status (such as low battery / damaged, peristaltic motor pump malfunctioning, overload or system damage, etc.) via the internet and via radio waves (to the server) to then be forwarded or picked up by the web application (browser) or mobile application (on smartphones), can also directly send SMS to certain numbers that have been recorded in the configuration (can be activated or disabled; list of numbers can be supplemented and subtracted from the app or via SMS as well, with a security code).

In terms of system development, Automatic Rain-Gauge provides API (Application Programming Interface) through https://api.hujan.id/ to facilitate application development on other platforms by any party (for example: web-based applications, Android mobile applications, iOS mobile applications,

Figure 11. Diagram of how automatic rain-gauge works

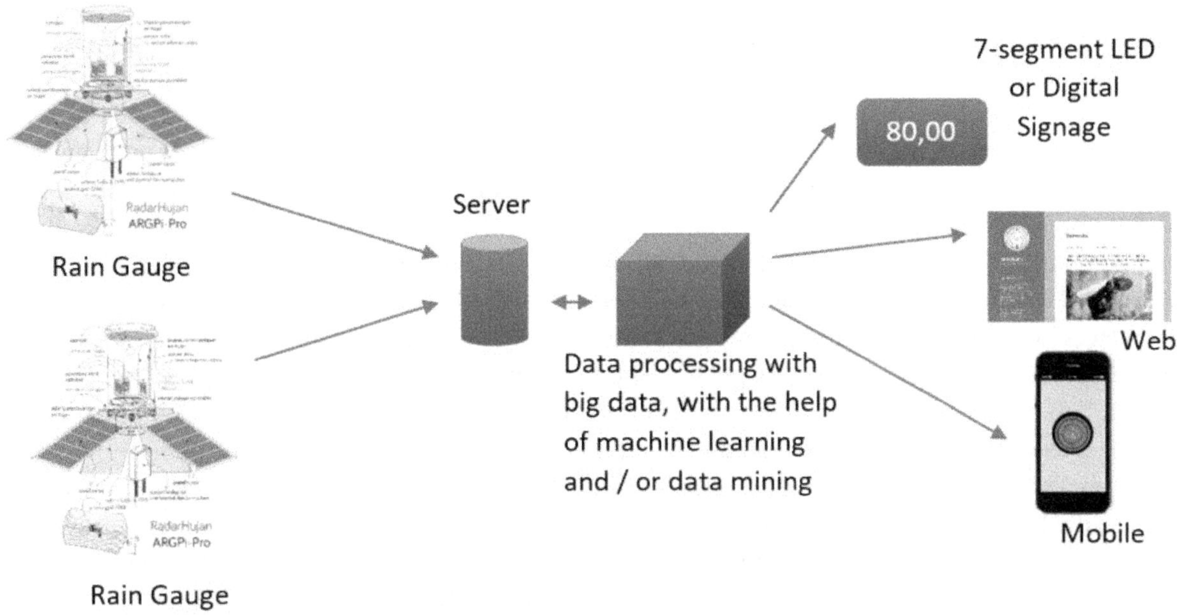

WindowsPhone mobile applications and so on). Meanwhile, this tool does not have mechanical parts that routinely move like in rain gauge tipping bucket type (the seesaw mechanical part is prone to jamming) so maintenance is less frequent. Although this tool has moving parts on the peristaltic motor pump, maintenance does not have to be done often. Firmware, operating system, scripts, applications can be updated over-the-air (via the internet, via radio waves and over wi-fi). It can also be manually during maintenance (by physically replacing MicroSD and flash drives).

Automatic Rain-Gauge has an autonomous power supply in the form of four units of 6V 3 watt solar panels and lithium gel batteries (the latest and best technology lithium batteries) with a total capacity of 40,000 mAh; So that this tool is able to turn on for at least 3 days without new energy intake. Solar panels can be replaced or combined with other natural power plants --- adapted to natural resources in --- locations such as: micro-hydro generators, wind-belt micro generators, thermo-electric and others.

The design of the tools and components used is selected that is weatherproof, as dustproof and waterproof as possible; So that overall, the tool will be safe if placed in a location with high humidity. The device input (analog / digital) supports the use of channels for rain breeding sensors. Automatic Rain-Gauge also has a storage medium to record / record data separate from the operating system used, which is in the form of a flash drive with a size of 15 GB (the size of the flash can be adjusted).

Measurements can be set within a certain time range, as early as once every 5 minutes for 24 hours. Meanwhile, for data delivery, it can be set within a certain time range. The fastest data transmission is once every 10 minutes (for 24 hours). Data transmission orders can be performed at any time. The last data sent is data within the last 10 minutes. In the case of data reading, the units of reading results are in "millimeters" or "mm" (can also be converted or adjusted); Also included are other data needed, for example: remaining voltage and battery power data, sunlight conditions (charging process), temperature inside the device, remaining credit or remaining internet quota.

CONCLUSION

IoT-based rain gauge systems that are planned and built effectively can detect state changes in tipping buckets that indicate a rise in rainfall and communicate data to cloud servers. With the system's ability to access the internet and send data as an IoT application, rainfall data and intensity can be collected remotely without requiring human labor, allowing the system to be distributed at test points and the data to be easily accessible to users in various locations for further processing because the data can be spread through the internet network.

Based on the results of the tests conducted, the accuracy of the tool is quite high, which is ± 2% (two percent) and will be further developed after being tested in the lab and in the field. Units of reading in "millimeters" or "mm" (can also be converted or adjusted); Also included are other data needed, for example: remaining voltage and battery power data, sunlight conditions (charging process), temperature inside the device, remaining credit or remaining internet quota. Measurements can be set within a certain time range, as early as once every 5 minutes (for 24 hours). The data transmission can be set within a certain time range, at the earliest for 10 minutes (for 24 hours). Data send orders can be performed at any time (the last data sent is data within the last 10 minutes). Meanwhile, firmware, operating system, scripts, applications can be updated over-the-air (via the internet, via radio waves and via wi-fi). It can also be manually during maintenance (by physically replacing MicroSD and flash drives).

IoT-based rain gauge is designed to be able to send notifications or special status (such as battery running low/damaged, peristaltic motor pump not working, system overload or damage, etc.) via the internet and via radio waves (to the server) to then be forwarded or taken by a web application (browser) or mobile application (on a smartphone), you can also directly send SMS to certain numbers that have been recorded in the configuration (can be activated or deactivated; the number list can be added and subtracted from the application or via SMS too, with security code).

REFERENCES

Aryastana, P., & Tasuku, T. (2010). Characteristic of Rainfall Pattern Before Flood Occur in Indonesia Based on Rainfall Data. *GSMaP, 7*, 1–4.

Chui, M., Löffler, M., & Roberts, R. (2015). The Internet of Things. *McKinsey Quarterly*.

Das, A., Sarma, M., Sarma, K., & Mastorakis, N. (2018). Design of an IoT based real time environment monitoring system using legacy sensors. *MATEC Web of Conference*, *210*, 1–4. doi:10.1051/matec-conf/201823700001

Daucik, K. (2011). Revised Release on the Pressure along the Melting and Sublimation Curves of Ordinary Water Substance. The International Association for the Properties of Water and Steam. IAPWS R14-08.

Emam, A. R., Mishra, B. K., Kumar, P., Masago, Y., & Fukushi, K. (2016). *Impact assessment of climate and land-use changes on flooding behavior in the upper ciliwung river*. Indonesia Water.

Ersue, M., Romascanu, D., Schoenwaelder, J., & Sehgal, A. (2015). *Management of Networks with Constrained Devices: Use Cases*. IETF Internet Draft. doi:10.17487/RFC7548

Frick, J., & Hegg, C. (2011). Can end-users' flood management decision making beimproved by information about forecast uncertainty? *Atmospheric Research*, *100*(2-3), 296–303. doi:10.1016/j.atmosres.2010.12.006

Gaitan, S., Calderoni, L., Palmieri, P., Veldhuis, M., Maio, D., & van Riemsdijk, M. B. (2014). From Sensing to Action: Quick and Reliable Access to Information in Cities Vulnerable to Heavy Rain. *IEEE Sensors Journal*, *14*(12), 4175–4184. doi:10.1109/JSEN.2014.2354980

Gillis, A. (2021). *What is internet of things (IoT)?* IOT Agenda.

Gubbi, J., Buyya, R., Marusic, S., & Palaniswami, M. (2013). Internet of Things (IoT): A vision, architectural elements, and future directions. *Future Generation Computer Systems*, *29*(7), 1645–1660. doi:10.1016/j.future.2013.01.010

Hart, J. K., & Martinez, K. (2015). Toward an environmental Internet of Things. *Earth and Space Science (Hoboken, N.J.)*, *2*(5), 194–200. doi:10.1002/2014EA000044

Hendricks, D. (2015). *The Trouble with the Internet of Things*. London Datastore. Greater London Authority.

Joseph, F. (2019). IoT Based Weather Monitoring System for Effective Analytics. *International Journal of Engineering and Advanced Technology*, *8*(4), 311–315.

Kusumastuti, D. I., Jokowinarno, D., Khotimah, S. N., & Dewi, C. (2017). The Use of Infiltration Wells to Reduce the Impacts of Land Use Changes on Flood Peaks. *An Indonesian Catchment Case Study*, *25*, 407–424.

Li, S., Wang, H., Xu, T., & Zhou, G. (2011). Application Study on Internet of Things in Environment Protection Field. *Lecture Notes in Electrical Engineering*, *133*, 99–106. doi:10.1007/978-3-642-25992-0_13

Maspo, N., Harun, A., Goto, M., Nawi, M., & Haron, N. (2019). Development of Internet of Thing (IoT) Technology for Flood Prediction and Early Warning System (EWS). *International Journal of Innovative Technology and Exploring Engineering*, *8*(4), 2019–2228.

Matese, A., Gennaro, S., & Zaldei, A. (2015). Agrometeorological monitoring: Low-cost and open-source – Is it possible? *Italian Journal of Agrometeorology*, (3), 81–88.

Narulita, I., & Ningrum, W. (2017). Extreme flood event analysis in Indonesia based on rainfall intensity and recharge capacity. *IOP Conf. Series: Earth and Environmental Science, 118*(2018), 012045. 10.1088/1755-1315/118/1/012045

Nilanjan, Hassanien, Bhatt, Ashour, & Satapathy. (2018). Internet of things and big data analytics toward next-generation intelligence. Academic Press.

Pathania, A., Kumar, P., Kesri, J., Priyanka, A. S., Mali, N., Singh, R., Chaturvedi, P., Uday, K.V., & Dutt, V. (2019). Reducing power consumption of weather stations for landslide monitoring. *Proceedings of 3rd International Conference on Information Technology in GeoEngineering 2019,* 1-16.

Sarkar, I., Pal, B., Datta, A., & Roy, S. (2018). WiFi based portable weather station for monitoring temperature, relative humidity, pressure, precipitation, wind speed and direction. *Proceedings of International Conference on ICT for Sustainable Development (ICT4SD) 2018.*

Shafiq, M., Gu, Z., Cheikhrouhou, O., Alhakami, W., & Hamam, H. (2022). The Rise of "Internet of Things": Review and Open Research Issues Related to Detection and Prevention of IoT-Based Security Attacks. *Wireless Communications and Mobile Computing*. doi:10.1155/2022/8669348

Sharma, D., Shukla, A., Bhondekar, A., Ghanshyam, C., & Ohja, A. (2016). A technical assessment of IOT for Indian agriculture sector. *IJCA Proceedings on National Symposium on Modern Information & Communication Technologies for Digital India 2016.*

Zehetbauer, T., Plöckinger, A., Emminger, C., & Çakmak, U. D. (2021). Mechanical Design and Performance Analyses of a Rubber-Based Peristaltic Micro-Dosing Pump. *Actuators*, *10*(8), 198. doi:10.3390/act10080198

Chapter 12
Empowering Safety by Embracing IoT for Leak Detection Excellence

Neha Bhati
https://orcid.org/0009-0008-0171-2786
AVN Innovations Pvt. Ltd., India

Ronak Duggar
AVN Innovations Pvt. Ltd., India

Abeer Saber
https://orcid.org/0000-0002-9261-0927
Damietta University, Egypt

ABSTRACT

Improvements in connectivity and data analysis enabled by the internet of things (IoT) are set to revolutionize various sectors, with a particular emphasis on making workplaces safer. Manual leak inspections, which can be both time-consuming and dangerous, are quickly being replaced by IoT-driven devices. These systems are more than just an improvement in technology; they usher in a new paradigm with their ability to monitor in real time, issue immediate alerts, and locate leaks with pinpoint accuracy. Because of the benefits that IoT provides, several sectors are making the switch from more traditional practices. Leak detection enabled by the internet of things represents a step toward safer, greener production. The promise of improved worker safety and environmental sustainability lies at the heart of the internet of things, which should be rapidly adopted by businesses.

INTRODUCTION

The fast expansion of IoT has revolutionized leak detection, delivering effective preventative measures for safer homes and workplaces. Networks of intelligent sensors strategically installed in pipelines,

DOI: 10.4018/979-8-3693-1194-3.ch012

storage tanks, and other vital places provide real-time monitoring and data recording for IoT-based leak detection systems. This information is sent to central computers, which are processed and analyzed quickly by advanced analytics and Artificial Intelligence (AI) algorithms. The system can detect leaks and irregularities using AI-driven analytics, allowing rapid steps to reduce dangers. Water waste, environmental harm, and infrastructure disruption may all be avoided with the help of IoT leak detection thanks to its real-time monitoring and early warning capabilities. As a result of these developments, IoT leak detection systems are widely used across many sectors, from improving water management in intelligent cities to strengthening safety protocols in manufacturing plants.

Leak detection in the IoT shows potential for future development and widespread use, which bodes well for all buildings' safety and sustainability (Banerjee & Banerjee, 2023).

Risk management has become increasingly important in today's fast-paced, linked society. The IoT has ushered in a new era of technology, revolutionizing many facets of modern life, including leak detection. The IoT has enabled digital and physical systems to merge through the years by connecting them without any hitches. Thanks to this networked infrastructure, we can now collect, analyze, and respond to data in real time, making the world more connected and innovative (Bedi et al., 2018).

Historically, Leak detection has been performed reactively and manually, resulting in delays and sometimes catastrophic consequences. However, things have changed drastically since the introduction of IoT, which allows for the 24*7 monitoring and proactive leak detection of vital infrastructure. Leak detection systems based on the IoT can track and report important metrics like temperature, pressure, and flow rate thanks to intelligent sensors and constant data transfer. This data is analyzed by centralized servers using advanced analytics and machine learning technologies, allowing for early detection and remediation of leaks.

Figure 1 illustrates how IoT, AI, and data analytics enhance leak detection and safety.

Figure 1. IoT system architecture for enhanced leak detection

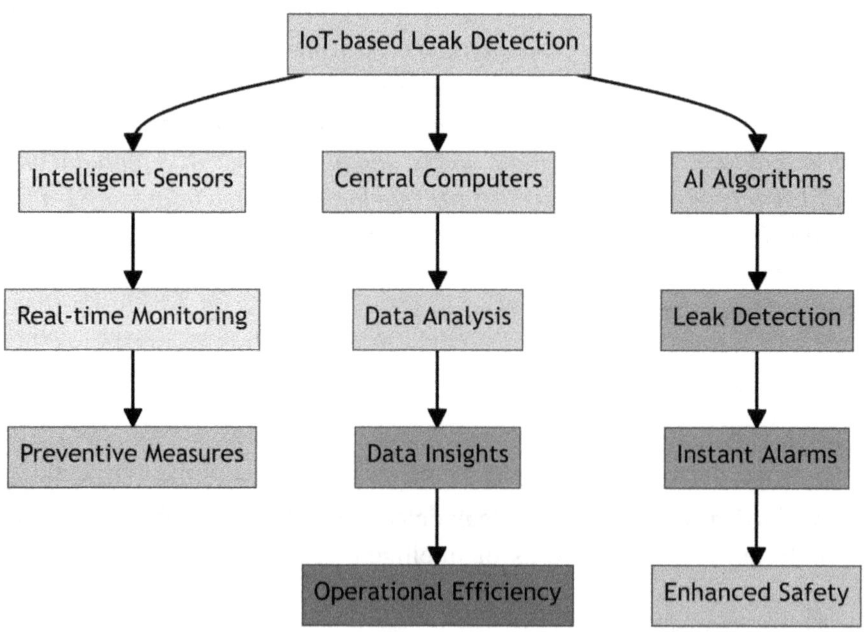

IoT leak detection using a mix of Data Analytics and AI has also revealed hitherto unexplored insights. We can spot patterns, trends, and outliers using IoT devices' massive amounts of data. Algorithms taught by machine learning may pick up on subtle behavioural changes that might otherwise go undetected, providing early warning of leaks. Leak detection is simplified because these AI-powered systems constantly learn from previous data, increasing their precision and prediction skills.

SYSTEM COMPONENTS AND ARCHITECTURE

The IoT-based leak detection system is a game-changing innovation that will affect several sectors' approaches to safety management. Each part of this complex system is essential to its overall function, and together, they form a sophisticated architecture. The sensors are the system's nerve centre and the first line of defense. These high-tech sensors are strategically placed throughout the infrastructure, from pipes to storage tanks, to keep tabs on real-time temperature, pressure, and flow rate.

The Importance of Sensors in Leak Detection

Any IoT-based leak detection system must begin with sensors. Their job is to record information about the surrounding environment so the system can look for abnormalities or leaks. In leak detection applications, sensors can monitor pressure, temperature, humidity, and fluid levels. They are built to perform accurately and dependably under various situations.

Some sensors can even do self-checks to guarantee they are continuously operating as intended (Elleuchi et al., 2019).

Data Collection Units for Digital Conversion

For more effective processing and transmission, the analogue data acquired by sensors must be transformed into digital data. The data-gathering units are crucial in this transition.

These devices are responsible for digitizing the analogue sensor readings, guaranteeing that the information is consistent and suitable for the IoT's communication protocols. Data conversion to digital format simplifies processing, storage, and transmission.

Communication Protocols Allowing for Convenient Data Exchange

Communication between sensors, data-collecting devices, and the central processing platform is crucial in an IoT-based leak detection system. Communication protocols allow this to happen. These protocols provide the norms and guidelines for exchanging data, allowing its transmission to proceed without a hitch. Message Queuing Telemetry Transport (MQTT), Constrained Application Protocol (CoAP), and LoRaWAN are typical communication protocols in IoT leak detection systems. Considerations like data volume, power consumption, and network coverage inform the decision of which communication protocol to choose.

Table 1 contrasts the features of popular communication protocols employed by IoT leak detection systems.

Table 1. Comparative analysis of IoT communication protocols

Criteria	MQTT	CoAP	LoRaWAN
Data Volume	High	Medium	Low
Power Consumption	Medium	Low	Very Low
Network Coverage	Wide	Medium	Very Wide

The System's Brain, the Central Data Processing Platform

The central data processing platform is the backbone of an IoT-based leak detection system. The digital data generated by the sensors is sent to the platform through the

data-collecting devices, where it is processed in real-time. It uses sophisticated analytics and AI algorithms to examine incoming data for trends, outliers, and security holes (Chhaya & Gupta, 1997). When the data from the sensors is compared against thresholds, the platform sends out instant notifications if anything out of the ordinary is found. Leaks can be found and fixed before they start since the platform can utilize previous data to train machine learning models, giving it predictive skills.

A few primary points are emphasized: Sensors, which gather information about the surrounding environment, and Data Collection Units, which are in charge of digitizing this information. Communication Protocols establish guidelines for sharing information and a centralized data processing platform, which does real-time analysis using sophisticated methods; these components work together to provide a complete picture.

Table 2 explains the primary parts of an IoT-based leak detection system, including their roles and examples of their use.

Table 2. IoT-based leak detection system components and architecture

Component	Functionality / Role	Examples / Details
Sensors	Detect environmental conditions and potential abnormalities.	Monitors pressure, temperature, humidity, and fluid levels—self-checking capability.
Units for Data Collection	Convert analogue data from sensors to digital format.	Responsible for digitization, ensuring data compatibility with IoT's communication protocols.
Communication Protocols	Facilitate the exchange of data between devices and platforms.	MQTT, CoAP, LoRaWAN. The choice is based on data volume, power consumption, and network coverage.
Central Data Processing	Processes and analyzes incoming sensor data. Predicts potential leaks.	It uses AI algorithms, real-time data analysis, and machine learning for predictions.
Platform User Interfaces and Alerts	Allows operators to visualize, understand, and act on the data provided by the system.	Web-based, mobile, or desktop dashboards. Customizable alert settings with notifications via email, text, or push.

User-Friendly Interfaces and Real-Time Alerts

The success of an IoT leak detection system depends on its operators having easy access to and understanding of the system's data. The dashboards themselves may be web-based, mobile, or desktop apps. The intuitive interface displays current sensor data, alarms, and historical patterns to give the user a complete picture of the system's health. The alert settings may be customized, and operators can choose to be notified through email, text message, or push notification. Because of this, they may take immediate action if a leak is suspected, reducing possible dangers and increasing overall security (Pointl & Fuchs-Hanusch, 2021).

For the reasons mentioned above, it is clear that the IoT-based leak detection system is a significant step forward in the field of safety management. Intelligent leak detection systems have several interconnected components, including sensors, data collectors, communication protocols, centralized data processing platforms, and user-friendly interfaces. The technology can detect possible breaches in real-time and allow rapid reactions by continually monitoring vital metrics and employing sophisticated analytics and artificial intelligence. Because of this, the IoT-based leak detection system is a potent instrument for protecting the environment, infrastructure, and human lives across a wide range of sectors and applications.

RECOGNIZING IOT-BASED LEAK DETECTION METHODS

This section delves into the systematic workings of such systems, explaining how anomaly detection algorithms are applied to the data acquired from intelligent sensors and how that information is then transformed into real-time alerts and messages. To ensure a thorough and effective leak detection strategy, we will also investigate the remote monitoring and management capabilities that enable operators to respond quickly to possible breaches (Ahmed et al., 2023).

Figure 2 represents the layered architecture of an IoT-based leak detection system, showing how data flows from sensors to remote monitoring.

Figure 2. Layered architecture of IoT-based leak detection

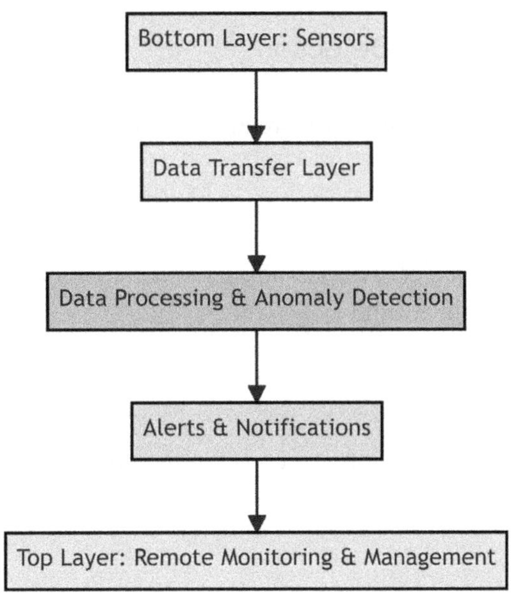

Smart Sensor Data Collection

Leak detection approaches that leverage the IoT rely on intelligent sensors installed at critical nodes. These sensors can detect various conditions, from temperature, pressure, humidity, and fluid levels. Sensors generate a continuous stream of data by gathering information from their respective settings. The leak detection system can monitor vital infrastructure in real-time with the help of intelligent sensors, giving an accurate picture of the system's state and quickly picking up on any irregularities (Kayastha et al., 2014).

Transfer of Information to the Primary Server

After intelligent sensors have gathered data, it must be sent to a centralized hub for further processing and analysis. Wireless communication technologies, including Wi-Fi, Bluetooth, Zigbee, and cellular networks, are commonly used to support this kind of data transfer. The distance between the sensors and the hub platform also plays a role in determining the optimal communication protocol (Kim et al., 2001).

Algorithms for Data Processing and Anomaly Detection

When information reaches the central hub, it is processed and analyzed thoroughly.

The large quantities of data are processed instantly by the central platform's artificial intelligence and complex algorithms. The system determines regular operating patterns by comparing them to past data and establishing thresholds. The anomaly detection algorithms will flag any behaviour that deviates from these norms as a possible security breach. Leaks may be detected with high sensitivity and accuracy thanks to these algorithms, which are taught to recognize even the smallest changes in sensor readings (Sunny et al., 2022).

Algorithm: Anomaly-Based Leak Detection

Objective: Detect anomalous patterns in sensor readings indicative of potential leaks using a moving average approach.

Inputs:

- – **S** = [s1, s2,.., sn]: A list of sensor readings over a given timeframe.
  - **s**$_i$: A sensor reading at time i.
  - **T**: Threshold value.
  - **w**: Window Size.

Output:

- – Alert signal if an anomaly score exceeds the threshold T.

Algorithm Steps:

```
Step 1: Data Preprocessing.
•         - Normalize the sensor readings to ensure all values fall within a
standard range. This facilitates consistent anomaly detection.
```

Step 2: Compute Moving Average.

- - For each sensor reading si (from i = w to n):
- *MAi = w1 Σ j = i − w + 1isj*
- **MA**$_i$: The moving average for the sensor reading at time the i.
- **w**: The window size.

Step 3: Calculate Anomaly Score.

- - For each sensor reading s_i:
- *ASi =| si − MAi |*

Step 4: Leak Detection.

- - For each sensor reading si:
- If *ASi > T*, an anomaly is detected.
- Trigger an alert indicating a potential leak at the corresponding time step.

Elements Used:

- - **S**: Represents the list of sensor readings.
- **s**$_i$: A specific sensor reading at time i.
- **T**: The threshold value used to determine if an anomaly score indicates a potential leak.
- **w**: The window size used to compute the moving average.
- **MA**$_i$: The moving average of the sensor readings for a specific time i.
- **AS**$_i$: The anomaly score for a specific sensor reading at time i. It represents the absolute difference between the sensor reading and its moving average.

Alerts and Notifications

The IoT-enabled leak detection system immediately sends alerts and notifications if the anomaly detection algorithms find a possible leak or strange behaviour. Email, text messages, and push notifications on mobile devices are some ways operators might get these alerts. The severity and location of a possible leak are included in the signals, along with other specifics regarding the observed anomaly. In addition, the system may activate local, auditory, or visible signs to notify onsite workers, who may then take appropriate action (Yadav et al., 2022).

Capabilities for Remote Monitoring and Management

The ability to remotely monitor and operate systems is a significant benefit of IoT-enabled leak detection solutions. Anywhere there is an internet connection, operators may log in to the central platform and see how things are doing with the system at any given moment. Operators may respond quickly to breaches without being physically there, thanks to remote access. It can access sensor data, analyze trends, and react swiftly to system warnings. In addition, operators may optimize efficiency and flexibility by configuring and personalizing the settings, such as alert levels and communication preferences, remotely (Mabrouki et al., 2021).

Figure 3. Comprehensive IoT-based leak detection process

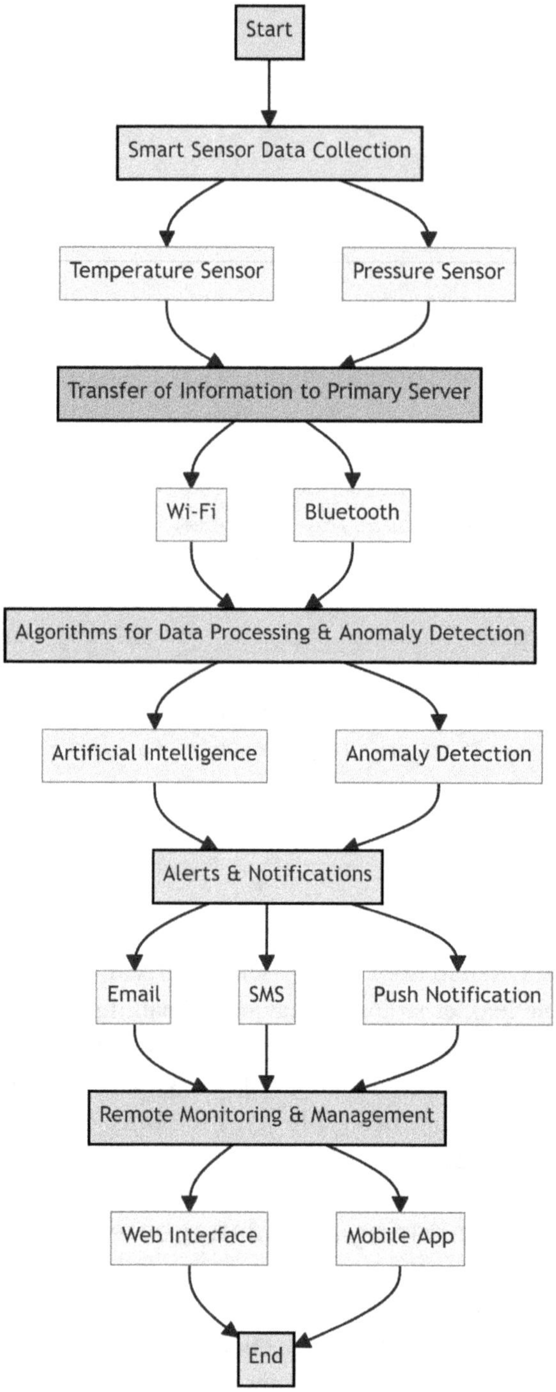

The flowchart in Figure 3 delves deeper into the IoT-based leak detection system, breaking down each component and sub-components for a more detailed understanding.

Figure 4. Timeline illustrating operator and system response actions after an alert is triggered

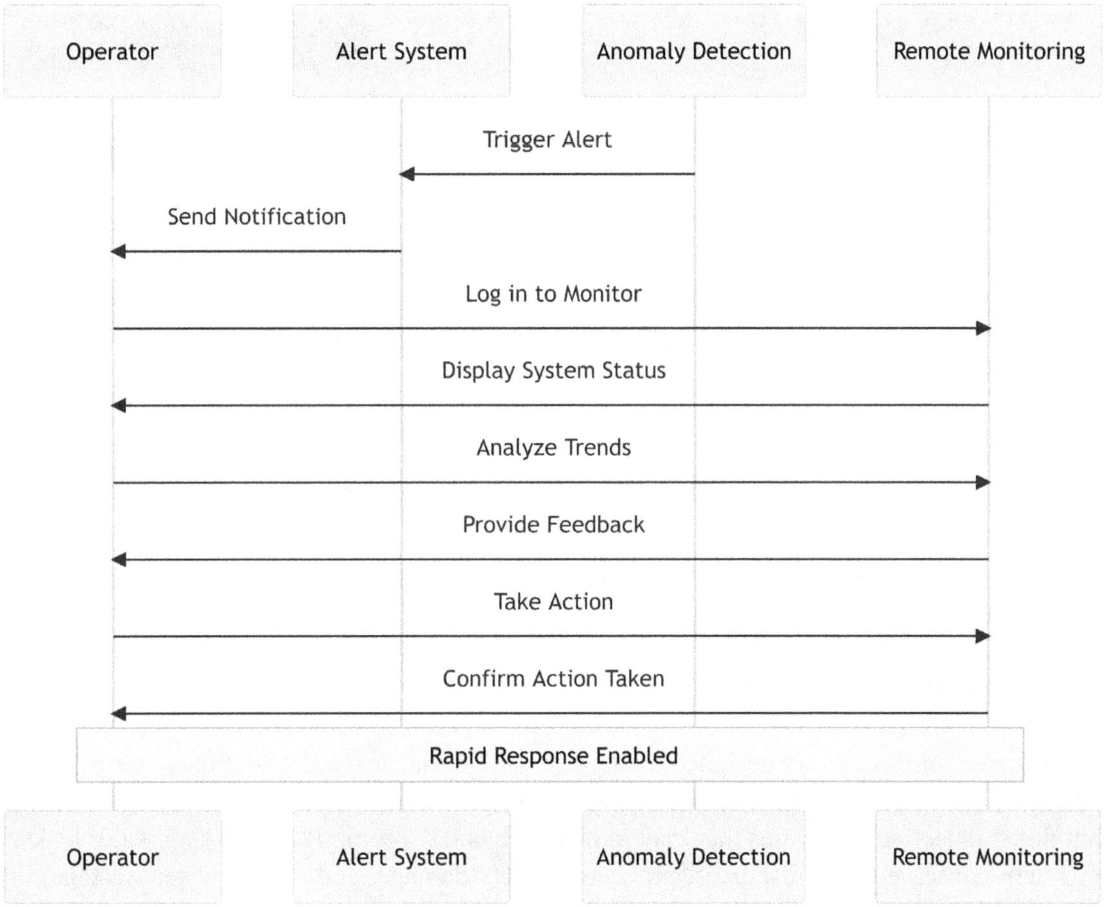

Using sophisticated sensors, sophisticated data processing, anomaly detection algorithms, real-time alerts, and remote monitoring capabilities, IoT-enabled leak detection solutions function as a single, unified process. This all-encompassing method enables businesses to see possible breaches early, allowing for prompt reactions and effective risk mitigation. These techniques, which use the IoT and leak detection, improve safety, lessen environmental impact, and maximize resource management across various settings and applications.

The timeline in Figure 5 illustrates the sequence of actions taken by operators and the system after an alert is triggered, emphasizing rapid and efficient response.

IOT-BASED LEAK DETECTION SYSTEMS

In the fast-evolving landscape of safety management, IoT-based leak detection systems have emerged as versatile and proactive solutions to prevent and mitigate leaks. These innovative systems cater to various industries and applications, offering specialized water, gas, chemical, and environmental leak detection

Figure 5. Evolution of benefits in IoT-based leak detection systems

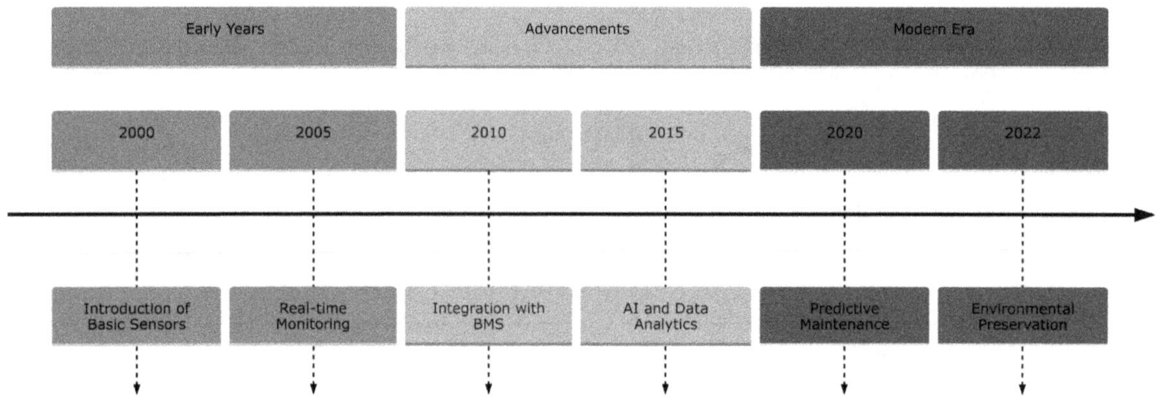

functionalities. Additionally, multi-purpose leak detection systems and those designed explicitly for infrastructure further expand the scope of IoT-based solutions. This section explores the diverse varieties of IoT-based leak detection systems, highlighting their unique features, applications, and benefits (Khasawneh et al., 2023).

Water Leak Detection Devices

Water leak detection devices are designed to identify and localize leaks in plumbing systems, water supply networks, and buildings. These devices utilize intelligent sensors that monitor water flow, pressure, and humidity, detecting anomalies that may indicate leaks. By promptly identifying water leaks, these systems help conserve water resources, prevent property damage, and reduce water wastage, making them invaluable assets for domestic and industrial settings (Chuang et al., 2019).

Gas Leak Detection Systems

Gas leak detection systems are crucial in industries dealing with natural gas, propane, or other hazardous gases. These systems employ specialized sensors that detect even minute gas leaks in pipelines, storage tanks, and industrial facilities. The IoT-based gas leak detection systems have real-time monitoring capabilities, enabling swift response and preventing potential hazards. These systems enhance safety in oil and gas, petrochemicals, and manufacturing industries, safeguarding personnel and assets from gas-related risks (Suma et al., 2019).

Chemical Leak Detection Systems

Chemical leak detection systems are tailored to detect and manage leaks of hazardous chemicals in industrial environments. These systems rely on advanced sensors specifically designed to detect the presence of specific substances and their concentration levels. By continuously monitoring chemical storage areas, processing plants, and transportation systems, IoT-based chemical leak detection systems minimize the risk of exposure to harmful substances, protect the environment, and comply with stringent safety regulations (Hou et al., 2021).

Multi-Purpose Leak Detection Systems

Multi-purpose leak detection systems offer a versatile solution capable of detecting different types of leaks in various scenarios. These systems can be configured with intelligent sensors for detecting water, gas, and chemical leaks. The flexibility makes them ideal for applications with multiple leaks within a facility or infrastructure. Multi-purpose leak detection systems are widely used in commercial buildings, hotels, hospitals, and educational institutions, providing comprehensive leak protection (Choudhary et al., 2023).

Environmental Leak Detection Systems

Environmental leak detection systems focus on monitoring and protecting natural ecosystems from potential leaks or spills. These systems are commonly used in industries handling petroleum products, chemicals, or hazardous materials near sensitive ecological areas. Equipped with advanced sensors, environmental leak detection systems can promptly detect leaks and spills, enabling swift containment and mitigating environmental damage.

These systems support sustainable practices and environmental stewardship by minimizing the ecological impact (Joseph et al., 2023).

Leak Detection Systems in Infrastructure

Leak detection systems in infrastructure are tailored to address the unique challenges of large-scale projects such as pipelines, dams, bridges, and underground utilities. These systems employ a network of intelligent sensors distributed throughout the infrastructure, continuously monitoring key parameters. In case of any anomalies or potential leaks, the IoT-based systems trigger real-time alerts, enabling proactive maintenance and reducing the risk of catastrophic failures. Leak detection systems in infrastructure enhance safety and extend the lifespan of critical assets, reducing repair and replacement costs.

In conclusion, the varieties of IoT-based leak detection systems offer a comprehensive and tailored approach to address specific leak detection needs across diverse industries and applications. These systems empower organizations to proactively safeguard their assets, environment, and personnel, from water and gas to chemical and environmental leak detection. The versatility of multi-purpose leak detection systems and the specialized focus of infrastructure-specific systems further demonstrate the adaptability and effectiveness of IoT-based solutions. As the demand for robust leak detection mechanisms grows, IoT-based systems will undoubtedly play an instrumental role in enhancing safety and sustainability across many industries worldwide (Khalil et al., 2021).

Table 3 provides the features, benefits, and references of several Internet of Things-based leak detection solutions.

Table 3. Comparison of IoT-based leak detection systems

Detection System Type	Applications	Key Features	Benefits
Water Leak	Plumbing, Water Supply, Buildings	Monitors Water Flow, Pressure, Humidity	Conserves Water, Prevents Damage
Gas Leak	Natural Gas, Propane Industries	Detects Minute Gas Leaks	Enhances Safety, Real-time Monitoring
Chemical Leak	Industrial Chemical Environments	Detects Specific Chemicals & Concentration	Minimizes Risk of Chemical Exposure
Multi- Purpose	Various Scenarios (Water, Gas, Chemical) Petroleum,	Versatile, Configurable Sensors	Comprehensive Leak Protection
Environ- mental	Chemical Industries near Ecological Areas	Protects Natural Ecosystems	Mitigates Environmental Damage Proactive
Infrastructure	Pipelines, Dams, Bridges	Monitors Large-Scale Projects	Maintenance, Extends Asset Lifespan

BENEFITS OF IOT-DRIVEN LEAK DETECTION SYSTEMS

IoT-based leak detection systems are revolutionizing how businesses of all stripes improve safety and output. By tapping into the enormous IoT intelligence network, these systems have the potential to collect and analyze data continually. The potential dangers linked with leakage can be significantly reduced thanks to this real-time analysis. Not only can early detections save money by preventing possible harm, but they are also crucial in protecting natural resources. There is a drastic drop in pollution, and conservation efforts are significantly boosted. IoT-based leak detection solutions are essential for organizations that want to improve their safety, security, and sustainability. Their interoperability with preexisting systems, scalability, and adaptability only add to their appeal (Heidari & Jamali, 2022).

The timeline in Figure 5 illustrates the progressive evolution of IoT-based leak detection systems, highlighting technological advancements, safety, efficiency, and environmental impact over time.

Early Detection and Threat Mitigation

The IoT's power lies in its exceptional early detection capabilities when finding leaks. As a result, businesses may take preventative action before problems escalate into significant disruptions that are difficult to recover from and can cost a lot to fix.

Enhanced Safety and Security

IoT-based leak detection systems provide noticeable improvements in security across a wide range of settings thanks to their constant monitoring and in-depth data analysis. To dramatically reduce mishaps, injuries, or possible danger to both personnel and the broader public, these systems must be able to identify potential leaks proactively.

Efficient Use of Available Resources

The IoT promise for efficient use of resources in leak detection should be noticed.

Organizations can save money by carefully monitoring how much water and other fluids are used. Not only does this save money, but it also promotes the responsible management of natural resources.

Environmental Preservation and Risk Reduction

IoT-based solutions are crucial to environmental protection because of how quickly they can identify and stop the leaking of dangerous chemicals. These devices support conservation initiatives by reducing the wastage of essential resources and the leakage of hazardous substances (Gao et al., 2020).

Real-Time Monitoring and Distant Access

Leak detection systems built on the IoT benefit significantly from real-time remote monitoring, one of its most distinctive advantages. This feature is beneficial since it allows stakeholders to monitor their infrastructure from anywhere and at any time. Such rapid access allows for prompt decision-making and intervention, which can help prevent disasters.

Data Analysis and Predictive Upkeep

Leak detection solutions in the IoT combine data analytics with artificial intelligence to yield insights that might radically alter how infrastructure is managed. Predictive maintenance solutions, made possible by the convergence of these technologies, proactively address potential problems before they even occur, dramatically cutting down on downtime and associated costs (Cornelius & Shanthini, 2023).

Building Management System (BMS Systematic Coordination)

Integrating IoT-based leak detection systems with BMSs ushers in a new era in facility management. This alliance simplifies the complex duties of managing a building's infrastructure. Facility managers may better assess possible risks by combining leak detection's real-time monitoring capabilities with the all-encompassing management of BMSs. By working together, these can better allocate resources and save money on repairs by preventing problems from getting out of hand. When IoT and BMSs are combined, infrastructure management becomes more comprehensive, efficient, and proactive (Tatari et al., 2022).

Adaptability/Flexibility

Leak detection technologies built on the IoT are highly scalable because of their adaptability. In addition, their flexibility makes them suitable for a wide variety of settings, from single-family homes to massive industrial complexes, and assures they will continue to be useful even if infrastructure standards change.

Reducing Environmental Pollutants

IoT-driven leak detection has far-reaching ecological effects beyond the immediate advantages. These innovations help clean up their environment by quickly locating and fixing leaks. Protecting the ecosystems by ensuring clean air, water, and soil is essential for health. In conclusion, many advantages of IoT-based leak detection systems highlight their importance in today's infrastructure management. Businesses with varying priorities might benefit from their services because they promise to improve security, increase efficiency, and promote environmental conservation (Narmilan & Puvanitha, 2020).

Obstacles and Design Considerations in IoT-Based Leak Detection Systems

Challenges for leak detection systems based on the IoT include overcoming false positives and negatives, guaranteeing dependable connectivity, optimizing power usage, resolving data security and privacy concerns, and more. Other significant difficulties include scalability, environmental awareness, interoperability, and latency reduction. Additional design considerations essential to ensure the efficacy and success of these systems are cost-effectiveness, integration with existing infrastructure, device maintenance, and user-friendly interfaces with clear notifications.

Privacy and Data Security Issues

Leak detection solutions based on the IoT place a premium on data security and privacy. Because of the sensitive nature of the data they gather, these systems might be the subject of an attack. Robust encryption solutions for protecting data during transmission and storage are necessary to alleviate this worry. In addition, authentication and access restrictions should be used to guarantee that only approved users may access the information. It is essential to adhere to data protection legislation and standards (Atlam & Wills, 2020).

Sensor and Algorithm Reliability and Precision

Leak detection systems rely heavily on their sensors' and algorithms' precision and dependability. The sensors must accurately report even the most negligible leakage.

Algorithms must be developed that can distinguish between genuine leaks and false positives. Maintaining a sensor's accuracy over time requires regular calibration and testing.

Furthermore, the overall dependability of the system may be improved by data-driven algorithmic enhancement and refinement (Yeong et al., 2021).

Reliability and Availability of the Network

Leak detection solutions built on the IoT require a stable network to function correctly.

For real-time data transmission and timely alarms in the event of leakage, stable network connectivity is essential. In the event of a network outage, it is important to have redundancy and failover procedures in place. In addition, connection issues may arise in rugged and isolated areas, necessitating careful attention to the system's design.

Energy Efficiency and Power Management

Many IoT leak detection devices run on battery power; thus, improving energy efficiency is crucial for extending battery life and decreasing maintenance needs. It may be accomplished using low-power modes and sensor wake-up procedures, both products of effective power management practices. Energy-collecting technology might be included to lessen the need for battery replacement further.

Compatibility With Preexisting Infrastructure

Existing water supply and distribution infrastructure must be seamlessly integrated with IoT leak detection devices for optimal implementation. Implementing standardized communication protocols to ensure hardware and software compatibility with the current infrastructure may be necessary. There should be as little downtime as possible throughout this integration.

Initial Capital Outlay and Benefits Evaluation

Investing in hardware, software, and installation is necessary to launch an IoT-based leak detection system. A thorough cost-benefit analysis must determine the system's financial sustainability and ROI. The initial expenditures can be rationalized, and broader adoption can be encouraged by identifying potential savings from leak prevention and early detection.

False Alarms and Signal Interference

False alarms and other signal interference can compromise the system's trustworthiness. Using redundant sensors and developing more robust algorithms can reduce false positives and negatives. To further guarantee precise and dependable data transfer, dealing with signal interference difficulties such as radio frequency interference is essential.

Compliance With Regulations and Ethical Concerns

Tremendous privacy, water management, and environmental standards apply to IoT-based leak detection devices. Compliance with these rules is crucial to avoid legal consequences and preserve ethical standards. Deployment that is both sustainable and responsible must also consider the environmental effect of the system's components and activities.

Ongoing Training and Support

IoT devices must be regularly serviced to ensure they operate at peak performance. Long-term dependability may be provided by establishing routine maintenance plans and processes. Training the people using and maintaining the system is also crucial to ensuring its optimal performance. Data analysis, algorithm revisions, and problem-solving techniques should all be part of any relevant training.

IOT-BASED LEAK DETECTION SYSTEM CASE STUDIES IN THE REAL WORLD

Leak detection systems built on the IoT are reliable and flexible in real-world settings.

These technologies have been vital in preventing leaks, decreasing water losses, improving safety, and decreasing property damage for water utility companies in smart cities and industrial sites for chemical processing. Real-time monitoring, early leak identification, and quick reaction measures are made possible via sensors, sophisticated algorithms, and cloud-based systems. Whether in office buildings for facilities management or oil and gas pipelines for monitoring, these systems play an essential role in safeguarding infrastructure, conserving the environment, and maximizing productivity in various sectors (Ramachandran et al., 2023).

The Discovery of a Water Leak in an Office Building

A leak in the building's water system caused expensive repairs and higher utility bills for this company. The facility management team addressed the issue by installing an IoT-based water leak monitoring system as a solution. Near critical points like water fixtures, supply lines, and other weak spots, they set up water flow sensors and moisture detectors. These sensors monitored the real-time water flow and humidity levels, sending that information to a control centre. The IoT system employed complex algorithms to examine the data for irregularities that might indicate the presence of leaks. When a leak was detected, the system immediately sent an SMS or email to the facility management staff (Sater & Hamza, 2021).

The water leak detection system based on the IoT helped the commercial building drastically cut down on property damage and water waste due to leaks. The proactive leak detection method also resulted in financial savings by fixing small leaks before they balloon into major problems.

Gas Leak Detection in an Industrial Environment: A Case Study

The potential for gas leaks in an industrial context where gas is stored and distributed was a significant safety issue for the employees and the surrounding community. The manufacturing plant used an IoT-based gas leak detection system to improve safety measures and forestall mishaps. Gas sensors were installed in the system at critical locations, such as near the gas tanks and pipes.

These gas sensors monitored the atmosphere for dangerous gases and relayed that information to a command centre in real time. The system used complex algorithms to examine the gas concentration readings and identify any unexpected changes that would suggest a leak. The technology triggered sirens and notifications when a gas leak was discovered, allowing for swift evacuation and separation of impacted regions.

The industrial site significantly increased safety standards and decreased the danger of potential gas-related mishaps after implementing the IoT-based gas leak detection system. The system's early detection and instant notifications made it possible to take rapid action, protecting the employees and the surrounding community.

Third Example: Environmental Leak Detection Along an Oil Pipeline

Detecting environmental leaks in an oil pipeline was a key concern to avoid oil spills and reduce ecological damage. In response, a network of sensors connected via the IoT was installed along the pipeline's length to monitor for leaks. A network of pressure sensors, flow meters, and sound sensors ensured pipeline safety.

The sensors constantly sent information about pressure, flow rates, and acoustic vibrations to a command centre. The data was analyzed by cutting-edge algorithms in search of vulnerabilities in the system. It is the system's job to detect and categorize deviations from the norm that may indicate a leak in the pipeline.

In the case of a leak or other abnormality, the system would quickly notify the pipeline operators, who in turn might begin emergency shutdown procedures to cut off the leaking piece of the pipeline. Oil spills were avoided, ecosystems were safeguarded, and regulatory compliance was kept up thanks to this proactive approach to environmental leak detection.

The oil pipeline firm proved its dedication to environmental stewardship and reduced the possibility of negative repercussions related to oil transportation by deploying an IoT-based environmental leak detection system. Because of the system's ability to detect problems early on, corrective action could be taken quickly, limiting ecological harm and the expense of cleaning it up.

IOT-BASED LEAK DETECTION SYSTEM DEVELOPMENTS AND TRENDS

Improvements in sensor technology, the incorporation of AI and ML, cloud-based platforms, wireless connection, and innovative city programs have all contributed to the rapid development of IoT-based leak detection systems in recent years. IoT and 5G networks working together have improved real-time monitoring capabilities, and energy harvesting solutions deal with low battery life issues. Data security and privacy improvements and work towards interoperability standards guarantee that leak detection is practical and trustworthy. These innovations and shifts make IoT-based leak detection systems increasingly useful for water-saving, environmental protection, and safety management (Kaur & Sharma, 2022).

Improvements in the Use of AI and ML in Everyday Life

In IoT-based leak detection systems, incorporating AI and ML has emerged as a distinguishing trend. Artificial intelligence and machine learning algorithms are used to make sense of the massive volumes of data generated by sensors. Leak detection accuracy is improved by fewer false alarms and better early warning capabilities because of these intelligent algorithms' ability to learn from prior data and adapt to shifting situations.

AI-driven leak detection systems are the most dependable and effective since they are continuously fine-tuned and improved.

Edge Computing for Faster Data Processing and Response

Edge computing has emerged as a beneficial alternative to keep up with the ever-increasing need for low latency and real-time data processing in IoT leak detection systems. Edge computing eliminates the need to send data to remote cloud servers and instead processes it where it is generated (at the sensors). That allows for quicker action in urgent leak detection instances and faster reaction times overall. IoT-based leak detection systems can benefit from edge computing to better detect and manage leaks.

Identifying IoT Leaks in Smart City Infrastructure

Smart cities have quickly deployed IoT-based leak detection systems to improve water management and sustainability. Leaks and water waste may be detected and prevented when these technologies are included in smart city infrastructure, including water supply networks, sewage systems, and public buildings. Authorities may optimize water distribution, minimize water losses, and lessen environmental consequences by incorporating IoT leak detection into intelligent city infrastructure. The information gathered by these technologies may also be used to develop and maintain infrastructure.

Adoption of Low-Power Wide-Area Network (LPWAN) Technologies

Because of their low power consumption and extended range, LPWAN technologies like LoRaWAN and NB-IoT have become more attractive for IoT-based leak detection systems. Even under harsh conditions, LPWAN allows for unhindered data transfer between sensors and a centralized monitoring hub. As a result of the LPWAN devices' low power consumption, battery-operated detectors may remain in service for much longer, cutting down on maintenance time and improving system dependability.

Integrating Multiple Sensors for Complete Tracking

IoT-based systems increasingly use multi-sensor fusion methods to detect leaks more thoroughly and precisely pinpoint their origins. The system can cross-validate information and identify leak locations by integrating data from many sensors (acoustic, pressure, and flow sensors, for example). In addition to improving the system's responsiveness, multi-sensor fusion makes it easier to tell the difference between natural leak occurrences and false alarms.

Figure 6 illustrates the interconnected components of an IoT-based leak detection system, highlighting data flow and functionalities in a smart city context.

IoT-based leak detection systems are making great strides forward thanks to the combination of AI and ML, edge computing, and LPWAN technologies. As innovative city efforts gain traction, these technologies become increasingly important in water management and infrastructure sustainability. IoT leak detection systems use multi-sensor fusion approaches to provide thorough and accurate leak monitoring, reducing water waste, environmental effects, and maintenance expenses.

CONCLUSION

Leak detection, in particular, has entered a new era of safety and efficiency because of the IoT's rapid development. The potential for radical change offered by IoT-based leak detection systems has been explored in this chapter. The advantages of these systems over more conventional approaches have been highlighted, especially for real-time monitoring, early warning capabilities, and data analytics. Leak detection may be made more proactive with the help of IoT systems thanks to their intelligent sensors, cutting-edge algorithms, and centralized data processing. Increased security, less waste, and better use of resources are just a few of the many advantages of IoT-driven leak detection. It is essential, however, to deal with the difficulties of introducing such cutting-edge systems. To ensure the effective deployment

Figure 6. IoT-based leak detection system architecture

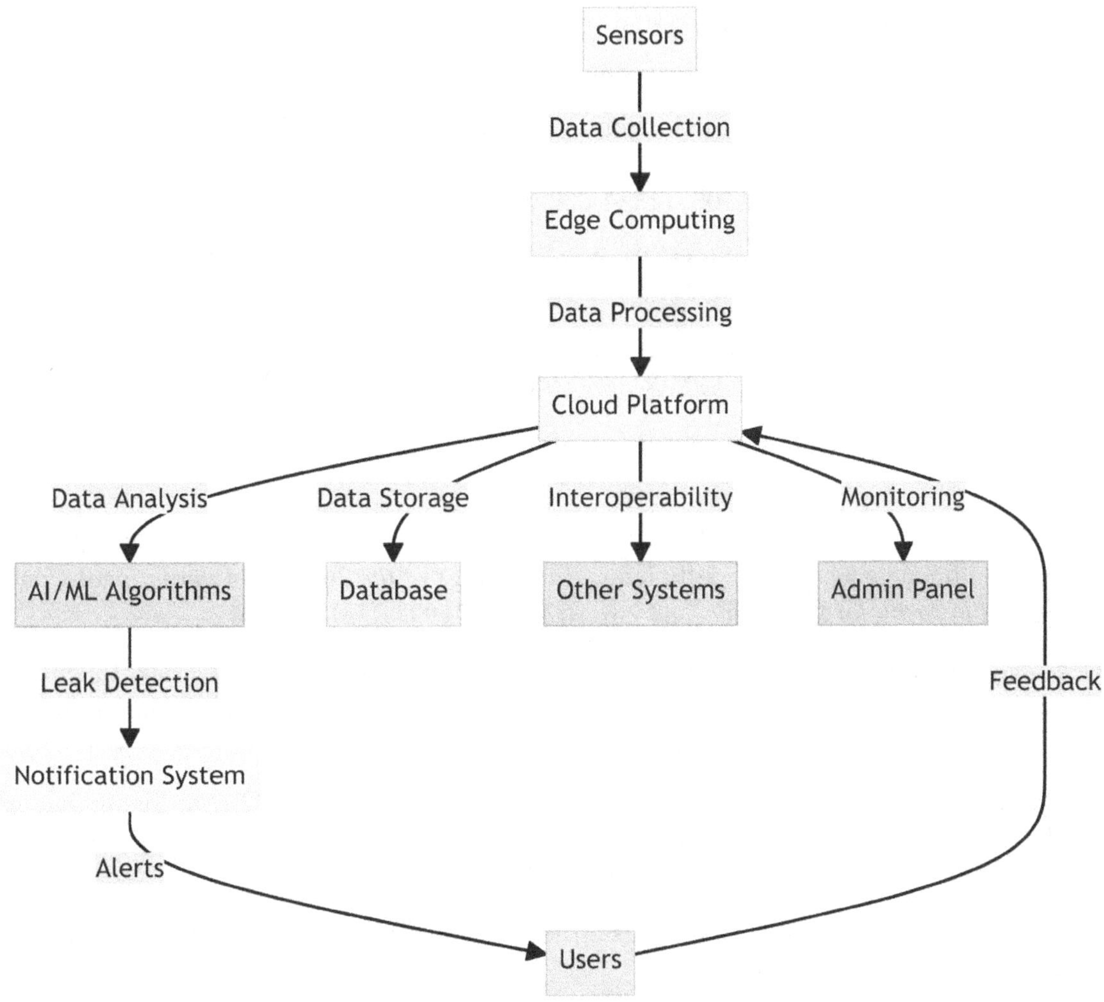

and operation of IoT-based leak detection systems, it is essential to carefully evaluate concerns related to data privacy, sensor reliability, and initial capital expenses.

There is hope that IoT system capabilities can be further expanded when new AI and ML techniques are developed and integrated with edge computing and LPWAN technology. The importance of the IoT in assuring safety and sustainability in our increasingly interconnected society cannot be emphasized.

IoT-based leak detection systems are a giant step toward a better, greener, and safer tomorrow. As technology advances, businesses and universities must work together to improve IoT systems for leak detection and other areas.

REFERENCES

Ahmed, S., Rahman, M. J., & Razzak, M. A. (2023). Design and Development of an IoT-Based LPG Gas Leakage Detector for Households and Industries. *IEEE World AI IoT Congress*, 762–767.

Banerjee, A., & Banerjee, C. (2023). A hybrid cellular automata-based model for leakage detection smart drip irrigation water pipeline structure using IoT sensors. *Innovations in Systems and Software Engineering*, *19*(1), 23–32. doi:10.100711334-022-00503-0

Bedi, G., Venayagamoorthy, G. K., Singh, R., Brooks, R. R., & Wang, K. C. (2018). Review of IoT (IoT) in electric power and energy systems. *IEEE IoT JWEnal, 5*(2), 847–870.

Chhaya, H. S., & Gupta, S. (1997). Performance modelling of asynchronous data transfer methods of IEEE 802.11 MAC protocol. *Wireless Networks*, *3*(3), 217–234. doi:10.1023/A:1019109301754

Choudhary, P., Botre, B. A., & Akbar, S. A. (2023). *Urban Water JWEnal*. doi:10.1080/157306 2X.2022.2164732

Chuang, W.-Y., Tsai, Y.-L., & Wang, L.-H. (2019). Leak detection in water distribution pipes based on CNN with mel frequency cepstral coefficients. *Proceedings of the 2019 3rd International Conference on Innovation in Artificial Intelligence*, 83–86. 10.1145/3319921.3319926

Elleuchi, M., Khelif, R., Kharrat, M., Aseeri, M., Obeid, A., & Abid, M. (2019). Water Pipeline, Monitoring and Leak Detection, using soil moisture Sensors: IoT-based solution. *16th International Multi-Conference on Systems, Signals & Devices (SSD)*, 772–775.

Gao, Q., Guo, S., Liu, X., Manogaran, G., Chilamkurti, N., & Kadry, S. (2020). Simulation analysis of supply chain risk management system based on IoT information platform. *Enterprise Information Systems*, *14*(9), 1354–1378. doi:10.1080/17517575.2019.1644671

Heidari, A., & Jamali, M. A. (2022). IoT intrusion detection systems: A comprehensive review and future directions. *Cluster Computing*, 1–28.

Hou, J., Gai, W., Cheng, W., & Deng, Y. (2021). Hazardous chemical leakage accidents and emergency evacuation response from 2009 to 2018 in China: A review. *Safety Science*, *135*, 105101–105101. doi:10.1016/j.ssci.2020.105101

Joseph, K., Sharma, A. K., van Staden, R. C., Wasantha, P. L. P., Cotton, J., & Small, S. (2023). Application of Software and Hardware-Based Technologies in Leaks and Burst Detection in Water Pipe Networks: A Literature Review. *Water (Basel)*, *15*(11), 2046. doi:10.3390/w15112046

Kaur, P., & Sharma, N. (2022). An IOT-Based Smart Home Security Prototype Using IFTTT Alert System. *2022 International Conference on Machine Learning, Big Data, Cloud and Parallel Computing (COM-IT-CON), 1*, 393–400. 10.1109/COM-IT-CON54601.2022.9850500

Kayastha, N., Niyato, D., Hossain, E., & Han, Z. (2014). Smart grid sensor data collection, communication, and networking: A tutorial. *Wireless Communications and Mobile Computing*, *14*(11), 1055–1087. doi:10.1002/wcm.2258

Khalil, M., Ismanto, I., & Akhsani, R. (2021). Development Of an LPG Leak Detection System Using Instant Messaging Infrastructure Based on IoT. *2021 International Conference on Electrical and Information Technology*, 147–150.

Khasawneh, M., & Azab, A., & Alrabaee, S. (2023). An IoT-based System Towards Detecting Oil Spills. *2023 International Conference on Intelligent Computing, Communication, Networking and Services (ICCNS)*, 197–202.

Kim, S.-T., Park, H.-J., & Kim, Y.-C. (2001). The load monitoring of the Web server using a mobile agent. *International Conferences on Info-Tech and Info-Net. Proceedings, 6*, 130–135.

Mabrouki, J., Azrwe, M., Fattah, G., Dhiba, D., & Hajjaji, S. (2021). Intelligent monitoring system for biogas detection based on the IoT: Mohammedia, Morocco city landfill case. *Big Data Mining and Analytics*, *4*(1), 10–17. doi:10.26599/BDMA.2020.9020017

Narmilan, A., & Puvanitha, N. (2020). Mitigation techniques for agricultural pollution by precision technologies focusing on the IoT (IoT): A review. *Agricultural Reviews (Karnal)*, *41*(3), 279–284. doi:10.18805/ag.R-151

Pointl, M., & Fuchs-Hanusch, D. (2021). Assessing the Potential of LPWAN Communication Technologies for Near Real-Time Leak Detection in Water Distribution Systems. *Sensors (Basel)*, *21*(1), 293. doi:10.339021010293 PMID:33406751

Ramachandran, B., Youssef, H. Y., Karunamurthi, J. V., Ebisi, F., Alnuaimi, A. N. A., & Najjar, J. (2023). IoT-based Test facility for a Water Transmission Line with Leak Simulation. *2023 International Conference on IT Innovation and Knowledge Discovery (ITIKD)*, 1–5. 10.1109/ITIKD56332.2023.10099751

Sater, R. A., & Hamza, A. B. (2021). A federated learning approach to anomaly detection in intelligent buildings. *ACM Transactions on IoT*, *2*(4), 1–23.

Suma, V., Shekar, R. R., & Akshay, K. A. (2019). Gas leakage detection based on IOT. *3rd International Conference on Electronics, Communication and Aerospace Technology (ICECA)*, 1312–1315.

Sunny, J. S., Patro, C. P. K., Karnani, K., Pingle, S. C., Lin, F., Anekoji, M., Jones, L. D., Kesari, S., & Ashili, S. (2022). Anomaly detection framework for wearables data: A perspective review on data concepts, data analysis algorithms, and prospects. *Sensors (Basel)*, *22*(3), 756–756. doi:10.339022030756 PMID:35161502

Tatari, M., Agarwal, P., Alam, M. A., & Ahmed, J. (2022). Review of intelligent building management system. *ICT Systems and Sustainability: Proceedings of ICT4SD 2021, 1*, 167–176.

Yadav, D., Pandey, A., Mishra, D., Bagchi, T., Mahapatra, A., Chandrasekhar, P., & Kumar, A. (2022). IoT-enabled smart dustbin with messaging alert system. *International Journal of Information Technology : an Official Journal of Bharati Vidyapeeth's Institute of Computer Applications and Management*, *14*(7), 3601–3609. doi:10.100741870-022-00947-4

Yeong, D. J., Velasco-Hernandez, G., Barry, J., & Walsh, J. (2021). Sensor and sensor fusion technology in autonomous vehicles: A review. *Sensors (Basel)*, *21*(6), 2140–2140. doi:10.339021062140 PMID:33803889

Chapter 13
Using Augmented Reality (AR) and the Internet of Things (IoT) to Improve Water Management Maintenance and Training

Muskan Sharma
Chandigarh University, India

Yash Mahajan
https://orcid.org/0009-0000-2530-0366
Chandigarh University, India

Abeer Saber
https://orcid.org/0000-0002-9261-0927
Damietta University, Egypt

ABSTRACT

The convergence of augmented reality (AR) and the internet of things (IoT) holds great promise for transforming water management by improving maintenance, troubleshooting, and training. This chapter explores the synergy between AR and IoT in water management, highlighting their potential to enhance professionals' efficiency and address critical challenges. AR overlays real-time data and interactive guidance onto the physical environment, streamlining maintenance, troubleshooting, and training. IoT provides real-time data and remote monitoring capabilities, facilitating proactive decision-making and predictive maintenance. The integration of AR and IoT offers a powerful toolkit to tackle water management issues, promising increased reliability and sustainability for water resources in a digitally augmented world.

DOI: 10.4018/979-8-3693-1194-3.ch013

1. INTRODUCTION

The Internet of Things (IoT) and Augmented Reality (AR) have just come together, and this combination has the potential to revolutionize many different sectors. This powerful combination has opened the way for cutting-edge maintenance, troubleshooting, and training solutions with enormous promise for the water management industry. This chapter's introduction looks into the exciting world of AR and IoT technologies and examines how they may be used to manage water resources. The main emphasis is on the ways in which these technical developments may enable technicians to increase their effectiveness and productivity. We want to solve urgent concerns and discover new possibilities for the efficient management of water resources by using the combination of AR and IoT.

This chapter focuses on how AR and the IoT might be useful tools for managing water resources. Our goal is to analyse the complex interplay between AR and IoT, offering insightful information on how technicians might use AR for training, maintenance, and troubleshooting in the context of water management. This in-depth analysis aims to provide readers a thorough knowledge of the potential benefits and practical uses of this cutting-edge fusion within the sector.

This chapter sets off on a fascinating trip via the meeting point of AR and the IoT within the context of water management. It conducts an exhaustive analysis of their mutually beneficial connection, shedding light on their crucial functions in upkeep, troubleshooting, and specialist training. By closely examining the manners in which these technologies smoothly interact, we reveal cutting-edge strategies that may completely alter current methods of water management. The insights gained from this chapter are crucial for both professionals and stakeholders since this transformation has the potential to enable the efficient and sustainable use of this priceless resource. This investigation of AR and IoT promises to change the face of water resource management as the boundaries between the real and digital worlds become more permeable.

2. AR IN WATER MANAGEMENT

2.1. Overview of AR

With the help of the cutting-edge technology known as AR, which combines digital data with the actual environment, water management procedures may be improved in a number of ways (Revolti et al., 2023). To overlay digital material over the real world, AR uses gadgets like smartphones, tablets, smart glasses, or specialised headsets. This substantially enhances how technicians see and interact with their surroundings. With the help of this technology, professionals can more easily grasp complicated systems and spot problems in water systems. AR apps may show real-time data like water flow rates and temperature by connecting with IoT sensors and data streams, enabling personnel to make educated judgements and take prompt action.

By superimposing instructions and animations over technicians' actions to guide them through processes (Rahman et al., 2021), AR significantly reduces mistakes and shortens downtime in maintenance and troubleshooting. Additionally, it makes it possible for remote support, which lets professionals direct on-site personnel through tasks through live video feeds. For water management technicians, AR-based training modules provide immersive learning experiences that support skill growth and knowledge acquisition. By decreasing downtime, increasing productivity, and minimising mistakes overall, the implementation of AR in water management may result in cost effectiveness, eventually optimising maintenance and operating procedures.

2.2. AR Applications in Water Management

The way technicians, engineers, and operators engage with water infrastructure is being revolutionised by AR, which has a wide range of applications in the area of water management. AR accelerates maintenance activities (Seeliger et al., 2023), speeds up training procedures, and improves decision-making by superimposing digital data onto the physical environment. Several crucial areas in the management of water are where AR is useful:

1. **Asset Visualization and Inspection:** AR can help with the visualisation of sewage systems, water treatment facilities, and subterranean water pipes. Real-time data on asset status, such as pipe corrosion levels, flow rates, or pressure readings, may be accessed by technicians using AR headsets. This capacity makes it possible to identify problems quickly, assisting with preventative maintenance and decreasing downtime.
2. **Remote Assistance:** Remote specialists may now direct field personnel in real time thanks to AR. Remote professionals may help with difficult issue diagnosis, repair work, or inspections using live video feeds and AR annotations. This saves time and costs since it expedites problem resolution and decreases the requirement for on-site physical presence.
3. **Training and Onboarding:** Water management staff may undergo immersive training using AR. Through engaging AR simulations, new personnel may learn about intricate water treatment procedures, equipment operation, and safety procedures. This strategy improves learning retention and lowers the danger of on-the-job training.
4. **Data Visualization:** On physical infrastructure, AR may immediately show real-time sensor data, analytics, and historical patterns (Bakhtiari et al., 2023). Operators are able to make defensible judgements based on the available facts thanks to this visual depiction. Operators of water treatment facilities, for example, may immediately react to departures from allowed levels by monitoring water quality metrics in real-time.
5. **Data Integration With IoT**: AR can access and show data from numerous sensors and devices installed across the water infrastructure when it is coupled with the IoT. The information shown in the AR environment is more accurate and relevant because to this connection.

AR and IoT integration in water management provide a potent toolkit for enhancing maintenance, troubleshooting, and training procedures. Water management experts may increase their effectiveness, lower operational risks, and guarantee the sustainable management of water resources by using AR applications (Proud et al., 2023). The water management business is set to change as a result of the convergence of AR and IoT, which will provide a more data-driven and responsive approach to solving the difficulties of maintaining and optimising water infrastructure.

3. IOT IN WATER MANAGEMENT

3.1. Introduction to IoT in Water Management

Recent years have witnessed a considerable development in the area of water management, partly due to the incorporation of cutting-edge technology. The IoT has emerged as one of them and is a potent technology that is revolutionising the way we monitor, manage, and regulate water resources (Gholizadeh et al., 2023). Overview of IOT in water management is illustrated in Fig. 1. This chapter examines the crucial part that IoT plays in water management and looks at how AR and IoT may work together to improve maintenance, troubleshooting, and training procedures.

Water is a valuable and limited resource that is essential for maintaining life and several businesses. This chapter's major focus is on the management of water resources utilizing AR and the IoT. This chapter aims to shed light on how technicians may efficiently use AR for training, maintenance, and troubleshooting in the context of water management by studying the synergies between AR and IoT. Readers will get comprehensive knowledge of the possible advantages and real-world uses of this cutting-edge combo in the field via this investigation.

Figure 1. Overview of IoT in water management

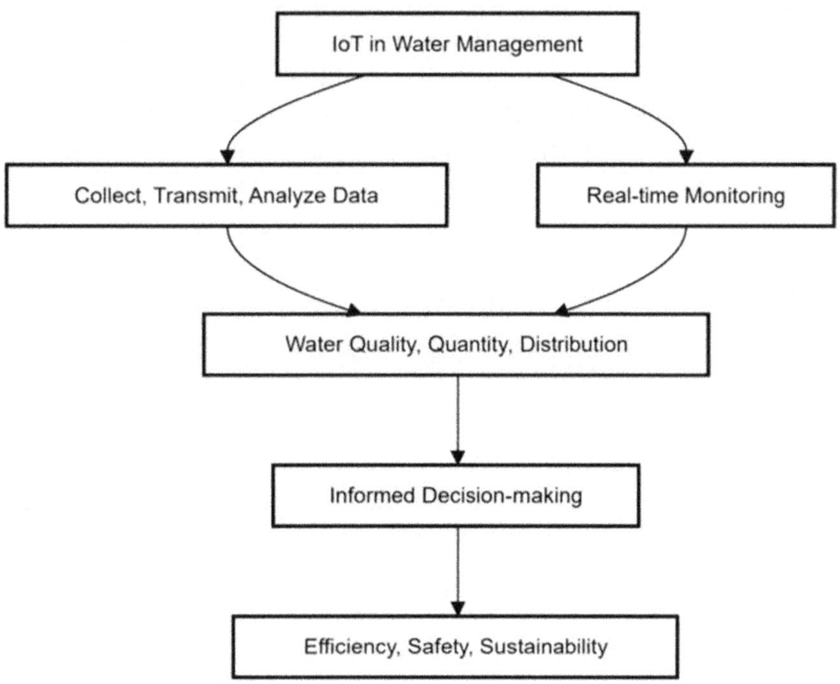

Two of the IoT's key advantages in water management are real-time data and the ability to conduct remote monitoring. Water quality, flow rates, temperature, pressure, and other variables are continually gathered by sensors positioned at different sites throughout the water infrastructure (Hadipour et al., 2020). This information is sent to a central hub, where it is immediately processed and examined. The information is then accessible to decision-makers through web-based dashboards or mobile apps, enabling them to act quickly on the information.

In this chapter, we'll look at how AR and IoT technologies work together to improve the skills of those in the water management industry. AR projects digital data over the physical environment to provide technicians real-time instructions and visualisations (Morchid et al., 2023). Technicians can efficiently do maintenance duties, diagnose problems, and get training by fusing AR with IoT data.

Water infrastructure may be made more effective, safe, and sustainable thanks to the interaction between AR and IoT. We will go further into the real-world uses of AR and IoT in the sections that follow, demonstrating how these tools enable personnel to operate water systems with accuracy and knowledge.

3.2. IoT Devices and Sensors in Water Management

The incorporation of IoT devices and sensors has ushered in a new era of efficiency, accuracy, and sustainability in the field of water management. These IoT gadgets have developed into crucial instruments for controlling, monitoring, and improving water-related operations (Heidary Dahooie et al., 2023). IoT devices are essential for assuring the dependability and sustainability of water systems, from assessing water quality to maintaining infrastructure.

Water quality is the most important factor to consider when managing water resources, and IoT devices with sensors are essential for regularly monitoring different water parameters. Real-time data from sensors' measurements of variables including pH levels, turbidity, dissolved oxygen, and chemical concentrations enables the quick identification of abnormalities or contamination occurrences. How iot sensors and devices may be employed in water management is shown in Fig. 2. This capacity to monitor water quality in real-time allows quick reactions to possible hazards, preserving human health and safeguarding natural ecosystems.

Figure 2. IoT devices and sensors in water management

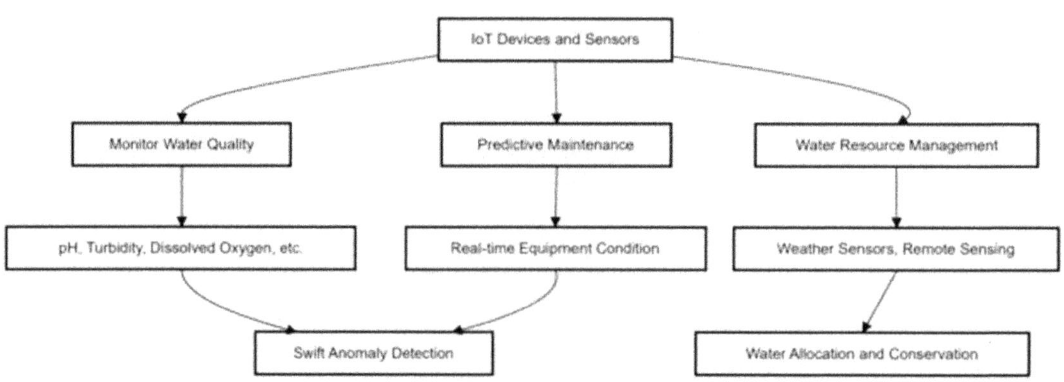

IoT devices are essential to water management's predictive maintenance strategies in addition to monitoring (Patel et al., 2023). Maintenance personnel may get real-time information about the state of equipment by using sensors installed on important water infrastructure elements like pumps and valves. After analysing this data, predictive analytics systems can determine when maintenance is required. Water utilities may decrease downtime, optimise maintenance schedules, and increase the lifetime of their infrastructure by using such predictive maintenance solutions, which can save money and enhance operational effectiveness.

IoT gadgets are essential for managing water resources (Ali et al., 2022). Water authorities may more accurately forecast and manage water supply by using weather sensors and data from remote sensing satellites, particularly in areas vulnerable to drought or severe weather events. This data-driven methodology enables more precise water allocation and conservation methods, enabling the effective and sustainable use of water resources.

In conclusion, the use of IoT devices and sensors in water management is revolutionising the sector by supplying real-time data, boosting infrastructure effectiveness, and encouraging sustainability. These tools help water industry experts make wise judgements, act quickly when problems arise, and streamline their processes (Sunku Mohan et al., 2023). The fusion of IoT and AR promises to boost the water management industry further as it continues to develop by giving personnel cutting-edge tools for training, maintenance, and troubleshooting. The future of water management will surely be shaped by this integration, which will increase its resilience and guarantee the provision of clean water for future generations.

3.3. The Role of IoT in Water Quality Monitoring

Monitoring and maintaining water quality, which is essential for guaranteeing the security of water supplies, is one of the key components of efficient water management. The use of IoT technology has a significant impact on how water quality monitoring procedures are conducted. IoT plays a significant role in monitoring water quality, as seen in Fig. 3. The capacity to gather real-time data from multiple places within water distribution networks, reservoirs, and treatment facilities is provided by IoT devices and sensors. This information is essential for monitoring and evaluating several aspects of water quality, including pH levels, turbidity, chlorine concentration, temperature, and others.

Systems for monitoring water quality that are IoT-enabled provide a number of benefits when used with AR applications. First and foremost, they provide technicians and water management experts a precise and current picture of the state of the network's water quality parameters. As a result, it will be easier for technicians to visualise and understand water quality data while they are on the job (Malkawi et al., 2023). AR overlays may be effortlessly included into these real-time data streams. IoT gadgets make it possible to monitor water quality metrics proactively and quickly identify abnormalities or deviations. AR systems with IoT integration may immediately notify technicians of any pollution or odd changes in water quality. This capacity to respond quickly is crucial for preventing crises and dealing with problems as soon as they arise.

Beyond the urgent demands for maintenance and troubleshooting, IoT plays a significant role in water quality monitoring. It also makes a substantial contribution to capacity-building and training initiatives. Technicians may simulate different water quality conditions using AR-powered training modules that link with IoT data. Modern water quality monitoring in water management is based on IOT technologies (Li et al., 2023). Numerous benefits, including as real-time data visualisation, anomaly detection, predictive analysis, and improved training chances, are provided by its integration with AR. Through

Figure 3. The role of IoT in water quality monitoring

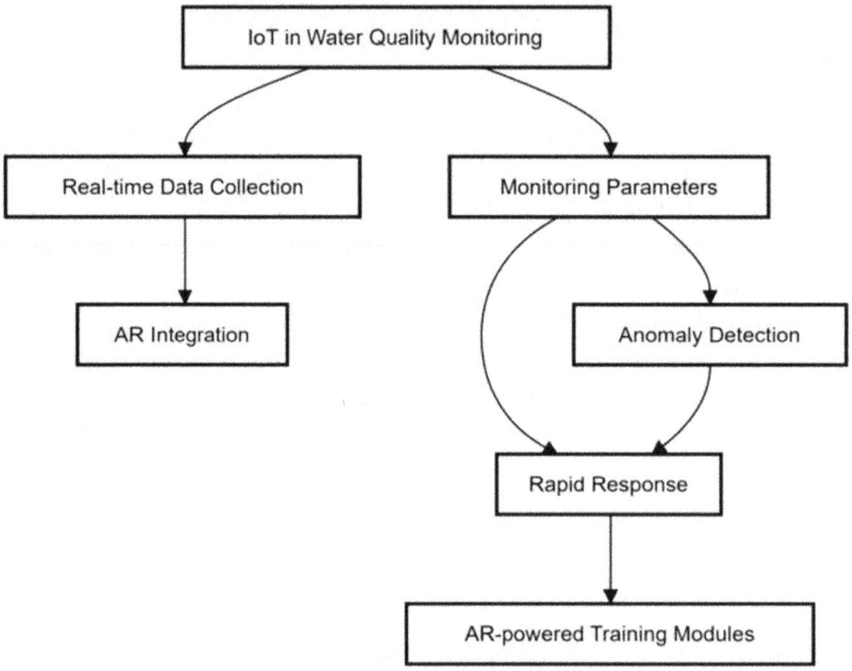

effective monitoring, maintenance, and training procedures, IoT and AR enable water management experts to assure the security and sustainability of water supplies.

4. SYNERGIES BETWEEN AR AND IOT

4.1. Combining AR and IoT for Water Management

Enhancing water management practises is made possible by the combination of AR and the IoT. The fusion of AR and IoT in the field of water management offers a special set of capabilities that may greatly enhance maintenance, troubleshooting, and training procedures.

Accessing real-time data from IoT sensors and devices right inside the AR environment is one of the main benefits of merging AR and IoT in water management. IoT sensors may provide crucial data including water flow rates, pressure measurements, temperature readings, and the condition of different water system components (Shaharuddin et al., 2023). Overview of IOT and AR integration in water management is given in Fig. 4. Field workers may get a quick understanding of the condition and operation of water infrastructure by superimposing this data on the technician's AR monitor. When abnormalities or problems develop, technicians can make wise judgements and respond quickly thanks to this real-time data integration.

Through the application of AR and IoT, maintenance tasks in water management systems may be considerably improved. Wearing AR headsets, technicians may get visual signals and instructions superimposed on actual equipment, leading them through step-by-step maintenance procedures (Yang

Figure 4. Integration of IoT and AR in water management

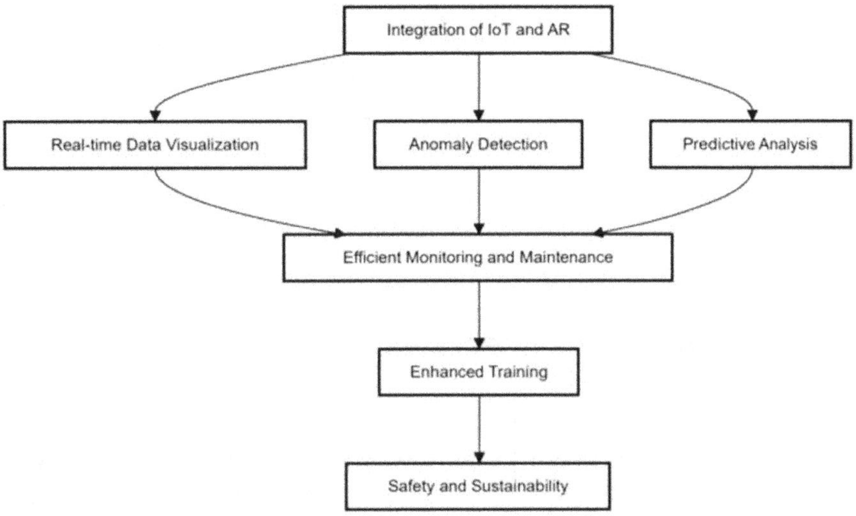

et al., 2020). IoT sensors may spot early indications of equipment wear or breakdowns and launch AR messages to immediately notify personnel.

Another use was combining AR and IoT is beneficial is troubleshooting. Technicians may use AR to access maintenance manuals, schematics, and other paperwork superimposed into their field of vision while dealing with complicated water management concerns. Additionally, AR can determine the precise location and type of an issue by using IoT data streams, enabling specialists to solve it quickly and effectively. This lessens the need for time-consuming trial-and-error troubleshooting, assuring effective issue resolution and reducing service interruptions.

In conclusion, the IoT and AR together create a potential new frontier in water management. Technicians can use real-time data, optimise maintenance processes, conduct efficient troubleshooting, and provide immersive training experiences thanks to the synergies between AR and IoT. The fusion of AR and IoT is positioned to play a crucial role in guaranteeing the sustainability and dependability of water infrastructure as water management continues to advance.

4.2. How AR Enhances IoT in Water Management

IoT applications may now be enhanced with the help of AR, especially in the field of water management. This section examines the mutually beneficial link between AR and the IoT and explains how AR may greatly improve IoT solutions for the upkeep, troubleshooting, and training elements of water management. When it comes to training, AR has the potential to completely transform how water management staff are instructed and equipped for their responsibilities. AR may be used to create training simulations that let students engage with digital water systems and practise maintenance or emergency response techniques (C.N. et al., 2023). This interactive, hands-on instruction not only quickens the learning process but also improves understanding and retention of important ideas.

The use of AR in IoT for water management goes beyond the technical. By placing real-time danger identification and safety instructions right in the technician's line of sight, it also improves safety. This guarantees that employees are informed of possible dangers and can take the necessary safety measures.

Water management may benefit from the combination of AR and IoT in a variety of ways, including better decision-making, greater cooperation, expedited training, and increased safety (Jahid et al., 2023). Water management experts may streamline their processes, save downtime, and guarantee the effective and sustainable use of water resources by using AR. Water management is about to undergo a revolution as a result of the convergence of AR and IoT, making it more effective, economical, and ecologically benign.

4.3. Case Studies of Successful Integrations

Numerous effective case studies have recently shown how AR and the IoT may work together seamlessly in the subject of water management. These examples demonstrate the real advantages that result from the marriage of these two cutting-edge technologies (Mahajan et al., 2023), giving technicians and experts the ability to improve maintenance, troubleshooting, and training procedures in the water management industry.

One such instance comes from a sizable municipal water treatment facility in a busy city. The factory has been dealing with a lot of equipment failures, which caused service interruptions and expensive repairs. Maintenance staff have access to real-time data on the state of different components, such as pumps and valves, by using AR headsets fitted with IoT sensors. Technicians were able to see problems and make targeted repairs since this data was overlaid into their field of vision (Chen et al., 2020). The end result was a considerable decrease in maintenance costs and downtime, which ultimately improved the water treatment system's dependability.

Another case included a water utility firm that wanted to improve the training it provided to staff members in charge of operating complex control systems. They unveiled training courses based on AR that made use of information from IoT sensors positioned across their network of water distribution pipes and reservoirs. Students might use AR glasses to see the network's architecture and interact with simulated elements. Through an interactive learning experience, technicians were not only better equipped to react quickly to real-world situations, but also increased the efficacy of training (Truong et al., 2023).

A case study from a small agricultural community in the country also demonstrated the potential of AR and IoT for remote monitoring and management of irrigation systems. IoT sensors installed in the fields sent vital information on crop health, weather patterns, and soil moisture levels. Farmers were able to access this information via AR programmes on their mobile devices or AR glasses, giving them real-time insights into when and how much to irrigate, eventually saving water and improving crop yields.

These case studies highlight how combining AR and IoT may revolutionise water management. They illustrate how the use of real-time data visualisation, remote monitoring, and immersive training may boost productivity, lower costs, and improve the capacity for decision-making in the water sector (Sanzana et al., 2022). The convergence of AR and IoT promises to revolutionise maintenance, troubleshooting, and training procedures as technology advances, enabling future water management systems that are robust and sustainable.

5. LEVERAGING AR FOR MAINTENANCE IN WATER MANAGEMENT

5.1. AR-Assisted Maintenance Procedures

In the sphere of water management, AR has emerged as a game-changing technology that offers creative fixes for upkeep operations. Technicians and maintenance staff may access a variety of real-time data, digital overlays, and interactive guidance by seamlessly integrating AR with the IoT, which greatly improves their capacity to carry out maintenance duties precisely and effectively.

Figure 5. AR-assisted maintenance procedures

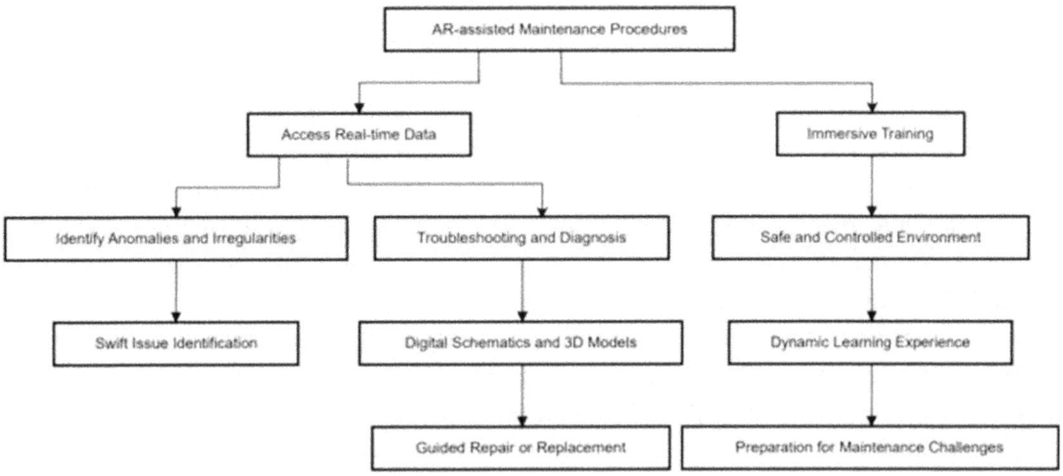

The ability to provide personnel immediate access to crucial information is one of the key benefits of AR in maintenance. Figure 5 illustrates how AR may help with maintenance procedures (Dzulkifli et al., 2021). Maintenance staff may monitor real-time data from IoT sensors distributed across the water management infrastructure using AR-enabled devices like smart glasses or tablets. Information on water flow rates, pressure readings, temperature, and other critical characteristics are all included in this data. With this knowledge, professionals may see any abnormalities or irregularities promptly and take immediate action to avert possible problems.

In addition, AR is essential for identifying and solving issues with water management systems. Technicians may use AR software to superimpose digital blueprints, 3D models, and detailed maintenance instructions onto actual equipment. This enhanced perspective helps professionals discover defective parts and directs them through the repair or replacement procedure.

A significant advantage of AR for sustaining water management systems is its potential for use in training (Huang et al., 2023). AR may be used to create immersive, interactive training courses. These modules allow technicians to practise maintenance procedures in a safe environment that replicates real-world circumstances. Due to the inclusion of IoT data, these training modules may also adapt to changing conditions, providing professionals with a dynamic learning experience that prepares them for a range of maintenance challenges.

In summary, the use of AR and the IoT to water management is revolutionising how maintenance is carried out. Employee accuracy and productivity are increased because to AR's access to real-time data, diagnostic tools, and remote collaboration platforms. Aside from that, AR-driven training modules prepare professionals for a range of maintenance scenarios, ensuring that they are well-prepared to address the evolving challenges of water management maintenance. As AR technology advances, its significance in optimising upkeep in water management will only grow.

5.2. Real-Time Data Integration Through IoT

When it comes to integrating real-time data, the IoT and AR integration have a lot of promise for water management. The way we gather and communicate data from diverse water management assets, such as pipes, pumps, and water treatment plants, has been revolutionised by IoT technology. Technicians and maintenance staff may access a variety of real-time information by leveraging the potential of IoT, enabling them to make wise choices and act quickly in urgent circumstances.

Monitoring asset performance and condition continually is one of the main benefits of integrating IoT into water management. Water infrastructure includes sensors and equipment that may collect information on variables including water flow rates, pressure, temperature, and quality. Then, this data is sent to a central platform so that it can be instantly analysed and visualized (Kumar et al., 2023). The presentation of this data to technicians and operators in an understandable and user-friendly way may be greatly aided by AR.

To receive a real-time flow of data from IoT sensors, technicians may put on AR glasses or utilise AR-enabled mobile devices via AR interfaces. For instance, they may check the flow rate in a specific pipeline section or keep an eye on a pump's condition from a distance. A useful layer of contextual information is provided by this real-time data integration, which may help with maintenance and troubleshooting chores. When doing preventative maintenance, it enables professionals to immediately spot abnormalities, possible problems, or deviations from regular operating conditions.

AR may also increase the effectiveness of maintenance activities by superimposing pertinent data onto the actual environment. AR allows technicians to superimpose equipment diagrams, maintenance checklists, and 3D models onto their field of vision. Consequently, combining AR and IoT for water management provides a potent method for integrating real-time data. It improves the procedures for maintenance, troubleshooting, and training by giving personnel instant access to crucial data and visualisations. The management and upkeep of water infrastructure might be revolutionised by this AR and IoT synergy, assuring its dependability, effectiveness, and sustainability.

5.3. Benefits and Efficiency Gains

Numerous advantages and efficiency improvements are realised when AR and the IoT are combined in the field of water management. The way specialists approach maintenance, troubleshooting, and training in the water management business is revolutionised by this combination.

The improvement of efficiency is one of the most important benefits of using AR for maintenance. Through AR-enabled devices like smart glasses or tablets, technicians may now instantly access vital information and instructions. The time needed to identify and fix problems is drastically shortened by this quick access to data, drawings, and processes. For instance, while repairing a broken pump station, workers may project digital data onto the actual equipment, speeding up the identification of damaged parts and streamlining the repair procedure (Sugiura et al., 2023). This results in less downtime and operational disturbances, which ultimately helps to save money and increase service dependability.

Additionally, AR and IoT allow for remote help and cooperation. With the ability to share live video feeds and augmented views of the equipment they are working on, technicians may communicate with specialists or coworkers located anywhere in the globe. This not only makes problem-solving quicker but also gives less experienced technicians a priceless learning opportunity. Training and skill development are made more efficient via real-time coaching and information exchange, providing a knowledgeable and adaptable workforce in the water management industry.

Analytics and data visualisation are important for predictive maintenance, and AR helps with both. The state of water infrastructure is regularly monitored by IoT sensors, and AR enables workers to meaningfully visualise this data. Technicians may see patterns and trends that can hint at future problems before they become serious by superimposing real-time sensor data onto physical equipment. By being proactive, you may reduce expensive breakdowns, increase asset longevity, and allocate resources more efficiently.

The decrease of human mistake is another benefit. AR places checklists and step-by-step instructions right in the technician's line of sight to guarantee that jobs are carried out accurately and consistently. This is especially important for water management, where accuracy and adherence to rules are essential for guaranteeing the provision of safe, high-quality water.

In conclusion, the use of AR and IoT in water management results in a number of advantages and productivity improvements. As the water management sector adopts these technologies, it will be able to manage crucial water infrastructure with improved levels of dependability, sustainability, and affordability.

6. AR FOR TRAINING IN WATER MANAGEMENT

6.1 Training Technicians With AR

A creative and useful way to improve the skill sets of technicians in the area of water management is to include AR into their training. Tech professionals may study information and practical skills more successfully thanks to AR technology's immersive and interactive learning environment.

Real-time visualisation is a crucial method for AR technician training. Trainees are able to examine complicated water management systems with pertinent data and instructional material superimposed on them using AR-enabled equipment like smart glasses or mobile apps. They can more easily identify possible problems and solutions since they have a greater understanding of the intricate nature of the systems they will be working on thanks to this real-time visualisation.

Hands-on simulation is another crucial method. Through the use of AR, personnel may practise maintenance and troubleshooting techniques in a secure setting. They may gain expertise and confidence in handling problems they could face in the real world while working on water management systems thanks to this hands-on training.

AR also enables remote cooperation and professional advice. During training exercises or actual fieldwork, technicians may use AR devices to connect with mentors or distant specialists who can provide real-time support and direction. This method boosts learning while also enhancing performance in the workplace.

In summary, incorporating AR technology into technician training for water management provides strategies including real-time visualisation, practical simulation, and remote collaboration, all of which contribute to more effective and efficient skill development in this crucial subject.

6.2. Interactive Learning Environment

In the context of AR for Training in Water Management, interactive learning environments are essential for improving technicians' expertise. These settings use the IoT and AR to create immersive teaching environments. The use of IoT sensors and data into AR simulations is a fundamental strategy. The ability to engage with real-time data from water management systems gives technicians a realistic and dynamic environment in which to practise troubleshooting and maintenance techniques.

AR overlays and annotations are yet another crucial method. Using this method, technicians may obtain important data and instructions while they work. Technicians can easily understand complicated water management systems and processes thanks to AR overlays that may provide step-by-step instructions, highlight important components, and give real-time feedback.

Gamification is also utilised to make technical training for technicians more interesting and enjoyable. This technique transforms training materials into entertaining games that promote participation and memory retention. Technicians may apply their skills to solve real-world water management issues in gamified AR settings.

IoT connection, AR overlays, and gamification are hence the basic strategies employed in Interactive Learning Environments in AR for Training in Water Management. These techniques improve technicians' ability to maintain and diagnose water management systems while facilitating their speedy acquisition of the necessary abilities and information.

6.3. Simulated Training Scenarios

One of the main applications of AR in water management is the development and usage of virtual training environments. Professionals may acquire vital skills and knowledge in these flexible settings, which provide a realistic and regulated environment. Students may visually engage with a variety of water management tools and systems using AR technology, imitating the challenges they could encounter in real-world situations. Through this immersion, they get a deeper grasp of intricate processes and are encouraged to engage in practical learning.

A variety of water management activities, such as valve maintenance, leak detection, or pump troubleshooting, may be included in simulated training situations in AR. With the use of digital overlays, technicians may get real-time feedback on their activities, visual signals, and step-by-step instructions. This participatory method reduces the hazards connected with on-the-job training while simultaneously quickening the learning curve.

Additionally, AR makes it possible to design adaptive scenarios that modify difficulty levels in accordance with student progress, resulting in a customised learning environment. It enables technicians to master water management methods, eventually resulting in more effective and efficient field maintenance procedures. The AR and IoT integration strengthen the skills of water management specialists thanks to these virtual training situations.

7. PRACTICAL IMPLEMENTATION CONSIDERATIONS

7.1. Technical Infrastructure Requirements

The effective use of this novel strategy depends on the technical infrastructure requirements for integrating AR and the IoT in water management. A strong and dependable IoT network must be created first. All of the sensors, devices, and AR-enabled equipment installed throughout the water management system should be seamlessly connected via this network, which ought to cover a significant region. Real-time data transfer and AR interactions need high bandwidth and low latency connections.

Additionally, the AR gadgets themselves are essential. AR headsets or smart glasses with sophisticated features like spatial mapping, gesture recognition, and accurate tracking are required for technicians. These gadgets have to interact with the IoT network and have excellent data transmission and reception capabilities. Additionally, interpreting the enormous quantity of data produced by IoT sensors and quickly providing AR content need a strong central computer infrastructure. In order to guarantee compatibility between various hardware and software components, the integration of AR and IoT requires well defined protocols and standards. To safeguard sensitive information and stop unauthorised access to the IoT and AR systems, security measures, such as encryption and authentication processes, are essential.

In conclusion, building the proper technological foundation for integrating AR with IoT in water management entails having a strong IoT network, top-notch AR devices, a strong computing infrastructure, adhering to industry standards, and having strong security measures. These elements are necessary for water management technicians to engage in effective maintenance, troubleshooting, and training procedures.

7.2 Integration Challenges and Solutions

1. Data Synchronization:
   Real-time synchronisation between AR interfaces and IoT sensor data might be difficult to achieve. Implement solid data pipelines that effectively transport sensor data to AR devices while maintaining low latency to solve this.
2. Device Compatibility:
   There may be concerns with interoperability between different AR devices and IoT systems. Standardise APIs and communication protocols to allow easy hardware platform interaction.
3. Security and Privacy:
   Data on water management must be protected at all costs. To protect IoT and AR systems and avoid unauthorised access or data breaches, use robust encryption and authentication techniques.
4. Network Reliability:
   It might be difficult to guarantee a dependable network connection at outlying water management facilities. Use failover and redundancy measures to keep connection up even under difficult circumstances.
5. Scalability:
   Scalability is crucial as water management systems grow. Consider scalability when designing IoT and AR systems so that additional sensors and AR devices may be added quickly.

6. User Training:
   For technicians to utilise AR interfaces successfully, training may be necessary. Create thorough training programmes to introduce them to AR technologies and their use with IoT data.
7. Maintenance and Updates:
   System performance depends on regular upkeep and upgrades. Automated software update procedures should be put in place to maintain both IoT and AR components.

Water management experts may fully use AR and IoT for maintenance, troubleshooting, and training reasons by overcoming these integration problems, resulting in efficient and effective operations.

7.3. Data Security and Privacy Concerns

When combining AR with IoT in water management, data security and privacy considerations are of utmost importance. It is crucial to use encryption methods to safeguard communication between IoT devices and AR apps in order to allay these worries. Access control procedures should also be put in place to prevent unauthorised employees from accessing sensitive data. To reduce vulnerabilities, regular software upgrades and patch management are essential. Techniques for data reduction and anonymization may also be used to lower the risk of privacy violations by making sure that only necessary data is gathered and disseminated. To maintain water management systems' confidentiality and integrity, compliance with relevant data protection standards is essential.

8. FUTURE TRENDS AND INNOVATIONS

The following are the next developments and trends that are directly connected to the use of AR and the IoT for upkeep and instruction in water management:

1. **Spatial Mapping and Object Recognition:** Future water management AR systems could make use of sophisticated spatial mapping and object identification methods. For more exact maintenance and troubleshooting, these technologies will allow AR devices to properly detect and interact with diverse water infrastructure components, such as pipelines, valves, and sensors.
2. **Real-time Data Integration:** The development of IoT sensors will lead to more real-time data streams from water systems. The ability of AR apps to include this data into the user's augmented view will continue to grow. When doing maintenance, technicians may then get instantaneous information on the state of the system, flow rates, and water quality.
3. **Machine Learning Integration:** Predictive maintenance is possible when machine learning algorithms are combined with AR and IoT. AR systems may provide maintenance professionals insights into possible problems before they become urgent, enabling proactive maintenance, by analysing previous data and present sensor inputs.
4. **Enhanced Remote Assistance:** Through AR glasses or headsets, specialists or managers may be able to remotely support on-site personnel in the future. Using real-time video feeds, annotations, and 3D modelling, technicians will be guided in this way, increasing productivity and decreasing downtime.

5. **Holographic Interfaces:** Technologists will be able to engage with intricate 3D models of water systems thanks to the development of increasingly sophisticated holographic interfaces. These interfaces will simplify the training process and improve component visualisation and comprehension.
6. **Multi-Modal Feedback:** Future AR systems could include voice control and multi-modal feedback techniques like haptic feedback. By enabling technicians to offer instructions in plain language and get haptic replies, this will improve the user experience and make system interactions more natural.
7. **Cloud-Based Collaboration:** Cloud-based solutions will make it easier for maintenance teams and specialists to collaborate. Access to shared information, documents, and real-time communication channels using AR devices will improve collaboration and knowledge sharing.
8. **Integration with Wearable IoT:** AR glasses or headsets might easily link with IoT devices as they become more wearable. When doing water system maintenance, for instance, wearable IoT sensors might broadcast vital signs or environmental data to AR devices, assuring the safety and wellbeing of the professionals.
9. **Environmental Sensing:** Environmental sensing capabilities in AR systems for water management may enable specialists to evaluate the effects of water infrastructure on the local ecology. For preserving ecological balance and abiding with environmental standards, this may be essential.
10. **Cybersecurity and Privacy:** To safeguard sensitive data and stop unauthorised access to crucial infrastructure, there will be a rising need for effective cybersecurity measures as AR and IoT are increasingly integrated into water management. This is an area where innovation will be crucial.

These next developments and trends show how AR and IoT technologies are still developing in the area of water management. They have the potential to enhance the procedures for maintenance, troubleshooting, and training, eventually resulting in more effective and sustainable management of water resources.

9. CONCLUSION

In conclusion, AR and the IoT integration have become a paradigm shift in the field of water management. We have examined the subtleties of how AR and IoT technologies work together to improve maintenance, troubleshooting, and training procedures in this important subject throughout this chapter.

Technicians may access a multitude of information at their fingertips using AR-enhanced maintenance, visualising crucial data points and historical records in real-time. This enables them to quickly make educated choices, leading to more effective maintenance practises that save costs and minimise downtime. It is impossible to exaggerate the value of AR in troubleshooting. AR overlays may help technicians visualise complicated water systems, spot abnormalities, and track the source of issues. This facilitates a greater knowledge of the water management infrastructure in addition to speeding up problem solutions.

Beyond maintenance, AR and IoT integration improves the effectiveness of water management training. With the use of AR simulations, students may interact with real-world environments and tools, promoting hands-on learning without the need for tangible resources. This expedites skill learning and guarantees that technicians are well-equipped to handle problems in the actual world.

In summary, the merging of AR and IoT technologies in water management provides a bright future in which procedures for maintenance, troubleshooting, and training are improved to new levels of efficacy and efficiency. The potential for additional innovation in water management is limitless as these technologies advance and become more widely available. Professionals in water management must adopt these innovations if they want to ensure a more sustainable and robust future for our essential water supplies.

REFERENCES

Ali, M. H., Al-Azzawi, W. K., Jaber, M., Abd, S. K., Alkhayyat, A., & Rasool, Z. I. (2022). Improving coal mine safety with IoT based Dynamic Sensor Information Control System. *Physics and Chemistry of the Earth Parts A/B/C*, *128*, 103225. doi:10.1016/j.pce.2022.103225

Bakhtiari, V., Piadeh, F., Behzadian, K., & Kapelan, Z. (2023). A critical review for the application of cutting-edge digital visualisation technologies for effective urban flood risk management. *Sustainable Cities and Society*, *99*, 104958. doi:10.1016/j.scs.2023.104958

Chen, K., Yang, J., Cheng, J. C. P., Chen, W., & Li, C. T. (2020). Transfer learning enhanced AR spatial registration for facility maintenance management. *Automation in Construction*, *113*, 103135. doi:10.1016/j.autcon.2020.103135

C.N., P., Vimala, H. S., & J., S. (2023). A systematic survey on content caching in ICN and ICN-IoT: Challenges, approaches and strategies. *Computer Networks*, *233*, 109896. doi:10.1016/j.comnet.2023.109896

Dzulkifli, N., Sarbini, N. N., Ibrahim, I. S., Abidin, N. I., Yahaya, F. M., & Nik Azizan, N. Z. (2021). Review on maintenance issues toward building maintenance management best practices. *Journal of Building Engineering*, *44*, 102985. doi:10.1016/j.jobe.2021.102985

Gholizadeh, A., Khiadani, M., Foroughi, M., Siuki, H. A., & Mehrfar, H. (2023). Wastewater treatment plants: The missing link in global One-Health surveillance and management of antibiotic resistance. *Journal of Infection and Public Health.* Advance online publication. doi:10.1016/j.jiph.2023.09.017 PMID:37865529

Hadipour, M., Derakhshandeh, J. F., & Shiran, M. A. (2020). An experimental setup of multi-intelligent control system (MICS) of water management using the IoT. *ISA Transactions*, *96*, 309–326. doi:10.1016/j.isatra.2019.06.026 PMID:31285060

Heidary Dahooie, J., Mohammadian, A., Qorbani, A. R., & Daim, T. (2023). A portfolio selection of IoT (IoTs) applications for the sustainable urban transportation: A novel hybrid multi criteria decision making approach. *Technology in Society*, *75*, 102366. doi:10.1016/j.techsoc.2023.102366

Huang, H., Zhao, L., & Wu, Y. (2023). An IoT and machine learning enhanced framework for real-time digital human modeling and motion simulation. *Computer Communications*, *212*, 78–89. Advance online publication. doi:10.1016/j.comcom.2023.09.024

Jahid, A., Alsharif, M. H., & Hall, T. J. (2023). The convergence of blockchain, IoT and 6G: Potential, opportunities, challenges and research roadmap. *Journal of Network and Computer Applications*, *217*, 103677. doi:10.1016/j.jnca.2023.103677

Kumar, D., Singh, R. K., Mishra, R., & Daim, T. U. (2023). Roadmap for integrating blockchain with IoT for sustainable and secured operations in logistics and supply chains: Decision making framework with case illustration. *Technological Forecasting and Social Change*, *196*, 122837. doi:10.1016/j.techfore.2023.122837

Li, W., Ye, Z., Wang, Y., Yang, H., Yang, S., Gong, Z., & Wang, L. (2023). Development of a distributed MR-IoT method for operations and maintenance of underground pipeline network. *Tunnelling and Underground Space Technology*, *133*, 104935. doi:10.1016/j.tust.2022.104935

Mahajan, Y., Shandilya, D., Batta, P., & Sharma, M. (2023). 3D Object 360-Degree Motion Detection Using Ultra-Frequency PIR Sensor. *2023 IEEE World Conference on Applied Intelligence and Computing (AIC)*, 614–619. 10.1109/AIC57670.2023.10263926

Malkawi, A., Ervin, S., Han, X., Chen, E. X., Lim, S., Ampanavos, S., & Howard, P. (2023). Design and applications of an IoT architecture for data-driven smart building operations and experimentation. *Energy and Building*, *295*, 113291. doi:10.1016/j.enbuild.2023.113291

Morchid, A., El Alami, R., Raezah, A. A., & Sabbar, Y. (2023). Applications of IoT and sensors technology to increase food security and agricultural Sustainability: Benefits and challenges. *Ain Shams Engineering Journal*, *102509*, 102509. Advance online publication. doi:10.1016/j.asej.2023.102509

Patel, S., Sutaria, S., Daga, R., Shah, M., & Prajapati, M. (2023). A systematic study on complementary metal-oxide semiconductor technology (CMOS) and IoT for radioactive leakage detection in nuclear plant. *Nuclear Analysis*, *2*(3), 100080. doi:10.1016/j.nucana.2023.100080

Proud, C., Fukai, S., Dunn, B., Dunn, T., & Mitchell, J. (2023). Effect of nitrogen management on grain yield of rice grown in a high-yielding environment under flooded and non-flooded conditions. *Crop and Environment*, *2*(1), 37–45. doi:10.1016/j.crope.2023.02.004

Rahman, A., Xi, M., Dabrowski, J. J., McCulloch, J., Arnold, S., Rana, M., George, A., & Adcock, M. (2021). An integrated framework of sensing, machine learning, and AR for aquaculture prawn farm management. *Aquacultural Engineering*, *95*, 102192. doi:10.1016/j.aquaeng.2021.102192

Revolti, A., Dallasega, P., Schulze, F., & Walder, A. (2023). AR to support the maintenance of urban-line infrastructures: A case study. *Procedia Computer Science*, *217*, 746–755. doi:10.1016/j.procs.2022.12.271

Sanzana, M. R., Maul, T., Wong, J. Y., Abdulrazic, M. O. M., & Yip, C.-C. (2022). Application of deep learning in facility management and maintenance for heating, ventilation, and air conditioning. *Automation in Construction*, *141*, 104445. doi:10.1016/j.autcon.2022.104445

Seeliger, A., Cheng, L., & Netland, T. (2023). AR for industrial quality inspection: An experiment assessing task performance and human factors. *Computers in Industry*, *151*, 103985. doi:10.1016/j.compind.2023.103985

Shaharuddin, S., Abdul Maulud, K. N., Syed Abdul Rahman, S. A. F., Che Ani, A. I., & Pradhan, B. (2023). The role of IoT sensor in smart building context for indoor fire hazard scenario: A systematic review of interdisciplinary articles. *Internet of Things : Engineering Cyber Physical Human Systems*, *22*, 100803. doi:10.1016/j.iot.2023.100803

Sugiura, T., Yamamura, K., Watanabe, Y., Yamakiri, S., & Nakano, N. (2023). Circuits and devices for standalone large-scale integration (LSI) chips and IoT applications: A review. *Chip (Würzburg)*, *2*(3), 100048. doi:10.1016/j.chip.2023.100048

Sunku Mohan, V., Sankaran, S., Nanda, P., & Achuthan, K. (2023). Enabling secure lightweight mobile Narrowband IoT (NB-IoT) applications using blockchain. *Journal of Network and Computer Applications*, *219*, 103723. doi:10.1016/j.jnca.2023.103723

Truong, Q. C., El Soueidy, C.-P., Hawchar, L., Li, Y., & Bastidas-Arteaga, E. (2023). Modelling two-dimensional chloride diffusion in repaired RC structures for sustainable maintenance management. *Structures*, *51*, 895–909. doi:10.1016/j.istruc.2023.03.088

Yang, C.-H., Lee, K.-C., & Li, S.-E. (2020). A mixed activity-based costing and resource constraint optimal decision model for IoT-oriented intelligent building management system portfolios. *Sustainable Cities and Society*, *60*, 102142. doi:10.1016/j.scs.2020.102142

Compilation of References

Aani, S. A., Bonny, T., Hasan, S. W., & Hilal, N. (2019). Can machine language and artificial intelligence revolutionize process automation for water treatment and desalination? *Desalination*, *458*, 84–96. doi:10.1016/j.desal.2019.02.005

Abbaszadeh, P., Alizadeh, F., & Arabi, M. (2020). Integration of artificial intelligence and physically-based hydrological model for drought prediction under different climatic regions. *The Science of the Total Environment*, *744*, 140664.

Abusukhon, A., & Altamimi, F. (2021). Water Preservation Using IoT: A proposed IoT System for Detecting Water Pipeline Leakage. In *2021 International Conference on Information Technology (ICIT)* (pp. 115–119). 10.1109/ICIT52682.2021.9491667

Adamowski, J. F. (2008). Peak Daily Water Demand Forecast Modeling Using Artificial Neural Networks. In Journal of Water Resources Planning and Management (Vol. 134, Issue 2, pp. 119–128). American Society of Civil Engineers (ASCE). doi:10.1061/(ASCE)0733-9496(2008)134:2(119)

Adams, S., & Acheampong, A. O. (2019). Reducing carbon emissions: The role of renewable energy and democracy. *Journal of Cleaner Production*, *240*, 118245. doi:10.1016/j.jclepro.2019.118245

Adnan, R. M., Mostafa, R. R., Elbeltagi, A., Yaseen, Z. M., Shahid, S., & Kisi, O. (2021). Development of new machine learning model for streamflow prediction: case studies in Pakistan. In Stochastic Environmental Research and Risk Assessment (Vol. 36, Issue 4, pp. 999–1033). Springer Science and Business Media LLC. doi:10.100700477-021-02111-z

Adnan, R. M. R., Mostafa, R., Kisi, O., Yaseen, Z. M., Shahid, S., & Zounemat-Kermani, M. (2021). Improving streamflow prediction using a new hybrid ELM model combined with hybrid particle swarm optimization and grey wolf optimization. In *Knowledge-Based Systems* (Vol. 230, p. 107379). Elsevier BV. doi:10.1016/j.knosys.2021.107379

Adu-Manu, K. S., Tapparello, C., Heinzelman, W., Katsriku, F. A., & Abdulai, J. (2017). Water quality monitoring using wireless sensor networks. *ACM Transactions on Sensor Networks*, *13*(1), 1–41. doi:10.1145/3005719

Afrifa, S., Zhang, T., Appiahene, P., & Varadarajan, V. (2022). Mathematical and machine learning models for groundwater level changes: A systematic review and bibliographic analysis. *Future Internet*, *14*(9), 259. doi:10.3390/fi14090259

Aguilera, H., Guardiola-Albert, C., Naranjo-Fernández, N., & Kohfahl, C. (2019). Towards flexible groundwater-level prediction for adaptive water management: using Facebook's Prophet forecasting approach. In Hydrological Sciences Journal (Vol. 64, Issue 12, pp. 1504–1518). Informa UK Limited. doi:10.1080/02626667.2019.1651933

Ahanger, T. A., & Aljumah, A. (2019). Internet of Things: A Comprehensive Study of Security Issues and Defense Mechanisms. In IEEE Access (Vol. 7, pp. 11020–11028). Institute of Electrical and Electronics Engineers (IEEE). doi:10.1109/ACCESS.2018.2876939

Ahansal, Y., Bouziani, M., Yaagoubi, R., Sebari, I., Sebari, K., & Kenny, L. (2022). Towards Smart Irrigation: A Literature Review on the Use of Geospatial Technologies and Machine Learning in the Management of Water Resources in Arboriculture. *Agronomy (Basel)*, *12*(2), 297. doi:10.3390/agronomy12020297

Ahmed, S., Rahman, M. J., & Razzak, M. A. (2023). Design and Development of an IoT-Based LPG Gas Leakage Detector for Households and Industries. *IEEE World AI IoT Congress*, 762–767.

Ahmed, A. N., Othman, F., Afan, H. A., Ibrahim, R. K., Fai, C. M., Hossain, S., Ehteram, M., & El-Shafie, A. (2019). Machine learning methods for better water quality prediction. *Journal of Hydrology (Amsterdam)*, *578*, 124084. doi:10.1016/j.jhydrol.2019.124084

Ahmed, U., Mumtaz, R., Anwar, H., Mumtaz, S., & Qamar, A. M. (2019). Water quality monitoring: From conventional to emerging technologies. *Water Science and Technology: Water Supply*, *20*(1), 28–45. doi:10.2166/ws.2019.144

Aivazidou, E., Banias, G., Lampridi, M., Vasileiadis, G., Anagnostis, A., Papageorgiou, E., & Bochtis, D. (2021). Smart Technologies for Sustainable Water Management: An Urban Analysis. *Sustainability (Basel)*, *13*(24), 13940. doi:10.3390u132413940

Akhtar, N., Syakir Ishak, M. I., Bhawani, S. A., & Umar, K. (2021). Various Natural and Anthropogenic Factors Responsible for Water Quality Degradation: A Review. *Water (Basel)*, *13*(19), 19. Advance online publication. doi:10.3390/w13192660

Akhtar, S., Fatima, R., Soomro, Z. A., Hussain, M., Ahmad, S. R., & Ramzan, H. S. (2019). Bacteriological quality assessment of water supply schemes (WSS) of Mianwali, Punjab, Pakistan. *Environmental Earth Sciences*, *78*(15), 458. Advance online publication. doi:10.100712665-019-8455-1

Aldhyani, T. H. H., Al-Yaari, M., Alkahtani, H., & Maashi, M. (2020). Water quality prediction using artificial intelligence algorithms. *Applied Bionics and Biomechanics*, *2020*, 1–12. doi:10.1155/2020/6659314 PMID:33456498

Al-Fuqaha, A., Guizani, M., Mohammadi, M., Aledhari, M., & Ayyash, M. (2015). Internet of Things: A survey on enabling technologies, protocols, and applications. *IEEE Communications Surveys and Tutorials*, *17*(4), 2347–2376. doi:10.1109/COMST.2015.2444095

Ali, A. E., Salem, W. M., Younes, S. M., & Kaid, M. (2020). Modeling climatic effect on physiochemical parameters and microorganisms of Stabilization Pond Performance. *Heliyon*, *6*(5), e04005. doi:10.1016/j.heliyon.2020.e04005 PMID:32478191

Ali, M. H., Al-Azzawi, W. K., Jaber, M., Abd, S. K., Alkhayyat, A., & Rasool, Z. I. (2022). Improving coal mine safety with IoT based Dynamic Sensor Information Control System. *Physics and Chemistry of the Earth Parts A/B/C*, *128*, 103225. doi:10.1016/j.pce.2022.103225

Ali, M. H., Popescu, I., Jonoski, A., & Solomatine, D. P. (2023). Remote Sensed and/or Global Datasets for Distributed Hydrological Modelling: A Review. *Remote Sensing (Basel)*, *15*(6), 1–43. doi:10.3390/rs15061642

Ali, M., & Qamar, A. M. (2013). Data analysis, quality indexing and prediction of water quality for the management of rawal watershed in Pakistan. In *Eighth International Conference on Digital Information Management (ICDIM 2013)* (pp. 108-113). IEEE. 10.1109/ICDIM.2013.6694009

Alloghani, M., Al-Jumeily, D., Mustafina, J., Hussain, A., & Aljaaf, A. J. (2019). A Systematic Review on Supervised and Unsupervised Machine Learning Algorithms for Data Science. In *Unsupervised and Semi-Supervised Learning* (pp. 3–21). Springer International Publishing. doi:10.1007/978-3-030-22475-2_1

AlMetwally, S. H., Hassan, M. K., & Mourad, M. (2020). Real time Internet of Things (IoT) based water quality management system. *Procedia CIRP*, *91*, 478–485. doi:10.1016/j.procir.2020.03.107

Almikaeel, W., Čubanová, L., & Šoltész, A. (2022). Hydrological Drought Forecasting Using Machine Learning—Gidra River Case Study. In Water (Vol. 14, Issue 3, p. 387). MDPI AG. doi:10.3390/w14030387

Alshammari, H. H. (2023). The IoT healthcare monitoring system based on MQTT protocol. *Alexandria Engineering Journal*, *69*, 275–287. doi:10.1016/j.aej.2023.01.065

Another Data Science Student's Blog—The 1cycle Policy. (n.d.). Available online: sgugger.github.io

Antunes, A., Andrade-Campos, A., Sardinha-Lourenço, A., & Oliveira, M. S. (2018). Short-term water demand forecasting using machine learning techniques. In Journal of Hydroinformatics (Vol. 20, Issue 6, pp. 1343–1366). IWA Publishing. doi:10.2166/hydro.2018.163

Antwi, P., Li, J., Meng, J., Deng, K., Quashie, F. K., Li, J., & Boadi, P. O. (2018). Feedforward neural network model estimating pollutant removal process within mesophilic upflow anaerobic sludge blanket bioreactor treating industrial starch processing wastewater. *Bioresource Technology*, *257*, 102–112. doi:10.1016/j.biortech.2018.02.071 PMID:29486407

Apa, A. D., Boenish, R., & Kleisner, K. (2023). *Effects of climate change and variability on large pelagic fish in the Northwest Atlantic Ocean : Implications for improving climate resilient management for pelagic longline fi sheries.* doi:10.3389/fmars.2023.1206911

Apaydin, H., Taghi Sattari, M., Falsafian, K., & Prasad, R. (2021). Artificial intelligence modelling integrated with Singular Spectral analysis and Seasonal-Trend decomposition using Loess approaches for streamflow predictions. In *Journal of Hydrology* (Vol. 600, p. 126506). Elsevier BV. doi:10.1016/j.jhydrol.2021.126506

Aprisal, Istijono, B., Ophiyandri, T., & Nurhamidah. (2019). A study of the quality of soil infiltration at the downstream of Kuranji River, Padang City. *Int. J. Geomate*, *16*, 16–20.

Arsene, D., Predescu, A., Pahonțu, B., Chiru, C. G., Apostol, E.-S., & Truică, C.-O. (2022). Advanced Strategies for Monitoring Water Consumption Patterns in Households Based on IoT and Machine Learning. *Water (Basel)*, *14*(14), 2187. doi:10.3390/w14142187

Aryastana, P., & Tasuku, T. (2010). Characteristic of Rainfall Pattern Before Flood Occur in Indonesia Based on Rainfall Data. *GSMaP, 7*, 1–4.

Ashiyani, N. (2015). *Adaptive Neuro Fuzzy Inference System (ANFIS) for prediction of groundwater quality index in Matar Taluka and Nadiad Taluka*. https://api.semanticscholar.org/CorpusID:40125977

Assessment, M. E. (2005). *Ecosystems and human Well-Being: Multiscale Assessments: Findings of the Sub-Global Assessments Working Group*. Island Press.

Atlam, H., Walters, R., & Wills, G. (2018). Fog Computing and the Internet of Things: A Review. In Big Data and Cognitive Computing (Vol. 2, Issue 2, p. 10). MDPI AG. doi:10.3390/bdcc2020010

Atzori, L., Iera, A., & Morabito, G. (2010). The Internet of Things: A survey. In Computer Networks (Vol. 54, Issue 15, pp. 2787–2805). Elsevier BV. doi:10.1016/j.comnet.2010.05.010

AWWA. (2021). *Buried No Longer: Confronting America's Water Infrastructure Challenge*. American Water Works Association.

Ay, M., & Özyıldırım, S. (2018). Artificial intelligence (AI) studies in water resources. *Natural and Engineering Sciences*, *3*(2), 187–195. doi:10.28978/nesciences.424674

Baird, R. B., Rice, E. W., & Eaton, A. (2017). *Standard Methods for the Examination of Water and Wastewater.*http://dspace.uniten.edu.my/handle/123456789/14241

Bakhtiari, V., Piadeh, F., Behzadian, K., & Kapelan, Z. (2023). A critical review for the application of cutting-edge digital visualisation technologies for effective urban flood risk management. *Sustainable Cities and Society*, *99*, 104958. doi:10.1016/j.scs.2023.104958

Balabin, R. M., & Lomakina, E. I. (2011). Support vector machine regression (SVR/LS-SVM)—An alternative to neural networks (ANN) for analytical chemistry? Comparison of nonlinear methods on near infrared (NIR) spectroscopy data. *Analyst*, *136*(8), 1703. doi:10.1039/c0an00387e PMID:21350755

Baldisserotto, B. (2011). Water pH and hardness affect growth of freshwater teleosts. *Brazilian Journal of Animal Science*, *40*(1), 138–144.

Ball, J. E., Anderson, D. T., & Chan, C. S. (2017). A comprehensive survey of deep learning in remote sensing: Theories, tools, and challenges for the community. *Journal of Applied Remote Sensing*, *11*(4), 042609–042609. doi:10.1117/1.JRS.11.042609

Banerjee, A., & Banerjee, C. (2023). A hybrid cellular automata-based model for leakage detection smart drip irrigation water pipeline structure using IoT sensors. *Innovations in Systems and Software Engineering*, *19*(1), 23–32. doi:10.100711334-022-00503-0

Barzegar, R., Aalami, M. T., & Adamowski, J. (2020). Short-term water quality variable prediction using a hybrid CNN–LSTM deep learning model. *Stochastic Environmental Research and Risk Assessment*, *34*(2), 415–433. doi:10.100700477-020-01776-2

BassineF. Z.EpuleT. E.KechchourA.ChehbouniA. (2023). *Recent applications of ML, remote sensing, and iot approaches in yield prediction: a critical review.* https://arxiv.org/abs/2306.04566

Baswaraju, S., Maheswari, V. U., Chennam, K. K., Thirumalraj, A., Kantipudi, M. P., & Aluvalu, R. (2023). Future Food Production Prediction Using AROA Based Hybrid Deep Learning Model in Agri-Sector. *Human-Centric Intelligent Systems*, 1-16.

Bata, M., Carriveau, R., & Ting, D. S.-K. (2020). Short-term water demand forecasting using hybrid supervised and unsupervised machine learning model. In Smart Water (Vol. 5, Issue 1). Springer Science and Business Media LLC. doi:10.118640713-020-00020-y

Bedi, G., Venayagamoorthy, G. K., Singh, R., Brooks, R. R., & Wang, K. C. (2018). Review of IoT (IoT) in electric power and energy systems. *IEEE IoT JWEnal, 5*(2), 847–870.

Bedi, S., Samal, A., Ray, C., & Snow, D. (2020). Comparative evaluation of machine learning models for groundwater quality assessment. *Environmental Monitoring and Assessment*, *192*(12), 776. Advance online publication. doi:10.100710661-020-08695-3 PMID:33219864

Begum, Y., Tanguturi, R. C., & Ahmed, B. (2022). Implementation of an Early Warning System using Internet of Things and Rainfall Threshold. In *2022 6th International Conference on Electronics, Communication and Aerospace Technology* (pp. 491–496). Academic Press.

Behera, B. B., Mohanty, R. K., & Pattanayak, B. K. (2022). Attack Detection and Mitigation in Industrial IoT: An Optimized Ensemble Approach. *Specialusis Ugdymas*, *1*(43), 879–905.

Behmel, S., Damour, M., Ludwig, R., & Rodriguez, M. J. (2016). Water quality monitoring strategies — A review and future perspectives. *The Science of the Total Environment*, *571*, 1312–1329. doi:10.1016/j.scitotenv.2016.06.235 PMID:27396312

Ben-Bouallegue, Z., Clare, M. C. A., Magnusson, L., Gascon, E., Maier-Gerber, M., Janousek, M., Rodwell, M., Pinault, F., Dramsch, J. S., Lang, S. T. K., Raoult, B., Rabier, F., Chevallier, M., Sandu, I., Dueben, P., Chantry, M., & Pappenberger, F. (2023). *The rise of data-driven weather forecasting* (Version 1). arXiv. doi:10.48550/ARXIV.2307.10128

Bhandari, N. S., & Nayal, K. (2008). Correlation Study on Physico-Chemical Parameters and Quality Assessment of Kosi River Water, Uttarakhand. *E-Journal of Chemistry*, *5*(2), 342–346. doi:10.1155/2008/140986

Bharadwaj, K. V., Dusarlapudi, K., Sudhakar, K., Polasi, V. S. K., Tiruvuri, C., & Raju, K. N. (2022). Novel IoT & Machine Learning Base Water Flow Monitoring System. In *2022 IEEE 2nd Mysore Sub Section International Conference (MysuruCon)* (pp. 1–5). IEEE.

Bhardwaj, A., Dagar, V., Khan, M. O., Aggarwal, A., Alvarado, R., Kumar, M., Irfan, M., & Proshad, R. (2022). Smart IoT and Machine Learning-based Framework for Water Quality Assessment and Device Component Monitoring. *Environmental Science and Pollution Research International*, *29*(30), 46018–46036. Advance online publication. doi:10.100711356-022-19014-3 PMID:35165843

Bhargava, A. (2016). Physico-Chemical Waste Water Treatment Technologies: An Overview. *International Journal of Scientific Research and Education*. https://doi.org/ doi:10.18535/ijsre/v4i05.05

Biamonte, J., Wittek, P., Pancotti, N., Rebentrost, P., Wiebe, N., & Lloyd, S. (2017). Quantum machine learning. *Nature*, *549*(7671), 195–202. doi:10.1038/nature23474 PMID:28905917

Bing, X., Yu, J., & Chen, J. (2017). A real-time sensor data-driven decision support system for water quality management in an industrial park. *Environmental Monitoring and Assessment*, *189*(12), 620. PMID:29124450

Bolick, M. M., Post, C. J., Naser, M. Z., Forghanparast, F., & Mikhailova, E. A. (2023). Evaluating Urban Stream Flooding with ML, LiDAR, and 3D Modeling. *Water (Basel)*, *15*(14), 1–25. doi:10.3390/w15142581

Bongards, M., Gaida, D., Trauer, O., & Wolf, C. (2014). Intelligent automation and IT for the optimization of renewable energy and wastewater treatment processes. *Energy, Sustainability and Society*, *4*(1), 19. Advance online publication. doi:10.118613705-014-0019-3

Bougadis, J., Adamowski, K., & Diduch, R. (2005). Short-term municipal water demand forecasting. In Hydrological Processes (Vol. 19, Issue 1, pp. 137–148). Wiley. doi:10.1002/hyp.5763

Bozorg-Haddad, O., Delpasand, M., & Loáiciga, H. A. (2021). Water quality, hygiene, and health. In Elsevier eBooks (pp. 217–257). doi:10.1016/B978-0-323-90567-1.00008-5

Brewster, C., Roussaki, I., Kalatzis, N., Doolin, K., & Ellis, K. (2017). IoT in Agriculture: Designing a Europe-Wide Large-Scale Pilot. In IEEE Communications Magazine (Vol. 55, Issue 9, pp. 26–33). Institute of Electrical and Electronics Engineers (IEEE). doi:10.1109/MCOM.2017.1600528

Broad, K., Pfaff, A., Taddei, R., Sankarasubramanian, A., Lall, U., & de Assis de Souza Filho, F. (2007). Climate, stream flow prediction and water management in northeast Brazil: societal trends and forecast value. In Climatic Change (Vol. 84, Issue 2, pp. 217–239). Springer Science and Business Media LLC. doi:10.100710584-007-9257-0

Burkhard, R., Deletić, A., & Craig, T. (2000). Techniques for water and wastewater management: A review of techniques and their integration in planning. *Urban Water*, *2*(3), 197–221. doi:10.1016/S1462-0758(00)00056-X

Butler, M. J., Yellen, B. C., Oyewumi, O., Ouimet, W., & Richardson, J. B. (2023). Accumulation and transport of nutrient and pollutant elements in riparian soils, sediments, and river waters across the Thames River Watershed, Connecticut, USA. *The Science of the Total Environment*, *899*(March), 165630. doi:10.1016/j.scitotenv.2023.165630 PMID:37467973

C.N., P., Vimala, H. S., & J., S. (2023). A systematic survey on content caching in ICN and ICN-IoT: Challenges, approaches and strategies. *Computer Networks*, *233*, 109896. doi:10.1016/j.comnet.2023.109896

Cabral, J. P., & Marques, C. J. (2006). Fecal Coliform Bacteria in Febros River (Northwest Portugal): Temporal Variation, Correlation with Water Parameters, and Species Identification. *Environmental Monitoring and Assessment*, *118*(1–3), 21–36. doi:10.100710661-006-0771-8 PMID:16897531

Chandra Sekhar, K., Venkatesh, B., Reddy, K. S., Giridhar, G., Nithin, K., & Eshwar, K. (2023). IoT-based Realtime Water Quality Management System using Arduino Microcontroller. *Turkish Journal of Computer and Mathematics Education*, *14*(02), 783–792.

Chandrasekaran, S., Khaparde, V., & Seshagiri Rao, G. (2020). Adoption of AI in Indian agriculture: Opportunities and challenges. *AI & Society*, *35*(4), 855–866.

Chen, B., Mu, X., Chen, P., Wang, B., Choi, J., Park, H., Xu, S., Wu, Y., & Yang, H. (2021). Machine learning-based inversion of water quality parameters in typical reach of the urban river by UAV multispectral data. *Ecological Indicators*, *133*, 108434. doi:10.1016/j.ecolind.2021.108434

Chen, F.-L., Yang, B.-C., Peng, S.-Y., & Lin, T.-C. (2020). Applying a deployment strategy and data analysis model for water quality continuous monitoring and management. *International Journal of Distributed Sensor Networks*, *16*(6), 1550147720929825. doi:10.1177/1550147720929825

Cheng, B., Solmaz, G., Cirillo, F., Kovacs, E., Terasawa, K., & Kitazawa, A. (2018). Easy Programming of IoT Services Over Cloud and Edges for Smart Cities. In IEEE Internet of Things Journal (Vol. 5, Issue 2, pp. 696–707). Institute of Electrical and Electronics Engineers (IEEE). doi:10.1109/JIOT.2017.2747214

Chen, K., Yang, J., Cheng, J. C. P., Chen, W., & Li, C. T. (2020). Transfer learning enhanced AR spatial registration for facility maintenance management. *Automation in Construction*, *113*, 103135. doi:10.1016/j.autcon.2020.103135

Chen, W. L., Wang, J. J., Liu, Y. F., Jin, M. G., Liang, X., Wang, Z. M., & Ferre, T. P. A. (2021). Using bromide data tracer and HYDRUS-1D to estimate groundwater recharge and evapotranspiration under film-mulched drip irrigation in an arid Inland Basin, Northwest China. *Hydrological Processes*, *35*(7), 14290. doi:10.1002/hyp.14290

Chhaya, H. S., & Gupta, S. (1997). Performance modelling of asynchronous data transfer methods of IEEE 802.11 MAC protocol. *Wireless Networks*, *3*(3), 217–234. doi:10.1023/A:1019109301754

Chinnappan, C. V., John William, A. D., Nidamanuri, S. K. C., Jayalakshmi, S., Bogani, R., Thanapal, P., Syed, S., Venkateswarlu, B., & Syed Masood, J. A. I. (2023). IoT-Enabled Chlorine Level Assessment and Prediction in Water Monitoring System Using ML. *Electronics (Basel)*, *12*(6), 1458. Advance online publication. doi:10.3390/electronics12061458

Chiu, M.-C., Yan, W.-M., Bhat, S. A., & Huang, N.-F. (2022). Development of smart aquaculture farm management system using IoT and AI-based surrogate models. *Journal of Agriculture and Food Research*, *9*, 100357. doi:10.1016/j.jafr.2022.100357

Cho, K., & Kim, Y. (2022). Improving streamflow prediction in the WRF-Hydro model with LSTM networks. In *Journal of Hydrology* (Vol. 605, p. 127297). Elsevier BV. doi:10.1016/j.jhydrol.2021.127297

Choubin, B., Abdolshahnejad, M., Moradi, E., Querol, X., Mosavi, A., Shamshirband, S., & Ghamisi, P. (2020). Spatial hazard assessment of the PM10 using machine learning models in Barcelona, Spain. *The Science of the Total Environment, 701*, 134474. doi:10.1016/j.scitotenv.2019.134474 PMID:31704408

Choudhary, P., Botre, B. A., & Akbar, S. A. (2023). *Urban Water JWEnal.* doi:10.1080/1573062X.2022.2164732

Chou, J.-S., Ho, C.-C., & Hoang, H.-S. (2018). Determining quality of water in reservoir using machine learning. In *Ecological Informatics* (Vol. 44, pp. 57–75). Elsevier BV. doi:10.1016/j.ecoinf.2018.01.005

Chuang, W.-Y., Tsai, Y.-L., & Wang, L.-H. (2019). Leak detection in water distribution pipes based on CNN with mel frequency cepstral coefficients. *Proceedings of the 2019 3rd International Conference on Innovation in Artificial Intelligence*, 83–86. 10.1145/3319921.3319926

Chui, M., Löffler, M., & Roberts, R. (2015). The Internet of Things. *McKinsey Quarterly.*

Collado-Fernández, M., González-SanJosé, M. L., & Pino-Navarro, R. (2000). Evaluation of turbidity: Correlation between Kerstez turbidimeter and nephelometric turbidimeter. *Food Chemistry, 71*(4), 563–566. doi:10.1016/S0308-8146(00)00212-0

Contreras, J. D., Meza, R., Siebe, C., Rodríguez-Dozál, S., López-Vidal, Y., Castillo-Rojas, G., Amieva, R. I., Solano-Gálvez, S. G., Mazarí-Hiriart, M., Silva-Magaña, M. A., Vázquez-Salvador, N., Rosas-Pérez, I., & Romero, L. M. (2017). Health risks from exposure to untreated wastewater used for irrigation in the Mezquital Valley, Mexico: A 25-year update. *Water Research, 123*, 834–850. doi:10.1016/j.watres.2017.06.058 PMID:28755783

Correll, D. L. (1998). The role of phosphorus in the eutrophication of receiving waters: A review. *Journal of Environmental Quality, 27*(2), 261–266. doi:10.2134/jeq1998.00472425002700020004x

Cosgrove, W. J., & Loucks, D. P. (2015). Water management: Current and future challenges and research directions. *Water Resources Research, 51*(6), 51. doi:10.1002/2014WR016869

Dadhich, S., Pathak, V., Mittal, R., & Doshi, R. (2021). Machine learning for weather forecasting. In *Machine Learning for Sustainable Development* (pp. 161–174). De Gruyter. doi:10.1515/9783110702514-010

Das, A., Sarma, M., Sarma, K., & Mastorakis, N. (2018). Design of an IoT based real time environment monitoring system using legacy sensors. *MATEC Web of Conference, 210*, 1–4. doi:10.1051/matecconf/201823700001

Dastres, R., & Soori, M. (2021). Artificial neural network systems. *International Journal of Imaging and Robotics, 21*(2), 13-25. https://www.researchgate.net/publication/350486076_Artificial_Neural_Network_Systems

Datta, B. (2019). Artificial Intelligence for Aquifer Remediation and Management: A Review. *The Science of the Total Environment, 670*, 550–566.

Daucik, K. (2011). Revised Release on the Pressure along the Melting and Sublimation Curves of Ordinary Water Substance. The International Association for the Properties of Water and Steam. IAPWS R14-08.

Davies-Colley, R. J., & Smith, D. G. (2001). Turbidity suspended sediment, and water clarity: A review. *Journal of the American Water Resources Association, 37*(5), 1085–1101. doi:10.1111/j.1752-1688.2001.tb03624.x

Deb, K., Pratap, A., Agarwal, S., & Meyarivan, T. A. M. T. (2002). A fast and elitist multiobjective genetic algorithm: NSGA-II. *IEEE Transactions on Evolutionary Computation, 6*(2), 182–197. doi:10.1109/4235.996017

Dehghanisanij, H., Emami, H., Emami, S., & Rezaverdinejad, V. (2022). A hybrid machine learning approach for estimating the water-use efficiency and yield in agriculture. In Scientific Reports (Vol. 12, Issue 1). Springer Science and Business Media LLC. doi:10.103841598-022-10844-2

Demirel, M. C. (2021). Streamflow Forecasting with Deep Learning: A Case Study in California's American River Basin. *Journal of Hydrology (Amsterdam)*, *598*, 126444.

Dey, S., Botta, S., Kallam, R., Angadala, R., & Andugala, J. (2021). Seasonal variation in water quality parameters of Gudlavalleru Engineering College pond. *Current Research in Green and Sustainable Chemistry*, *4*, 100058. doi:10.1016/j.crgsc.2021.100058

Di Baldassarre, G., Sivapalan, M., Rusca, M., Cudennec, C., Garcia, M., Kreibich, H., Konar, M., Mondino, E., Mård, J., Pande, S., Sanderson, M. R., Tian, F., Viglione, A., Wei, J., Wei, Y., Yu, D. J., Srinivasan, V., & Blöschl, G. (2019). Sociohydrology: Scientific challenges in addressing the sustainable development goals. *Water Resources Research*, *55*(8), 6327–6355. doi:10.1029/2018WR023901 PMID:32742038

Di Lecce, V., Menoni, S., Mancini, L., & Masseroli, M. (2018). Hydro informatics: Data integration and knowledge discovery for smarter water management. *Environmental Modelling & Software*, *101*, 1–4.

Diaconu, D. C., Koutalakis, P. D., Gkiatas, G. T., Dascalu, G. V., & Zaimes, G. N. (2023). River Sand and Gravel Mining Monitoring Using Remote Sensing and UAVs. *Sustainability (Basel)*, *15*(3), 1944. doi:10.3390u15031944

Dietz, T., Ostrom, E., & Stern, P. C. (2003). The struggle to govern the commons. *Science, 302*(5652), 1907-1912.

Dinku, T., Ceccato, P., Grover-Kopec, E., Lemma, M., Connor, S. J., & Ropelewski, C. F. (2018). Validation of satellite rainfall products over East Africa's complex topography. *International Journal of Remote Sensing*, *29*(18), 6577–6600.

Dirisu, C., Mafiana, M., Dirisu, G. B., & Amodu, R. (2016). Level of Ph in drinking water of an oil and gas producing community and perceived biological and health. *ResearchGate*. https://www.researchgate.net/publication/332012834_LEVEL_OF_pH_IN_DRINKING_WATER_OF_AN_OIL_AND_GAS_PRODUCING_COMMUNITY_AND_PERCEIVED_BIOLOGICAL_AND_HEALTH_IMPLICATIONS

Dolatabadi, M., Mehrabpour, M., Esfandyari, M., Hossein, A., & Davoudi, M. (2018). Modeling of simultaneous adsorption of dye and metal ion by sawdust from aqueous solution using of ANN and ANFIS. *Chemometrics and Intelligent Laboratory Systems*, *181*, 72–78. doi:10.1016/j.chemolab.2018.07.012

Drakaki, K.-K., Sakki, G.-K., Tsoukalas, I., Kossieris, P., & Efstratiadis, A. (2022). Day-ahead energy production in small hydropower plants: uncertainty-aware forecasts through effective coupling of knowledge and data. In *Advances in Geosciences* (Vol. 56, pp. 155–162). Copernicus GmbH. doi:10.5194/adgeo-56-155-2022

Dudgeon, D., Arthington, A. H., Gessner, M. O., Kawabata, Z. I., Knowler, D. J., Lévêque, C., Naiman, R. J., Prieur-Richard, A.-H., Soto, D., Stiassny, M. L. J., & Sullivan, C. A. (2006). Freshwater biodiversity: Importance, threats, status and conservation challenges. *Biological Reviews of the Cambridge Philosophical Society*, *81*(2), 163–182. doi:10.1017/S1464793105006950 PMID:16336747

Duerr, I., Merrill, H. R., Wang, C., Bai, R., Boyer, M., Dukes, M. D., & Bliznyuk, N. (2018). Forecasting urban household water demand with statistical and machine learning methods using large space-time data: A Comparative study. In Environmental Modelling & Software (Vol. 102, pp. 29–38). Elsevier BV doi:10.1016/j.envsoft.2018.01.002

Dutta, D., Deka, L., & Mandal, D. (2019). Multi-objective optimization in real-time reservoir operation using multi-agent reinforcement learning. *Journal of Hydrology (Amsterdam)*, *574*, 554–565.

Dwivedi, Y. K., Hughes, L., Ismagilova, E., Aarts, G., Coombs, C., Crick, T., Duan, Y., Dwivedi, R., Edwards, J. S., Eirug, A., Galanos, V., Ilavarasan, P. V., Janssen, M., Jones, P., Kar, A. K., Kizgin, H., Kronemann, B., Lal, B., Lucini, B., ... Williams, M. D. (2021). Artificial Intelligence (AI): Multidisciplinary perspectives on emerging challenges, opportunities, and agenda for research, practice and policy. *International Journal of Information Management*, *57*, 101994. doi:10.1016/j.ijinfomgt.2019.08.002

Dzulkifli, N., Sarbini, N. N., Ibrahim, I. S., Abidin, N. I., Yahaya, F. M., & Nik Azizan, N. Z. (2021). Review on maintenance issues toward building maintenance management best practices. *Journal of Building Engineering*, *44*, 102985. doi:10.1016/j.jobe.2021.102985

Elbeltagi, A., Aslam, M. R., Malik, A., Mehdinejadiani, B., Srivastava, A., Bhatia, A. S., & Deng, J. (2020). The impact of climate changes on the water footprint of wheat and maize production in the Nile Delta, Egypt. *The Science of the Total Environment*, *743*, 140770. doi:10.1016/j.scitotenv.2020.140770 PMID:32679501

Elbeltagi, A., Srivastava, A., Deng, J., Li, Z., Raza, A., Khadke, L., Yu, Z., & El-Rawy, M. (2023). Forecasting vapor pressure deficit for agricultural water management using machine learning in semi-arid environments. *Agricultural Water Management*, *283*, 108302. doi:10.1016/j.agwat.2023.108302

Elleuchi, M., Khelif, R., Kharrat, M., Aseeri, M., Obeid, A., & Abid, M. (2019). Water Pipeline, Monitoring and Leak Detection, using soil moisture Sensors: IoT-based solution. *16th International Multi-Conference on Systems, Signals & Devices (SSD)*, 772–775.

El-Magd, S. A., Ismael, I. S., El-Sabri, M. S., Abdo, M. S., & Farhat, H. I. (2023). Integrated machine learning–based model and WQI for groundwater quality assessment: ML, geospatial, and hydroindex approaches. *Environmental Science and Pollution Research International*, *30*(18), 53862–53875. doi:10.100711356-023-25938-1 PMID:36864333

Elshaboury, N., & Marzouk, M. (2022). Prioritizing water distribution pipelines rehabilitation using ML algorithms. *Soft Computing*, *26*(11), 5179–5193. doi:10.100700500-022-06970-8

Emam, A. R., Mishra, B. K., Kumar, P., Masago, Y., & Fukushi, K. (2016). *Impact assessment of climate and land-use changes on flooding behavior in the upper ciliwung river*. Indonesia Water.

Emami, M., Ahmadi, A., Daccache, A., Nazif, S., Mousavi, S.-F., & Karami, H. (2022). County-Level Irrigation Water Demand Estimation Using Machine Learning: Case Study of California. In Water (Vol. 14, Issue 12, p. 1937). MDPI AG. doi:10.3390/w14121937

EPA. (2020). *National Water Quality Inventory Report to Congress*. United States Environmental Protection Agency.

EPA. (2021). *Water Quality Monitoring and Assessment*. United States Environmental Protection Agency.

Eregno, F. E., Tryland, I., Tjomsland, T., Myrmel, M., Robertson, L. J., & Heistad, A. (2016). Quantitative microbial risk assessment combined with hydrodynamic modeling to estimate the public health risk associated with bathing after rainfall events. *The Science of the Total Environment*, *548–549*, 270–279. doi:10.1016/j.scitotenv.2016.01.034 PMID:26802355

Ersue, M., Romascanu, D., Schoenwaelder, J., & Sehgal, A. (2015). *Management of Networks with Constrained Devices: Use Cases*. IETF Internet Draft. doi:10.17487/RFC7548

ESA. (2023). *European Space Agency - Earth Observation*. [Online]. Retrieved on 07 September 2023 from https://www.esa.int/Applications/Observing_the_Earth

Feng, Z., Shi, P., Yang, T., Niu, W., Zhou, J., & Cheng, C. (2022). Parallel cooperation search algorithm and artificial intelligence method for streamflow time series forecasting. In *Journal of Hydrology* (Vol. 606, p. 127434). Elsevier BV. doi:10.1016/j.jhydrol.2022.127434

Figueiredo, I., Esteves, P., & Cabrita, P. (2021). Water wise - A digital water solution for smart cities and water management entities. *Procedia Computer Science*, *181*(2019), 897–904. doi:10.1016/j.procs.2021.01.245

Fishman, C. (2011). *The big Thirst: The Secret Life and Turbulent Future of Water*. Simon and Schuster.

Foreman, W. T., Williams, T. L., Furlong, E. T., Hemmerle, D. M., Stetson, S., Jha, V. K., Noriega, M. C., Decess, J. A., Reed-Parker, C., & Sandstrom, M. W. (2021). Comparison of detection limits estimated using single- and multiconcentration spike-based and blank-based procedures. *Talanta, 228*, 122139. doi:10.1016/j.talanta.2021.122139 PMID:33773706

Fox, I. (2019). Institutions for water management in a changing world. In *Water In A Developing World* (pp. 9–24). Routledge. https://digitalrepository.unm.edu/cgi/viewcontent.cgi?article=3454&context=nrj

Frick, J., & Hegg, C. (2011). Can end-users' flood management decision making beimproved by information about forecast uncertainty? *Atmospheric Research, 100*(2-3), 296–303. doi:10.1016/j.atmosres.2010.12.006

Fu, Z., Cheng, J., Yang, M., & Batista, J. R. (2018). Prediction of industrial wastewater quality parameters based on wavelet denoised ANFIS model. In *2018 IEEE 8th Annual Computing and Communication Workshop and Conference (CCWC)* (pp. 301-306). IEEE. 10.1109/CCWC.2018.8301761

Fu, G., Jin, Y., Sun, S., Yuan, Z., & Butler, D. (2022). The role of deep learning in urban water management: A critical review. *Water Research, 223*, 118973–118973. doi:10.1016/j.watres.2022.118973 PMID:35988335

Gaitan, S., Calderoni, L., Palmieri, P., Veldhuis, M., Maio, D., & van Riemsdijk, M. B. (2014). From Sensing to Action: Quick and Reliable Access to Information in Cities Vulnerable to Heavy Rain. *IEEE Sensors Journal, 14*(12), 4175–4184. doi:10.1109/JSEN.2014.2354980

Gao, Q., Guo, S., Liu, X., Manogaran, G., Chilamkurti, N., & Kadry, S. (2020). Simulation analysis of supply chain risk management system based on IoT information platform. *Enterprise Information Systems, 14*(9), 1354–1378. doi:10.1080/17517575.2019.1644671

Gedda, G., Balakrishnan, K., Devi, R. U., Shah, K. J., Gandhi, V., Gandh, V., & Shah, K. L. (2021). Introduction to conventional wastewater treatment technologies: Limitations and recent advances. *Mater. Res. Found, 91*, 1–36. doi:10.21741/9781644901151-1

Geetha, S., & Gouthami, S. (2016). Internet of Things enabled real-time water quality monitoring system. *Smart Water, 2*(1), 1. Advance online publication. doi:10.118640713-017-0005-y

Ghangrekar, M. M. (2022). Aerobic Wastewater Treatment Systems. In *Wastewater to Water: Principles, Technologies and Engineering Design* (pp. 395–474). Springer Nature Singapore. doi:10.1007/978-981-19-4048-4_10

Ghobadi, F., & Kang, D. (2023). Application of Machine Learning in Water Resources Management: A Systematic Literature Review. In Water (Vol. 15, Issue 4, p. 620). MDPI AG. doi:10.3390/w15040620

Gholami, V., Khaleghi, M. R., & Sebghati, M. (2016). A method of groundwater quality assessment based on fuzzy network-CANFIS and geographic information system (GIS). *Applied Water Science, 7*(7), 3633–3647. doi:10.100713201-016-0508-y

Gholizadeh, A., Khiadani, M., Foroughi, M., Siuki, H. A., & Mehrfar, H. (2023). Wastewater treatment plants: The missing link in global One-Health surveillance and management of antibiotic resistance. *Journal of Infection and Public Health*. Advance online publication. doi:10.1016/j.jiph.2023.09.017 PMID:37865529

GhoochaniS.KhorramM.NazemiN.ClassificationS.QualityD. W.ScholarG. (2023). *Uncovering Top-Tier ML Classifier for Drinking Water Quality Detection.* doi:10.20944/preprints202308.1636.v1

Ghorai, P., & Ghosh, D. (2022). Sustainable Approach for Insoluble Phosphate Recycling from Wastewater Effluents. In Springer eBooks (pp. 77–86). doi:10.1007/978-3-030-94148-2_7

Ghosh, D., Chaudhary, S., & Dhara, S. (2023). Prospects and Potentials of Microbial Applications on Heavy-Metal Removal from Wastewater. *Metal Organic Frameworks for Wastewater Contaminant Removal,* 177–201. doi:10.1002/9783527841523.ch8

Ghosh, D., Debnath, S., & Das, S. (2022). Microbial electrochemical platform: A sustainable workhorse for improving wastewater treatment and desalination. In Elsevier eBooks (pp. 239–268). doi:10.1016/B978-0-323-90765-1.00014-9

Gikas, J., & Grant, M. M. (2013). Mobile computing devices in higher education: Student perspectives on learning with cellphones, smartphones & social media. *The Internet and Higher Education*, *19*, 18–26. doi:10.1016/j.iheduc.2013.06.002

Gillis, A. (2021). *What is internet of things (IoT)?* IOT Agenda.

Gino Sophia, S. G., Ceronmani Sharmila, V., Suchitra, S., Sudalai Muthu, T., & Pavithra, B. (2020). Water management using genetic algorithm-based machine learning. In Soft Computing (Vol. 24, Issue 22, pp. 17153–17165). Springer Science and Business Media LLC. doi:10.100700500-020-05009-0

Gleick, P. H. (2018). Transitions to freshwater sustainability. *Proceedings of the National Academy of Sciences of the United States of America*, *115*(36), 8863–8871. doi:10.1073/pnas.1808893115 PMID:30127019

Głomb, P., Cholewa, M., Koral, W., Madej, A., & Romaszewski, M. (2023). Detection of emergent leaks using ML approaches. *Water Science and Technology: Water Supply*, *23*(6), 2371–2386. doi:10.2166/ws.2023.118

Godini, K., Azarian, G., Kimiaei, A., Drăgoi, E. N., & Curteanu, S. (2021). Modeling of a real industrial wastewater treatment plant based on aerated lagoon using a neuro-evolutive technique. *Process Safety and Environmental Protection*, *148*, 114–124. doi:10.1016/j.psep.2020.09.057

Gong, C., Wang, W., Zhang, Z., Wang, H., Luo, J., & Brunner, P. (2020). Comparison of field methods for estimating evaporation from bare soil using lysimeters in a semi-arid area. *Journal of Hydrology (Amsterdam)*, *590*, 125–334. doi:10.1016/j.jhydrol.2020.125334

Gong, J., Guo, X., Yan, X., & Hu, C. (2023). Review of Urban Drinking Water Contamination Source Identification Methods. *Energies*, *16*(2), 705. Advance online publication. doi:10.3390/en16020705

Grbčić, L., Kranjčević, L., & Družeta, S. (2021). ML and simulation-optimization coupling for water distribution network contamination source detection. *Sensors (Basel)*, *21*(4), 1–25. doi:10.339021041157 PMID:33562175

Grizzetti, B., Bouraoui, F., Billen, G., Van Grinsven, H., Cardoso, A. C., Thieu, V., Garnier, J., Curtis, C., Howarth, R. W., & Johnes, P. J. (2011). Nitrogen as a threat to European water quality. In Cambridge University Press eBooks (pp. 379–404). doi:10.1017/CBO9780511976988.020

Gubbi, J., Buyya, R., Marusic, S., & Palaniswami, M. (2013). Internet of Things (IoT): A vision, architectural elements, and future directions. *Future Generation Computer Systems*, *29*(7), 1645–1660. doi:10.1016/j.future.2013.01.010

Gude, V. G., Rumbos, P., & Mattson, J. E. (2019). Smart meters for enhanced urban water supply management: A review. *Journal of Water Resources Planning and Management*, *145*(9), 04019037.

Guo, H., Jeong, K., Lim, J., Jo, J., Kim, Y. M., Park, J., Kim, J. H., & Cho, K. H. (2015). Prediction of effluent concentration in a wastewater treatment plant using machine learning models. *Journal of Environmental Sciences (China)*, *32*, 90–101. doi:10.1016/j.jes.2015.01.007 PMID:26040735

Gupta, A. D., Pandey, P., Feijóo, A., Yaseen, Z. M., & Bokde, N. D. (2020). Smart Water Technology for Efficient Water Resource Management: A Review. *Energies*, *13*(23), 6268. doi:10.3390/en13236268

Gupta, H. V., Mohtar, R. H., & Pande, S. (2019). The Water Energy Food Nexus: An integrated assessment framework for policy analysis. *Environmental Science & Policy*, *93*, 101–110.

Gupta, J., Pathak, S., & Kumar, G. (2022). Deep Learning (CNN) and Transfer Learning: A Review. *Journal of Physics: Conference Series*, *2273*(1), 012029. doi:10.1088/1742-6596/2273/1/012029

Gupta, V. K., Ali, I., Saleh, T. A., Nayak, A., & Agarwal, S. (2012). Chemical treatment technologies for waste-water recycling—An overview. *RSC Advances*, *2*(16), 6380. doi:10.1039/c2ra20340e

Guyon, I., & Elisseeff, A. (2003). An introduction to variable and feature selection. *Journal of Machine Learning Research*, *3*(Mar), 1157–1182.

Hadipour, M., Derakhshandeh, J. F., & Shiran, M. A. (2019). An experimental setup of a multi-intelligent control system (MICS) of water management using the Internet of Things (IoT). *ISA Transactions*. Advance online publication. doi:10.1016/j.isatra.2019.06.026 PMID:31285060

Hafezparast, M. (2021). Monitoring groundwater level changes of Mianrahan aquifer with GRACE satellite data. *Iranian Journal of Irrigation and Drainage*, *15*(2), 428–443.

Ha, L. T., Bastiaanssen, W. G., Simons, G. W., & Poortinga, A. (2023). A New Framework of 17 Hydrological Ecosystem Services (HESS17) for Supporting River Basin Planning and Environmental Monitoring. *Sustainability (Basel)*, *15*(7), 6182. doi:10.3390u15076182

Hamadou, W. S., Sulieman, A. M. E., Alshammari, N., Snoussi, M., Alanazi, N. A., Alshammary, A., & Al-Azmi, M. (2023). Water quality assessment of the surface and groundwater from Wadi Al-Adairey, Hail, Saudi Arabia. *Sustainable Water Resources Management*, *9*(5), 144. Advance online publication. doi:10.100740899-023-00923-1

Han, B., & Basu, A. (2023). Level Sets Guided by SoDEF-Fitting Energy for River Channel Detection in SAR Images. *Remote Sensing (Basel)*, *15*(13), 3251. doi:10.3390/rs15133251

Han, S. (2019). A Machine Learning Approach to Drought Prediction in the Context of the U.S. Drought Monitor. *Environmental Monitoring and Assessment*, *191*(5), 317. PMID:31041530

Han, Y., Zou, Z., & Wang, H. (2010). Adaptive neuro fuzzy inference system for classification of water quality status. *Journal of Environmental Sciences (China)*, *22*(12), 1891–1896. doi:10.1016/S1001-0742(09)60335-1 PMID:21462706

Hao, H., Wang, Y., & Shi, B. (2019). NaLa(CO3)2 hybridized with Fe3O4 for efficient phosphate removal: Synthesis and adsorption mechanistic study. *Water Research*, *155*, 1–11. doi:10.1016/j.watres.2019.01.049 PMID:30826591

Hao, Z., Yang, D., Zhao, J., Wang, X., Xu, J., & Li, Z. (2018). Bayesian network-based multi-objective reservoir operation with multi-scenario simulation considering hydrological uncertainty. *Water Resources Management*, *32*(5), 1767–1783.

Harmarneh, S. H., Hani, R. B., & Yaseen, Z. M. (2020). Predicting the water quality index using ensemble-based machine learning models. *Journal of Water Process Engineering*, *37*, 101442.

Hart, J. K., & Martinez, K. (2015). Toward an environmental Internet of Things. *Earth and Space Science (Hoboken, N.J.)*, *2*(5), 194–200. doi:10.1002/2014EA000044

Hasan, M. K., Habib, A. A., Islam, S., Balfaqih, M., Alfawaz, K. M., & Singh, D. (2023). Smart Grid Communication Networks for Electric Vehicles Empowering Distributed Energy Generation: Constraints, Challenges, and Recommendations. *Energies*, *16*(3), 1140. Advance online publication. doi:10.3390/en16031140

Hayashi, M. (2004). Temperature-Electrical conductivity relation of water for environmental monitoring and geophysical data inversion. *Environmental Monitoring and Assessment*, *96*(1–3), 119–128. doi:10.1023/B:EMAS.0000031719.83065.68 PMID:15327152

Hayder, G., Kurniawan, I., & Mustafa, H. M. (2021). Implementation of ML methods for monitoring and predicting water quality parameters. *Biointerface Research in Applied Chemistry*, *11*(2), 9285–9295. doi:10.33263/BRIAC112.92859295

He, C., Liu, Z., Wu, J., Pan, X., Fang, Z., Li, J., & Bryan, B. A. (2021). Future global urban water scarcity and potential solutions. *Nature Communications*, *12*(1), 4667. doi:10.103841467-021-25026-3 PMID:34344898

Heidari, A., & Jamali, M. A. (2022). IoT intrusion detection systems: A comprehensive review and future directions. *Cluster Computing*, 1–28.

Heidary Dahooie, J., Mohammadian, A., Qorbani, A. R., & Daim, T. (2023). A portfolio selection of IoT (IoTs) applications for the sustainable urban transportation: A novel hybrid multi criteria decision making approach. *Technology in Society*, *75*, 102366. doi:10.1016/j.techsoc.2023.102366

Hejazi, M., Edmonds, J., Clarke, L., Kyle, P., Davies, E., Chaturvedi, V., Wise, M., Patel, P., Eom, J., Calvin, K., Moss, R., & Kim, S. (2014). Long-term global water projections using six socioeconomic scenarios in an integrated assessment modelling framework. *Technological Forecasting and Social Change*, *81*, 205–226. doi:10.1016/j.techfore.2013.05.006

Hemdan, E. E.-D., Essa, Y. M., Shouman, M., El-Sayed, A., & Moustafa, A. N. (2023). An efficient IoT based smart water quality monitoring system. *Multimedia Tools and Applications*, *82*(19), 28827–28851. doi:10.100711042-023-14504-z

Hendricks, D. (2015). *The Trouble with the Internet of Things*. London Datastore. Greater London Authority.

Herman, J. D., Zeff, H. B., Lamontagne, J. R., Reed, P. M., & Characklis, G. W. (2021). Balancing water allocation under deep uncertainty and evolving infrastructure networks: Lessons from the California Water System. *Environmental Research Letters*, *16*(1), 014006.

Hernández-Del-Olmo, F., Gaudioso, E., & Nevado, A. (2012). Autonomous Adaptive and Active Tuning Up of the Dissolved Oxygen Setpoint in a Wastewater Treatment Plant Using Reinforcement Learning. *IEEE Transactions on Systems, Man, and Cybernetics. Part C, Applications and Reviews*, *42*(5), 768–774. doi:10.1109/TSMCC.2011.2162401

HESS. (2023). *Hydrology and Earth System Sciences Journal*. https://www.hydrol-earth-syst-sci.net/

Hewage, P., Behera, A., Trovati, M., Pereira, E., Ghahremani, M., Palmieri, F., & Liu, Y. (2020). Temporal convolutional neural (TCN) network for an effective weather forecasting using time-series data from the local weather station. In Soft Computing (Vol. 24, Issue 21, pp. 16453–16482). Springer Science and Business Media LLC. doi:10.100700500-020-04954-0

He, Y., Qiu, H. J., Song, J. X., Zhao, Y., Zhang, L. M., Hu, S., & Hu, Y. Y. (2019). Quantitative contribution of climate change and human activities to runoff changes in the Bahe River watershed of the Qinling Mountains, China. *Sustainable Cities and Society*, *51*, 101729. doi:10.1016/j.scs.2019.101729

Hou, D., He, H., Huang, P., Zhang, G., & Loaiciga, H. (2013). Detection of water-quality contamination events based on multi-sensor fusion using an extented Dempster–Shafer method. *Measurement Science & Technology*, *24*(5), 055801. doi:10.1088/0957-0233/24/5/055801

Hou, J., Gai, W., Cheng, W., & Deng, Y. (2021). Hazardous chemical leakage accidents and emergency evacuation response from 2009 to 2018 in China: A review. *Safety Science*, *135*, 105101–105101. doi:10.1016/j.ssci.2020.105101

Huang, H., Zhao, L., & Wu, Y. (2023). An IoT and machine learning enhanced framework for real-time digital human modeling and motion simulation. *Computer Communications*, *212*, 78–89. Advance online publication. doi:10.1016/j.comcom.2023.09.024

Huang, P. C., & Lee, K. T. (2020). Refinement of the channel response system by considering time-varying parameters for flood prediction. *Hydrological Processes*, *34*(21), 4097–4111. doi:10.1002/hyp.13868

Huang, Y., Chen, H., Liu, B., Huang, K., Wu, Z., & Yan, K. (2023). Radar Technology for River Flow Monitoring: Assessment of the Current Status and Future Challenges. *Water (Basel)*, *15*(10), 1904. doi:10.3390/w15101904

Hughes, D. A. (2019). A simple approach to estimating channel transmission losses in large South African river basins. *Journal of Hydrology. Regional Studies*, *25*, 100619. doi:10.1016/j.ejrh.2019.100619

Hussain, A., & Naaz, S. (2020). Prediction of diabetes mellitus: Comparative study of various machine learning models. In Advances in intelligent systems and computing (pp. 103–115). doi:10.1007/978-981-15-5148-2_10

İçağa, Y. (2007). Fuzzy evaluation of water quality classification. *Ecological Indicators*, *7*(3), 710–718. doi:10.1016/j.ecolind.2006.08.002

Ighalo, J. O., Adeniyi, A. G., & Marques, G. (2020). Internet of Things for Water Quality Monitoring and Assessment: A Comprehensive Review. *Artificial Intelligence for Sustainable Development: Theory, Practice and Future Applications*, 245–259. doi:10.1007/978-3-030-51920-9_13

Ikram, R. M. A., Ewees, A. A., Parmar, K. S., Yaseen, Z. M., Shahid, S., & Kisi, O. (2022). The viability of extended marine predators algorithm-based artificial neural networks for streamflow prediction. In *Applied Soft Computing* (Vol. 131, p. 109739). Elsevier BV. doi:10.1016/j.asoc.2022.109739

Inoue, J., Yamagata, Y., Chen, Y., Poskitt, C. M., & Sun, J. (2017). Anomaly Detection for a Water Treatment System Using Unsupervised Machine Learning. In *2017 IEEE International Conference on Data Mining Workshops (ICDMW). 2017 IEEE International Conference on Data Mining Workshops (ICDMW)*. IEEE. 10.1109/ICDMW.2017.149

IPCC. (2021). *The Physical Science Basis. Intergovernmental Panel on Climate Change*. IPCC.

ISO. (2017). *ISO 5725-1:1994. Accuracy (trueness and precision) of measurement methods and results - Part 1: General principles and definitions*. International Organization for Standardization.

ISRIC World Soil Information. (2023) https://www.isric.org/

Issoufou Ousmane, B., Nazoumou, Y., Favreau, G., Abdou Babaye, M. S., Abdou Mahaman, R., Boucher, M., Issoufa, I., Lawson, F. M. A., Vouillamoz, J.-M., Legchenko, A., & Taylor, R. G. (2023). Changes in aquifer properties along a seasonal river channel of the Niger River basin: Identifying potential recharge pathways in a dryland environment. *Journal of African Earth Sciences*, *197*, 104742. doi:10.1016/j.jafrearsci.2022.104742

Jahid, A., Alsharif, M. H., & Hall, T. J. (2023). The convergence of blockchain, IoT and 6G: Potential, opportunities, challenges and research roadmap. *Journal of Network and Computer Applications*, *217*, 103677. doi:10.1016/j.jnca.2023.103677

Jain, R., Thakur, A., Kaur, P., Kim, K., & Devi, P. (2020). Advances in imaging-assisted sensing techniques for heavy metals in water: Trends, challenges, and opportunities. *Trends in Analytical Chemistry*, *123*, 115758. doi:10.1016/j.trac.2019.115758

Jakaria, A. H. M., Hossain, M. M., & Rahman, M. A. (2020). *Smart Weather Forecasting Using Machine Learning:A Case Study in Tennessee* (Version 1). arXiv. doi:10.48550/ARXIV.2008.10789

Jakaria, A. H. M., Tennessee, T. M. H., & Rahman, M. A. (2020). *Smart weather forecasting using machine learning: a case study in*. Research Gate.

Jaseena, K. U., & Kovoor, B. C. (2022). Deterministic weather forecasting models based on intelligent predictors: A survey. In Journal of King Saud University - Computer and Information Sciences (Vol. 34, Issue 6, pp. 3393–3412). Elsevier BV. doi:10.1016/j.jksuci.2020.09.009

Javaid, M., Haleem, A., Singh, R. P., Suman, R., & Gonzalez, E. S. (2022). Understanding the adoption of Industry 4.0 technologies in improving environmental sustainability. *Sustainable Operations and Computers, 3*, 203–217. doi:10.1016/j.susoc.2022.01.008

Jeihouni, M., Toomanian, A., & Mansourian, A. (2019). Decision Tree-Based Data Mining and Rule Induction for Identifying High Quality Groundwater Zones to Water Supply Management: A Novel Hybrid Use of Data Mining and GIS. *Water Resources Management, 34*(1), 139–154. doi:10.100711269-019-02447-w

Jing, L., Chen, B., & Zhang, B. (2014). Modeling of UV-Induced photodegradation of naphthalene in marine oily wastewater by artificial neural networks. *Water, Air, and Soil Pollution, 225*(4), 1906. Advance online publication. doi:10.100711270-014-1906-0

Jin, J., Li, Z., & Xu, C. Y. (2021). Reservoir operation with improved hydrological forecasts using reinforcement learning: A case study in China. *Environmental Modelling & Software, 137*, 104915.

Joseph, F. (2019). IoT Based Weather Monitoring System for Effective Analytics. *International Journal of Engineering and Advanced Technology, 8*(4), 311–315.

Joseph, K., Sharma, A. K., van Staden, R. C., Wasantha, P. L. P., Cotton, J., & Small, S. (2023). Application of Software and Hardware-Based Technologies in Leaks and Burst Detection in Water Pipe Networks: A Literature Review. *Water (Basel), 15*(11), 2046. doi:10.3390/w15112046

Jury, W. A., & Vaux, H. J. (2007). The emerging global water crisis: managing scarcity and conflict between water users. In Advances in Agronomy (pp. 1–76). doi:10.1016/S0065-2113(07)95001-4

K, A., Ajith, V., K, N., & Ramesh, M. V. (2022). Design of IoT and ML enabled Framework for Water Quality Monitoring. In *2022 3rd International Conference on Electronics and Sustainable Communication Systems (ICESC)* (pp. 452–460). Academic Press.

Kallis, G., Kiparsky, M., & Norgaard, R. B. (2015). Collaborative governance and adaptive management: Lessons from California's CALFED Water Program. *Environmental Science & Policy, 55*, 1–12.

Kamaruidzaman, N. S., & Rahmat, S. N. (2020). Water Monitoring System Embedded with Internet of Things (IoT) Device: A Review. *IOP Conference Series. Earth and Environmental Science, 498*(1), 012068. doi:10.1088/1755-1315/498/1/012068

Kamienski, C., Soininen, J.-P., Taumberger, M., Dantas, R., Toscano, A., Salmon Cinotti, T., Filev Maia, R., & Torre Neto, A. (2019). Smart Water Management Platform: IoT-Based Precision Irrigation for Agriculture. In Sensors (Vol. 19, Issue 2, p. 276). MDPI AG. doi:10.339019020276

Kanade, P., & Prasad, J. P. (2021). Arduino-based Machine Learning and IoT Smart Irrigation System. *International Journal of Soft Computing and Engineering, 10*(4), 1–5. doi:10.35940/ijsce.D3481.0310421

Kang, G., Gao, J. Z., & Xie, G. (2017). Data-driven water quality analysis and prediction: A survey. *Proceedings - 3rd IEEE International Conference on Big Data Computing Service and Applications, BigDataService 2017*, 224–232. 10.1109/BigDataService.2017.40

Kaur, M., & Aron, R. (2022a). A Novel Load Balancing Technique for Smart Application in a Fog Computing Environment. *International Journal of Grid and High Performance Computing, 14*(1), 1–19. doi:10.4018/IJGHPC.301583

Kaur, M., & Aron, R. (2022b). An Energy-Efficient Load Balancing Approach for Fog Environment Using Scientific Workflow Applications. *Lecture Notes in Electrical Engineering, 903*(September), 165–174. doi:10.1007/978-981-19-2281-7_16

Kaur, P., & Sharma, N. (2022). An IOT-Based Smart Home Security Prototype Using IFTTT Alert System. *2022 International Conference on Machine Learning, Big Data, Cloud and Parallel Computing (COM-IT-CON), 1*, 393–400. 10.1109/COM-IT-CON54601.2022.9850500

Kayastha, N., Niyato, D., Hossain, E., & Han, Z. (2014). Smart grid sensor data collection, communication, and networking: A tutorial. *Wireless Communications and Mobile Computing, 14*(11), 1055–1087. doi:10.1002/wcm.2258

Keller, A. A., Garner, K., Rao, N., Knipping, E., & Thomas, J. (2023). Hydrological models for climate-based assessments at the watershed scale: A critical review of existing hydrologic and water quality models. *Science of the Total Environment, 867*, 161209. doi:10.1016/j.scitotenv.2022.161209

Kemper, M., Veenman, C., Blaak, H., & Schets, F. M. (2023). A membrane filtration method for the enumeration of *Escherichia coli* in bathing water and other waters with high levels of background bacteria. *Journal of Water and Health, 21*(8), 995–1003. doi:10.2166/wh.2023.004 PMID:37632376

Kendon, E. J., Stratton, R. A., Tucker, S., Marsham, J. H., Berthou, S., Rowell, D. P., & Senior, C. A. (2019). Enhanced future changes in wet and dry extremes over Africa at convection-permitting scale. *Nature Communications, 10*(1), 1794. doi:10.103841467-019-09776-9 PMID:31015416

Kesari, K. K., Soni, R., Jamal, Q. M. S., Tripathi, P., Lal, J. A., Jha, N. K., Siddiqui, M. H., Kumar, P., Tripathi, V., & Ruokolainen, J. (2021). Wastewater Treatment and Reuse: A Review of its Applications and Health Implications. *Water, Air, and Soil Pollution, 232*(5), 208. Advance online publication. doi:10.100711270-021-05154-8

Khalil, M., Ismanto, I., & Akhsani, R. (2021). Development Of an LPG Leak Detection System Using Instant Messaging Infrastructure Based on IoT. *2021 International Conference on Electrical and Information Technology*, 147–150.

Khalilpourazari, S., & Hashemi Doulabi, H. (2022). Designing a hybrid reinforcement learning based algorithm with application in prediction of the COVID-19 pandemic in Quebec. *Annals of Operations Research, 312*(2), 1261–1305. doi:10.100710479-020-03871-7 PMID:33424076

Khan, J., Lee, E., Balobaid, A. S., & Kim, K. (2023). A Comprehensive Review of Conventional, Machine Leaning, and Deep Learning Models for Groundwater Level (GWL) Forecasting. *Applied Sciences (Basel, Switzerland), 13*(4), 2743. Advance online publication. doi:10.3390/app13042743

Khan, S., Saeed, S., Khan, S. A., & Abrar, M. (2023). Impact of Bacteriological Water Quality on Water Borne Diseases and its Health Costs among Students of the Institutions. *Journal of Social Sciences Review, 3*(1), 510–518. doi:10.54183/jssr.v3i1.164

Khasawneh, M., & Azab, A., & Alrabaee, S. (2023). An IoT-based System Towards Detecting Oil Spills. *2023 International Conference on Intelligent Computing, Communication, Networking and Services (ICCNS)*, 197–202.

Khatoon, N. (2013). Correlation Study For the Assessment of Water Quality and Its Parameters of Ganga River, Kanpur, Uttar Pradesh, India. *IOSR Journal of Applied Chemistry, 5*(3), 80–90. doi:10.9790/5736-0538090

Khosravi, K., Golkarian, A., & Tiefenbacher, J. P. (2022). Using Optimized Deep Learning to Predict Daily Streamflow: A Comparison to Common Machine Learning Algorithms. In Water Resources Management (Vol. 36, Issue 2, pp. 699–716). Springer Science and Business Media LLC. doi:10.100711269-021-03051-7

Kılıçaslan, Y., Tuna, G., Gezer, G., Gülez, K., Arkoç, O., & Potirakis, S. M. (2014). ANN-Based estimation of groundwater quality using a wireless water quality network. *International Journal of Distributed Sensor Networks*, *10*(4), 458329. doi:10.1155/2014/458329

Kim, S.-T., Park, H.-J., & Kim, Y.-C. (2001). The load monitoring of the Web server using a mobile agent. *International Conferences on Info-Tech and Info-Net. Proceedings, 6*, 130–135.

Kirankumar, P., Keertana, G., Sivarao, S. U., Vijaykumar, B., & Shah, S. C. (2021). Smart Monitoring and Water Quality Management in Aquaculture using IOT and ML. In *2021 IEEE International Conference on Intelligent Systems, Smart and Green Technologies (ICISSGT)* (pp. 32–36). 10.1109/ICISSGT52025.2021.00018

Koditala, N. K., & Pandey, P. S. (2018). Water Quality Monitoring System Using IoT and Machine Learning. In *2018 International Conference on Research in Intelligent and Computing in Engineering (RICE)* (pp. 1–5). 10.1109/RICE.2018.8509050

Kohnhäuser, F., Meier, D., Patzer, F., & Finster, S. (2021). On the Security of IIoT Deployments: An Investigation of Secure Provisioning Solutions for OPC UA. *IEEE Access : Practical Innovations, Open Solutions*, *9*, 99299–99311. doi:10.1109/ACCESS.2021.3096062

Kolenčík, M., & Šimanský, V. (2016). Application of various methodological approaches for assessment of soil micromorphology due to VESTA program applicable to prediction of the soil structures formation. *Acta Fytotechnica et Zootechnica*, *19*(2), 68–73. doi:10.15414/afz.2016.19.02.68-73

Kolli, S., Ranjani, M., Kavitha, P., Daniel, D. A. P., & Chandramauli, A. (2023). Prediction of water quality parameters by IoT and machine learning. In *2023 International Conference on Computer Communication and Informatics (ICCCI)* (pp. 1–5). 10.1109/ICCCI56745.2023.10128475

Krishnan, S. R., Nallakaruppan, M. K., Chengoden, R., Koppu, S., Iyapparaja, M., Sadhasivam, J., & Sethuraman, S. (2022). Smart Water Resource Management Using Artificial Intelligence—A Review. *Sustainability (Basel)*, *14*(20), 13384. doi:10.3390u142013384

Kroll, C., Warchold, A., & Pradhan, P. (2019). Sustainable Development Goals (SDGs): Are we successful in turning trade-offs into synergies? *Palgrave Communications*, *5*(1), 140. doi:10.105741599-019-0335-5

Kshirsagar, P. R., Manoharan, H., Selvarajan, S., Althubiti, S. A., Alenezi, F., Srivastava, G., & Lin, J. C.-W. (2022). A Radical Safety Measure for Identifying Environmental Changes Using Machine Learning Algorithms. *Electronics (Basel)*, *11*(13), 1950. doi:10.3390/electronics11131950

Kumar, N., Reddy, N. S., & Ashreetha, B. (2023). Technologies for Comprehensive Information Security in the IoT. In *2023 International Conference for Advancement in Technology (ICONAT)* (pp. 1–5). Academic Press.

Kumar, T. M. V., Firoz, C. M., Bimal, P., Harikumar, P. S., & Sankaran, P. (2020). Smart water management for smart Kozhikode metropolitan area. *Advances in 21st Century Human Settlements*, 241–306. doi:10.1007/978-981-13-6822-6_7

Kumar, B. S., Ramalingam, S., Balamurugan, S., Soumiya, S., & Yogeswari, S. (2022). Water Management and Control Systems for Smart City using IoT and Artificial Intelligence. In *2022 International Conference on Edge Computing and Applications (ICECAA)* (pp. 653–657). 10.1109/ICECAA55415.2022.9936166

Kumar, C. P. (2018). Water Resources Issues and Management in India. *The Journal of Scientific and Engineering Research*, *5*(9), 2394–2630.

Kumar, D., Singh, R. K., Mishra, R., & Daim, T. U. (2023). Roadmap for integrating blockchain with IoT for sustainable and secured operations in logistics and supply chains: Decision making framework with case illustration. *Technological Forecasting and Social Change*, *196*, 122837. doi:10.1016/j.techfore.2023.122837

Kumar, J. S. (2022). *Smart Weather Prediction Using Machine Learning*. Jibendu Kumar Mantri.

Kusumastuti, D. I., Jokowinarno, D., Khotimah, S. N., & Dewi, C. (2017). The Use of Infiltration Wells to Reduce the Impacts of Land Use Changes on Flood Peaks. *An Indonesian Catchment Case Study*, *25*, 407–424.

Lakshmi, N. S., Ajimunnisa, P., Prasanna, V. L., Yugasravani, T., & RaviTeja, M. (2021). Prediction of weather forecasting by using machine learning. *International Journal of Innovative Research in Computer Science & Technology*, *9*(4). Advance online publication. doi:10.21276/ijircst.2021.9.4.7

Lambrechts, H. A., Paparrizos, S., Brongersma, R., Kroeze, C., Ludwig, F., & Stoof, C. R. (2023). Governing wildfire in a global change context: Lessons from water management in the Netherlands. *Fire Ecology*, *19*(1), 6. Advance online publication. doi:10.118642408-023-00166-7

Lan, Y., Lee, B. J., Wei, Y., & Zhang, C. (2020). Artificial neural networks for optimizing regional-scale water resource allocation: A framework and case study. *Environmental Modelling & Software*, *131*, 104779.

LaTour, J., Weldon, E., Dupré, D. H., & Halfar, T. M. (2006). *Water Resources Data for Illinois - Water Year 2005 (Includes historical data)*. doi:10.3133/wdrIL051

Li, S., Wang, H., Xu, T., & Zhou, G. (2011). Application Study on Internet of Things in Environment Protection Field. *Lecture Notes in Electrical Engineering, 133*, 99–106. doi:10.1007/978-3-642-25992-0_13

Liakos, K. G., Busato, P., Moshou, D., & Pearson, S., & Bochtis, D. (2018). Machine Learning in Agriculture: A Review. *Sensors (Basel)*, 18–18. PMID:30110960

Liang, J., Yi, Y., Li, X., Yuan, Y., Yang, S., Li, X., Zhu, Z., Lei, M., Meng, Q., & Zhai, Y. (2021). Detecting changes in water level caused by climate, land cover and dam construction in interconnected river-lake systems. *The Science of the Total Environment*, *788*, 147692. doi:10.1016/j.scitotenv.2021.147692 PMID:34022570

Li, J., Luo, G., He, L., Jing, X., & Lyu, J. (2017). Analytical Approaches for Determining Chemical oxygen Demand in water bodies: A review. *Critical Reviews in Analytical Chemistry*, *48*(1), 47–65. doi:10.1080/10408347.2017.1370670 PMID:28857621

Lindle, J., Villholth, K. G., Ebrahim, G. Y., Sorensen, J. P. R., Taylor, R. G., & Jensen, K. H. (2023). Groundwater recharge influenced by ephemeral river flow and land use in the semiarid Limpopo Province of South Africa. *Hydrogeology Journal*, 1–16. doi:10.100710040-023-02682-x

Liu, F. T., Ting, K. M., & Zhou, Z. H. (2008, December). Isolation forest. In *2008 eighth IEEE International Conference on Data Mining* (pp. 413-422). IEEE. 10.1109/ICDM.2008.17

Liu, J., Yuan, X., Zeng, J., Jiao, Y., Li, Y., Zhong, L., & Yao, L. (2022). Ensemble streamflow forecasting over a cascade reservoir catchment with integrated hydrometeorological modeling and machine learning. In Hydrology and Earth System Sciences (Vol. 26, Issue 2, pp. 265–278). Copernicus GmbH. doi:10.5194/hess-26-265-2022

Liu, N., Cheng, S., Wang, X., Li, Z., Zheng, L., Lyu, Y., Ao, X., & Wu, H. (2022). Characterization of microplastics in the septic tank via laser direct infrared spectroscopy. *Water Research*, *226*, 119293. doi:10.1016/j.watres.2022.119293 PMID:36323216

Liu, Y., Hou, G., Huang, F., Qin, H., Wang, B., & Yi, L. (2022). Directed graph deep neural network for multi-step daily streamflow forecasting. In *Journal of Hydrology* (Vol. 607, p. 127515). Elsevier BV. doi:10.1016/j.jhydrol.2022.127515

Li, W., Fang, H., Qin, G., Tan, X., Huang, Z., Zeng, F., Du, H., & Li, S. (2020). Concentration estimation of dissolved oxygen in Pearl River Basin using input variable selection and machine learning techniques. *The Science of the Total Environment*, *731*, 139099. doi:10.1016/j.scitotenv.2020.139099 PMID:32434098

Li, W., Ye, Z., Wang, Y., Yang, H., Yang, S., Gong, Z., & Wang, L. (2023). Development of a distributed MR-IoT method for operations and maintenance of underground pipeline network. *Tunnelling and Underground Space Technology*, *133*, 104935. doi:10.1016/j.tust.2022.104935

Lofrano, G., & Brown, J. (2010). Wastewater management through the ages: A history of mankind. *The Science of the Total Environment*, *408*(22), 5254–5264. doi:10.1016/j.scitotenv.2010.07.062 PMID:20817263

López-Riquelme, J. A., Pavón-Pulido, N., Navarro-Hellín, H., Soto-Valles, F., & Torres-Sánchez, R. (2017). A software architecture based on FIWARE cloud for Precision Agriculture. In *Agricultural Water Management* (Vol. 183, pp. 123–135). Elsevier BV. doi:10.1016/j.agwat.2016.10.020

Lowe, M., Qin, R., & Mao, X. (2022). A Review on Machine Learning, Artificial Intelligence, and Smart Technology in Water Treatment and Monitoring. *Water (Basel)*, *14*(9), 1384. doi:10.3390/w14091384

Lumb, A., Sharma, T. C., & Bibeault, J. (2011). A review of genesis and evolution of Water Quality Index (WQI) and some future directions. *Water Quality, Exposure, and Health*, *3*(1), 11–24. doi:10.100712403-011-0040-0

Ly, Q. V., Nguyen, X. C., Lê, N. C., Truong, T., Hoang, H. H., Park, T. J., Maqbool, T., Pyo, J., Cho, K. H., Lee, K., & Hur, J. (2021). Application of Machine Learning for eutrophication analysis and algal bloom prediction in an urban river: A 10-year study of the Han River, South Korea. *The Science of the Total Environment*, *797*, 149040. doi:10.1016/j.scitotenv.2021.149040 PMID:34311376

Mabrouki, J., Azrwe, M., Fattah, G., Dhiba, D., & Hajjaji, S. (2021). Intelligent monitoring system for biogas detection based on the IoT: Mohammedia, Morocco city landfill case. *Big Data Mining and Analytics*, *4*(1), 10–17. doi:10.26599/BDMA.2020.9020017

Madani, A., Hagage, M., & Elbeih, S. F. (2022). Random Forest and Logistic Regression algorithms for prediction of groundwater contamination using ammonia concentration. *Arabian Journal of Geosciences*, *15*(20), 1619. Advance online publication. doi:10.100712517-022-10872-2

Madani, K. (2019). Water Resources Allocation: A Comprehensive Review. *Water Resources Management*, *33*(9), 3183–3213.

Mahajan, Y., Shandilya, D., Batta, P., & Sharma, M. (2023). 3D Object 360-Degree Motion Detection Using Ultra-Frequency PIR Sensor. *2023 IEEE World Conference on Applied Intelligence and Computing (AIC)*, 614–619. 10.1109/AIC57670.2023.10263926

Malkawi, A., Ervin, S., Han, X., Chen, E. X., Lim, S., Ampanavos, S., & Howard, P. (2023). Design and applications of an IoT architecture for data-driven smart building operations and experimentation. *Energy and Building*, *295*, 113291. doi:10.1016/j.enbuild.2023.113291

Malviya, A., & Jaspal, D. (2021). Artificial intelligence as an upcoming technology in wastewater treatment: A comprehensive review. *Environmental Technology Reviews*, *10*(1), 177–187. doi:10.1080/21622515.2021.1913242

Mamandipoor, B., Majd, M., Sheikhalishahi, S., Modena, C., & Osmani, V. (2020). Monitoring and detecting faults in wastewater treatment plants using deep learning. *Environmental Monitoring and Assessment*, *192*(2), 148. Advance online publication. doi:10.100710661-020-8064-1 PMID:31997006

Manjakkal, L., Mitra, S., Petillot, Y. R., Shutler, J., Scott, E. M., Willander, M., & Dahiya, R. (2021). Connected Sensors, Innovative Sensor Deployment, and Intelligent Data Analysis for Online Water Quality Monitoring. *IEEE Internet of Things Journal*, *8*(18), 13805–13824. doi:10.1109/JIOT.2021.3081772

Mannina, G., Rebouças, T. F., Cosenza, A., Sànchez–Marrè, M., & Gibert, K. (2019). Decision support systems (DSS) for wastewater treatment plants – A review of the state of the art. *Bioresource Technology*, *290*, 121814. doi:10.1016/j.biortech.2019.121814 PMID:31351688

Manny, L. (2023). Socio-technical challenges towards data-driven and integrated urban water management: A socio-technical network approach. *Sustainable Cities and Society, 90*, 104360. doi:10.1016/j.scs.2022.104360

Manu, D. S., & Thalla, A. K. (2017). Artificial intelligence models for predicting the performance of biological wastewater treatment plant in the removal of Kjeldahl Nitrogen from wastewater. *Applied Water Science*, *7*(7), 3783–3791. doi:10.100713201-017-0526-4

Marques dos Santos, M., Caixia, L., & Snyder, S. A. (2023a). Evaluation of wastewater-based epidemiology of COVID-19 approaches in Singapore's 'closed-system' scenario: A long-term country-wide assessment. *Water Research, 244*, 120406. doi:10.1016/j.watres.2023.120406

Marshall, S. (2011). The water crisis in Kenya: Causes, effects and solutions. *Global Majority E-Journal*, *2*(1), 31–45.

Martínez, R. F., Vela, N., Aatik, A. E., Murray, E., Roche, P. C., & Navarro, J. M. (2020). On the Use of an IoT Integrated System for Water Quality Monitoring and Management in Wastewater Treatment Plants. *Water (Basel)*, *12*(4), 1096. doi:10.3390/w12041096

Masood, A., Tariq, M. U. R., Hashmi, M. Z. U. R., Waseem, M., Sarwar, M. K., Ali, W., Farooq, R., Almazroui, M., & Ng, A. W. M. (2022). An Overview of Groundwater Monitoring through Point-to Satellite-Based Techniques. *Water (Basel)*, *14*(4), 565. doi:10.3390/w14040565

Maspo, N., Harun, A., Goto, M., Nawi, M., & Haron, N. (2019). Development of Internet of Thing (IoT) Technology for Flood Prediction and Early Warning System (EWS). *International Journal of Innovative Technology and Exploring Engineering*, *8*(4), 2019–2228.

Ma, T., Sun, S., Fu, G., Hall, J. W., Ni, Y., He, L., Yi, J., Zhao, N., Du, Y., Pei, T., Cheng, W., Song, C., Fang, C., & Zhou, C. (2020). Pollution exacerbates China's water scarcity and its regional inequality. *Nature Communications*, *11*(1), 650. doi:10.103841467-020-14532-5 PMID:32005847

Matese, A., Gennaro, S., & Zaldei, A. (2015). Agrometeorological monitoring: Low-cost and open-source – Is it possible? *Italian Journal of Agrometeorology*, (3), 81–88.

Miro, M. E., Groves, D., Tincher, B., Syme, J., Tanverakul, S., & Catt, D. (2021). Adaptive water management in the face of uncertainty: Integrating machine learning, groundwater modeling and robust decision making. In *Climate Risk Management* (Vol. 34, p. 100383). Elsevier BV. doi:10.1016/j.crm.2021.100383

Mishra, S., & Tyagi, A. K. (2022). The Role of ML Techniques in IoT-Based Cloud Applications. *IoT*, (February), 105–135. doi:10.1007/978-3-030-87059-1_4

Mitsch, W. J., Zhang, L., Stefanik, K. C., Nahlik, A. M., Anderson, C. J., Bernal, B., Hernandez, M., & Song, K. (2012). Creating wetlands: Primary succession, water quality changes, and self-design over 15 years. *Bioscience*, *62*(3), 237–250. doi:10.1525/bio.2012.62.3.5

Mohamed, E. S., Belal, A. A., Abd-Elmabod, S. K., El-Shirbeny, M. A., Gad, A., & Zahran, M. B. (2021). Smart farming for improving agricultural management. *The Egyptian Journal of Remote Sensing and Space Sciences*, *24*(3), 971–981. Advance online publication. doi:10.1016/j.ejrs.2021.08.007

Mohammadpour, R., Shaharuddin, S., Chang, C. K., Zakaria, N. A., Ghani, A. A., & Chan, N. W. (2014). Prediction of water quality index in constructed wetlands using support vector machine. *Environmental Science and Pollution Research International*, *22*(8), 6208–6219. doi:10.100711356-014-3806-7 PMID:25408070

Mohapatra, J. B., Jha, P., Jha, M. K., & Biswal, S. (2021). Efficacy of machine learning techniques in predicting groundwater fluctuations in agro-ecological zones of India. *The Science of the Total Environment*, *785*, 147319. doi:10.1016/j.scitotenv.2021.147319 PMID:33957597

Mondejar, M. E., Avtar, R., Diaz, H. L. B., Dubey, R. K., Esteban, J., Gómez-Morales, A., Hallam, B., Mbungu, N. T., Okolo, C. C., Prasad, K. A., She, Q., & Garcia-Segura, S. (2021). Digitalization to Achieve Sustainable Development Goals: Steps Towards a Smart Green Planet. *The Science of the Total Environment*, *794*, 148539. doi:10.1016/j.scitotenv.2021.148539 PMID:34323742

Morabito, R., Kjallman, J., & Komu, M. (2015). Hypervisors vs. Lightweight Virtualization: A Performance Comparison. In *2015 IEEE International Conference on Cloud Engineering. 2015 IEEE International Conference on Cloud Engineering (IC2E)*. IEEE. 10.1109/IC2E.2015.74

Morchid, A., El Alami, R., Raezah, A. A., & Sabbar, Y. (2023). Applications of IoT and sensors technology to increase food security and agricultural Sustainability: Benefits and challenges. *Ain Shams Engineering Journal*, *102509*, 102509. Advance online publication. doi:10.1016/j.asej.2023.102509

Moubayed, A., Shami, A., & Ibrahim, A. (2022). *Intelligent Transportation Systems Orchestration: Lessons Learned & Potential Opportunities*. arXiv preprint arXiv:2205.14040.

Mpala, S. C., Gagnon, A. S., Mansell, M. G., & Hussey, S. W. (2020). Modelling the water level of the alluvial aquifer of an ephemeral river in south-western Zimbabwe. *Hydrological Sciences Journal*, *65*(8), 1399–1415. doi:10.1080/02626667.2020.1750615

Musa, A. A., Malami, S. I., Alanazi, F., Ounaies, W., Alshammari, M., & Haruna, S. I. (2023). Sustainable Traffic Management for Smart Cities Using Internet-of-Things-Oriented Intelligent Transportation Systems (ITS): Challenges and Recommendations. *Sustainability (Basel)*, *15*(13), 1–15. doi:10.3390u15139859

Musa, M. H., & Idrus, S. (2021). Physical and Biological Treatment Technologies of slaughterhouse Wastewater: A review. *Sustainability (Basel)*, *13*(9), 4656. doi:10.3390u13094656

Nadkarni, S., Kriechbaumer, F., Rothenberger, M., & Christodoulidou, N. (2020). The path to the Hotel of Things: IoT and Big Data converging in hospitality. *Journal of Hospitality and Tourism Technology*, *11*(1), 93–107. doi:10.1108/JHTT-12-2018-0120

Nagajayanthi, B. (2021). Decades of Internet of Things Towards Twenty-first Century: A Research-Based Introspective. In Wireless Personal Communications (Vol. 123, Issue 4, pp. 3661–3697). Springer Science and Business Media LLC. doi:10.100711277-021-09308-z

Narendran, S., Pradeep, P., & Ramesh, M. V. (2017). An Internet of Things (IoT) based sustainable water management. *2017 IEEE Global Humanitarian Technology Conference (GHTC)*. 10.1109/GHTC.2017.8239320

Narmilan, A., & Puvanitha, N. (2020). Mitigation techniques for agricultural pollution by precision technologies focusing on the IoT (IoT): A review. *Agricultural Reviews (Karnal), 41*(3), 279–284. doi:10.18805/ag.R-151

Narulita, I., & Ningrum, W. (2017). Extreme flood event analysis in Indonesia based on rainfall intensity and recharge capacity. *IOP Conf. Series: Earth and Environmental Science, 118*(2018), 012045. 10.1088/1755-1315/118/1/012045

Nasrabadi, T., Ruegner, H., Sirdari, Z. Z., Schwientek, M., & Grathwohl, P. (2016). Using total suspended solids (TSS) and turbidity as proxies for evaluation of metal transport in river water. *Applied Geochemistry, 68*, 1–9. doi:10.1016/j.apgeochem.2016.03.003

Nasseri, M., Mahdavi, M., & Shahcheraghi, H. (2015). Multi-objective particle swarm optimization for real-time operation of reservoir systems considering fuzzy operation rules. *Water Resources Management, 29*(4), 1069–1087.

Nathan, N. S., Saravanane, R., & Sundararajan, T. (2017). Application of ANN and MLR models on groundwater quality using CWQI at Lawspet, Puducherry in India. *Journal of Geoscience and Environment Protection, 05*(03), 99–124. doi:10.4236/gep.2017.53008

Nayar, R. (2020). Assessment of Water Quality Index and Monitoring of Pollutants by Physico-Chemical Analysis in Water Bodies: A review. *International Journal of Engineering Research & Technology (Ahmedabad), V9*(01). Advance online publication. doi:10.17577/IJERTV9IS010046

Neary, D. G., Ice, G. G., & Jackson, C. R. (2009). Linkages between forest soils and water quality and quantity. *Forest Ecology and Management, 258*(10), 2269–2281. doi:10.1016/j.foreco.2009.05.027

Ngoma, H., Wen, W., Ojara, M., & Ayugi, B. (2021). Assessing current and future spatiotemporal precipitation variability and trends over Uganda, East Africa, based on CHIRPS and regional climate model datasets. *Meteorology and Atmospheric Physics*, 1–21.

Nguyen, D. H., Le, X. H., Anh, D. T., Kim, S.-H., & Bae, D.-H. (2022). Hourly streamflow forecasting using a Bayesian additive regression tree model hybridized with a genetic algorithm. In *Journal of Hydrology* (Vol. 606, p. 127445). Elsevier BV. doi:10.1016/j.jhydrol.2022.127445

Nilanjan, Hassanien, Bhatt, Ashour, & Satapathy. (2018). Internet of things and big data analytics toward next-generation intelligence. Academic Press.

Norouzi, H., & Moghaddam, A. A. (2020). Groundwater quality assessment using random forest method based on groundwater quality indices (case study: Miandoab plain aquifer, NW of Iran). *Arabian Journal of Geosciences, 13*(18), 912. Advance online publication. doi:10.100712517-020-05904-8

Nourani, V., Elkiran, G., & Abba, S. I. (2018). Wastewater treatment plant performance analysis using artificial intelligence – an ensemble approach. *Water Science and Technology, 78*(10), 2064–2076. doi:10.2166/wst.2018.477 PMID:30629534

Nourbakhsh, Z., Mehrdadi, N., Moharamnejad, N., Hassani, A. H., & Yousefi, H. (2015). Evaluating the suitability of different parameters for qualitative analysis of groundwater based on analytical hierarchy process. *Desalination and Water Treatment, 57*(28), 13175–13182. doi:10.1080/19443994.2015.1056837

Nova, K. (2023). AI-Enabled Water Management Systems: An Analysis of System Components and Interdependencies for Water Conservation. *Eigenpub Review of Science and Technology, 7*(1), 105-124. https://studies.eigenpub.com/index.php/erst/article/download/12/11

Nova, K. (2023). AI-Enabled Water Management Systems: An Analysis of System Components and Interdependencies for Water Conservation. *Eigenpub Review of Science and Technology, 7*(1), 105–124. https://studies.eigenpub.com/index.php/erst/article/view/12

Nunes, J. V., Da Silva, M. W. B., Couto, G. H., Bordin, E. R., Ramsdorf, W. A., Flôr, I. C., Vicente, V. A., De Almeida, J. D., Celinski, F., & Xavier, C. R. (2021). Microbiological diversity in an aerated lagoon treating kraft effluent. *BioResources*, *16*(3), 5203–5219. doi:10.15376/biores.16.3.5203-5219

Nzila, A., Al-Ayoubi, S., Al-Gharabli, S., & Sayadi, S. (2020). Nanotechnology applications for the removal of biological and chemical contaminants from water. *Environmental Technology & Innovation*, *17*, 100589.

Omambia, A., Maake, B., & Wambua, A. (2022, May). Water quality monitoring using IoT & machine learning. In 2022 IST-Africa Conference (IST-Africa) (pp. 1-8). IEEE. doi:10.23919/IST-Africa56635.2022.9845590

Omambia, B., Maake, A., & Wambua. (2022). Water Quality Monitoring Using IoT & Machine Learning. In *2022 IST-Africa Conference* (pp. 1–8). Academic Press.

Optoelectronics, S. (2023).. . *Semiconductor Optoelectronics*, *42*(1), 200–211.

Orr, I., Mazari, K., Shukle, J. T., Li, R., & Filippelli, G. M. (2023). The impact of combined sewer outflows on urban water quality: Spatiotemporal patterns of fecal coliform in indianapolis. *Environmental Pollution*, *327*, 121531. doi:10.1016/j.envpol.2023.121531 PMID:37004861

Oruganti, R. K., Biji, A. P., Lanuyanger, T., Show, P. L., Sriariyanun, M., Upadhyayula, V. K., Gadhamshetty, V., & Bhattacharyya, D. (2023). Artificial intelligence and machine learning tools for high-performance microalgal wastewater treatment and algal biorefinery: A critical review. *The Science of the Total Environment*, *876*, 162797. doi:10.1016/j.scitotenv.2023.162797 PMID:36907394

Pai, S., Shruthi, M., Naveen, B., & K. (2020). Internet of Things: A Survey on Devices, Ecosystem, Components and Communication Protocols. *2020 4th International Conference on Electronics, Communication and Aerospace Technology (ICECA)*, 611–616.

Paillex, A., Enters, T., Angelini, C., & Bruder, A. (2017). Machine learning and ecological modelling for river habitat management. *Ecological Modelling*, *346*, 42–54.

Palanikkumar, P. A., Mary, A. Y., & Begum, D. G. (2023). A Novel IoT Framework and Device Architecture for Efficient Smart city Implementation. In *2023 7th International Conference on Trends in Electronics and Informatics (ICOEI)* (pp. 420–426). Academic Press.

Palermo, S. A., Maiolo, M., Brusco, A. C., Turco, M., Pirouz, B., Greco, E., Spezzano, G., & Piro, P. (2022). Smart Technologies for Water Resource Management: An Overview. *Sensors (Basel)*, *22*(16), 16. Advance online publication. doi:10.339022166225 PMID:36015982

Pandey, G., Pathak, N., & Chatterjee, D. (2019). Implementation of artificial intelligence for urban water supply system: A case study of Singapore. *Procedia Computer Science*, *152*, 136–143.

Park, K., Jung, Y., Seong, Y., & Lee, S. (2022). Development of Deep Learning Models to Improve the Accuracy of Water Levels Time Series Prediction through Multivariate Hydrological Data. In Water (Vol. 14, Issue 3, p. 469). MDPI AG. doi:10.3390/w14030469

Park, J., Kim, K. T., & Lee, W. H. (2020). Recent Advances in Information and Communications Technology (ICT) and Sensor Technology for Monitoring Water Quality. *Water (Basel)*, *12*(2), 2. Advance online publication. doi:10.3390/w12020510

Parwal, M. (2015). A Review Paper on Water Resource Management. *International Journal of New Technology and Research*, *1*(2), 2454–4116.

Patel, S., Sutaria, S., Daga, R., Shah, M., & Prajapati, M. (2023). A systematic study on complementary metal-oxide semiconductor technology (CMOS) and IoT for radioactive leakage detection in nuclear plant. *Nuclear Analysis*, *2*(3), 100080. doi:10.1016/j.nucana.2023.100080

Pathania, A., Kumar, P., Kesri, J., Priyanka, A. S., Mali, N., Singh, R., Chaturvedi, P., Uday, K.V., & Dutt, V. (2019). Reducing power consumption of weather stations for landslide monitoring. *Proceedings of 3rd International Conference on Information Technology in GeoEngineering 2019,* 1-16.

Patil, P. N. (2012). Physico-chemical parameters for testing of water -A review. *ResearchGate*. https://www.researchgate.net/publication/344323551_Physico-chemical_parameters_for_testing_of_water_-A_review

Peng, L., Wu, H., Gao, M., Yi, H., Xiong, Q., Yang, L., & Cheng, S. (2022). TLT: Recurrent fine-tuning transfer learning for water quality long-term prediction. In Water Research (Vol. 225, p. 119171). Elsevier BV. doi:10.1016/j.watres.2022.119171

Perumal, B., Nagaraj, P., Raja, S. E., Sunthari, S., Keerthana, S., & Muthukumar, M. V. (2022). Municipality Water Management System using IoT. In *2022 International Conference on Automation, Computing and Renewable Systems (ICACRS)* (pp. 317–320). 10.1109/ICACRS55517.2022.10029181

Piemontese, L., Kamugisha, R., Tukahirwa, J., Tengberg, A., Pedde, S., & Jaramillo, F. (2021). Barriers to scaling sustainable land and water management in Uganda: A cross-scale archetype approach. *Ecology and Society*, *26*(3), art6. Advance online publication. doi:10.5751/ES-12531-260306

Pijanowski, B. C., Brown, D. G., Shellito, B. A., & Manik, G. A. (2002). Using neural networks and GIS to forecast land use changes: A land transformation model. *Computers, Environment and Urban Systems*, *26*(6), 553–575. doi:10.1016/S0198-9715(01)00015-1

Poff, N. L., Brown, C. M., Grantham, T. E., Matthews, J. H., Palmer, M. A., Spence, C. M., Wilby, R. L., Haasnoot, M., Mendoza, G. F., Dominique, K. C., & Baeza, A. (2016). Sustainable water management under future uncertainty with eco-engineering decision scaling. *Nature Climate Change*, *6*(1), 25–34. doi:10.1038/nclimate2765

Pointl, M., & Fuchs-Hanusch, D. (2021). Assessing the Potential of LPWAN Communication Technologies for Near Real-Time Leak Detection in Water Distribution Systems. *Sensors (Basel)*, *21*(1), 293. doi:10.339021010293 PMID:33406751

Pras, A., & Mamane, H. (2023). Nowcasting of fecal coliform presence using an artificial neural network. *Environmental Pollution*, *326*, 121484. doi:10.1016/j.envpol.2023.121484 PMID:36958657

Prasad, K. D. V. (2023). A novel RF-SMOTE model to enhance the definite apprehensions for IoT security attacks. *Journal of Discrete Mathematical Sciences and Cryptography*, *26*(3), 861–873. doi:10.47974/JDMSC-1766

Proud, C., Fukai, S., Dunn, B., Dunn, T., & Mitchell, J. (2023). Effect of nitrogen management on grain yield of rice grown in a high-yielding environment under flooded and non-flooded conditions. *Crop and Environment*, *2*(1), 37–45. doi:10.1016/j.crope.2023.02.004

Puterman, E., Weiss, J., Lin, J., Schilf, S., Slusher, A. L., Johansen, K. L., & Epel, E. S. (2018). Aerobic exercise lengthens telomeres and reduces stress in family caregivers: A randomized controlled trial Richter Award Paper 2018. *Psychoneuroendocrinology*, *98*, 245–252. doi:10.1016/j.psyneuen.2018.08.002 PMID:30266522

Qin, X., Gao, F., & Chen, G. (2012). Wastewater quality monitoring system using sensor fusion and machine learning techniques. *Water Research*, *46*(4), 1133–1144. doi:10.1016/j.watres.2011.12.005 PMID:22200261

Queensland, G. (2019). *Environmental protection (water) policy 2009-monitoring and sampling manual physical and chemical assessment*. Academic Press.

Rahman, M., Bepery, C., Hossain, M. J., & Islam, M. M. (2020). Internet of Things (IoT) based water quality monitoring system. *ResearchGate*. https://www.researchgate.net/publication/344167317_Internet_of_Things_IoT_Based_Water_Quality_Monitoring_System

Rahman, A., Xi, M., Dabrowski, J. J., McCulloch, J., Arnold, S., Rana, M., George, A., & Adcock, M. (2021). An integrated framework of sensing, machine learning, and AR for aquaculture prawn farm management. *Aquacultural Engineering*, *95*, 102192. doi:10.1016/j.aquaeng.2021.102192

Ramachandran, B., Youssef, H. Y., Karunamurthi, J. V., Ebisi, F., Alnuaimi, A. N. A., & Najjar, J. (2023). IoT-based Test facility for a Water Transmission Line with Leak Simulation. *2023 International Conference on IT Innovation and Knowledge Discovery (ITIKD)*, 1–5. 10.1109/ITIKD56332.2023.10099751

Ramesha, G. N. (2020). Internet of things: Internet revolution, impact, technology road map and features. *Adv. Math. Sci. J*, *9*(7), 4405–4414. doi:10.37418/amsj.9.7.11

Ramesha, J., Kudari, M., & Samal, A. (2021). Smart Agriculture and Smart Farming using IoT Technology. *Journal of Physics: Conference Series*, *2089*(1), 12038–12038. doi:10.1088/1742-6596/2089/1/012038

Rasp, S., Dueben, P. D., Scher, S., Weyn, J. A., Mouatadid, S., & Thuerey, N. (2020). WeatherBench: A Benchmark Data Set for Data-Driven Weather Forecasting. In Journal of Advances in Modeling Earth Systems (Vol. 12, Issue 11). American Geophysical Union (AGU). doi:10.1029/2020MS002203

Raut, J. (2021). A Review on Weather Forecasting using Machine Learning and Deep Learning Techniques. *International Advanced Research Journal in Science, Engineering and Technology*, *8*, 5–5.

Rebouh, S., Bouhedda, M., & Hanini, S. (2015). Neuro-fuzzy modeling of Cu(II) and Cr(VI) adsorption from aqueous solution by wheat straw. *Desalination and Water Treatment*, *57*(14), 6515–6530. doi:10.1080/19443994.2015.1009171

Reddy, S., P, & S, P. (2022). Data Analytics and Cloud-Based Platform for Internet of Things Applications in Smart Cities. *2022 International Conference on Industry 4.0 Technology (I4Tech)*, 1–6.

Remesan, R., & Mathew, J. (2014). Machine Learning and Artificial Intelligence-Based approaches. In Springer eBooks (pp. 71–110). doi:10.1007/978-3-319-09235-5_4

Revolti, A., Dallasega, P., Schulze, F., & Walder, A. (2023). AR to support the maintenance of urban-line infrastructures: A case study. *Procedia Computer Science*, *217*, 746–755. doi:10.1016/j.procs.2022.12.271

Ridwan, W. M., Sapitang, M., Aziz, A., Kushiar, K. F., Ahmed, A. N., & El-Shafie, A. (2021). Rainfall forecasting model using machine learning methods: Case study Terengganu, Malaysia. In Ain Shams Engineering Journal (Vol. 12, Issue 2, pp. 1651–1663). Elsevier BV. doi:10.1016/j.asej.2020.09.011

Rizal, N. N. M., Hayder, G., Mnzool, M., Elnaim, B. M. E., Mohammed, A. O. Y., & Khayyat, M. M. (2022). Comparison between Regression Models, Support Vector Machine (SVM), and Artificial Neural Network (ANN) in River Water Quality Prediction. *Processes (Basel, Switzerland)*, *10*(8), 1652. doi:10.3390/pr10081652

Rizal, N. N. M., Hayder, G., & Yussof, S. (2023). *River Water Quality Prediction and Analysis*. In G. H. A. Salih & R. A. Saeed (Eds.), *Deep Learning Predictive Models Approach BT - Sustainability Challenges and Delivering Practical Engineering Solutions* (pp. 25–29). Springer International Publishing.

Roffia, L., Azzoni, P., Aguzzi, C., Viola, F., Antoniazzi, F., & Salmon Cinotti, T. (2018). Dynamic Linked Data: A SPARQL Event Processing Architecture. In Future Internet (Vol. 10, Issue 4, p. 36). MDPI AG. doi:10.3390/fi10040036

Rosa, L., Chiarelli, D. D., Rulli, M. C., Dell'Angelo, J., & D'Odorico, P. (2020). Global agricultural economic water scarcity. In Science Advances (Vol. 6, Issue 18). American Association for the Advancement of Science (AAAS). doi:10.1126ciadv.aaz6031

Rosati, R., Romeo, L., Cecchini, G., Tonetto, F., Viti, P., Mancini, A., & Frontoni, E. (2023). From knowledge-based to big data analytic model: A novel IoT and ML based decision support system for predictive maintenance in Industry 4.0. *Journal of Intelligent Manufacturing*, *34*(1), 107–121. doi:10.100710845-022-01960-x

Roy, S. K., Misra, S., Raghuwanshi, N. S., & Das, S. K. (2020). AgriSens: IoT-based dynamic irrigation scheduling system for water management of irrigated crops. *IEEE Internet of Things Journal*, *8*(6), 5023–5030. doi:10.1109/JIOT.2020.3036126

Rozos, E. (2019). Machine Learning, Urban Water Resources Management and Operating Policy. In Resources (Vol. 8, Issue 4, p. 173). MDPI AG. doi:10.3390/resources8040173

Rusydi, A. F. (2018). Correlation between conductivity and total dissolved solid in various type of water: A review. *IOP Conference Series, 118*, 012019. 10.1088/1755-1315/118/1/012019

Ryu, J. H. (2022). UAS-based real-time water quality monitoring, sampling, and visualization platform (UASWQP). *HardwareX, 11*, e00277. doi:10.1016/j.ohx.2022.e00277 PMID:35509896

Sacre Regis, M. D., Mouhamed, L., Kouakou, K., Adeline, B., Arona, D., Koffi Claude A, K., ... Issiaka, S. (2020). Using the CHIRPS Dataset to Investigate Historical Changes in Precipitation Extremes in West Africa. *Climate (Basel)*, *8*(7), 84. doi:10.3390/cli8070084

Sagan, V., Peterson, K. T., Maimaitijiang, M., Sidike, P., Sloan, J., Greeling, B. A., Maalouf, S., & Adams, C. (2020). Monitoring inland water quality using remote sensing: Potential and limitations of spectral indices, bio-optical simulations, machine learning, and cloud computing. *Earth-Science Reviews*, *205*, 103187. doi:10.1016/j.earscirev.2020.103187

Sahoo, S. K., & Goswami, S. S. (2024). Theoretical framework for assessing the economic and environmental impact of water pollution: A detailed study on sustainable development of India. *Journal of Future Sustainability*, *4*(1), 23–34. doi:10.5267/j.jfs.2024.1.003

Sakizadeh, M. (2015). Artificial intelligence for the prediction of water quality index in groundwater systems. *Modeling Earth Systems and Environment*, *2*(1), 8. Advance online publication. doi:10.100740808-015-0063-9

Salam, A., & Salam, A. (2020). Internet of things in water management and treatment. *Internet of Things for Sustainable Community Development: Wireless Communications. Sensory Systems*, 273–298. doi:10.1007/978-3-030-35291-2_9

Salcedo-Sanz, S., Pérez-Aracil, J., Ascenso, G., Ser, J. D., Casillas-Pérez, D., Kadow, C., ... Castelletti, A. (2023, August). Analysis, characterization, prediction, and attribution of extreme atmospheric events with machine learning and deep learning techniques: A review. *Theoretical and Applied Climatology*. Advance online publication. doi:10.100700704-023-04571-5

Salehahmadi, Z. (2018). Artificial Neural Network and Support Vector Machine for Prediction of Water Quality Parameters. *Environmental Monitoring and Assessment*, *190*(9), 534. PMID:30128706

Samudro, G., & Mangkoedihardjo, S. (2010). Review on BOD, COD and BOD/COD ratio: a triangle zone for toxic, biodegradable and stable levels. *ResearchGate*. https://www.researchgate.net/publication/228497615_Review_on_BOD_COD_and_BODCOD_ratio_a_triangle_zone_for_toxic_biodegradable_and_stable_levels

Sanzana, M. R., Maul, T., Wong, J. Y., Abdulrazic, M. O. M., & Yip, C.-C. (2022). Application of deep learning in facility management and maintenance for heating, ventilation, and air conditioning. *Automation in Construction*, *141*, 104445. doi:10.1016/j.autcon.2022.104445

Sarkar, I., Pal, B., Datta, A., & Roy, S. (2018). WiFi based portable weather station for monitoring temperature, relative humidity, pressure, precipitation, wind speed and direction. *Proceedings of International Conference on ICT for Sustainable Development (ICT4SD) 2018.*

Sarmas, E., Spiliotis, E., Marinakis, V., Tzanes, G., Kaldellis, J. K., & Doukas, H. (2022). ML-based energy management of water pumping systems for the application of peak shaving in small-scale islands. *Sustainable Cities and Society*, *82*, 103873. doi:10.1016/j.scs.2022.103873

Sater, R. A., & Hamza, A. B. (2021). A federated learning approach to anomaly detection in intelligent buildings. *ACM Transactions on IoT*, *2*(4), 1–23.

Sattar Hanoon, M., Najah Ahmed, A., Razzaq, A., Oudah, A. Y., Alkhayyat, A., Huang, F. Y., Kumar, P., & El-Shafie, A. (2023). Prediction of hydropower generation via machine learning algorithms at three Gorges Dam, China. In Ain Shams Engineering Journal (Vol. 14, Issue 4, p. 101919). Elsevier BV. doi:10.1016/j.asej.2022.101919

Sawunyama, L., Oyewo, O. A., Seheri, N., Onjefu, S. A., & Onwudiwe, D. C. (2023). Metal oxide functionalized ceramic membranes for the removal of pharmaceuticals in wastewater. *Surfaces and Interfaces*, *38*, 102787. doi:10.1016/j.surfin.2023.102787

Saxena, R., Hardainiyan, S., Singh, N., & Rai, P. K. (2022). Prospects of microbes in mitigations of environmental degradation in the river ecosystem. In *Ecological Significance of River Ecosystems* (pp. 429–454). Elsevier. doi:10.1016/B978-0-323-85045-2.00003-0

Scanlon, B. R., Jolly, I., Sophocleous, M., & Zhang, L. (2007). Global impacts of conversions from natural to agricultural ecosystems on water resources: Quantity versus quality. *Water Resources Research*, *43*(3), 2006WR005486. Advance online publication. doi:10.1029/2006WR005486

Schmidt, T., Harris, P., Lee, S., & McCabe, B. K. (2019). Investigating the impact of seasonal temperature variation on biogas production from covered anaerobic lagoons treating slaughterhouse wastewater using lab scale studies. *Journal of Environmental Chemical Engineering*, *7*(3), 103077. doi:10.1016/j.jece.2019.103077

Seelen, L. M. S., Flaim, G., Jennings, E., & De Senerpont Domis, L. N. (2019). Saving water for the future: Public awareness of water usage and water quality. *Journal of Environmental Management*, *242*, 246–257. doi:10.1016/j.jenvman.2019.04.047 PMID:31048230

Seeliger, A., Cheng, L., & Netland, T. (2023). AR for industrial quality inspection: An experiment assessing task performance and human factors. *Computers in Industry*, *151*, 103985. doi:10.1016/j.compind.2023.103985

Shafiq, M., Gu, Z., Cheikhrouhou, O., Alhakami, W., & Hamam, H. (2022). The Rise of "Internet of Things": Review and Open Research Issues Related to Detection and Prevention of IoT-Based Security Attacks. *Wireless Communications and Mobile Computing*. doi:10.1155/2022/8669348

Shaharuddin, S., Abdul Maulud, K. N., Syed Abdul Rahman, S. A. F., Che Ani, A. I., & Pradhan, B. (2023). The role of IoT sensor in smart building context for indoor fire hazard scenario: A systematic review of interdisciplinary articles. *Internet of Things : Engineering Cyber Physical Human Systems*, *22*, 100803. doi:10.1016/j.iot.2023.100803

Sharghi, E., Nourani, V., Zhang, Y., & Ghaneei, P. (2022). Conjunction of cluster ensemble-model ensemble techniques for spatiotemporal assessment of groundwater depletion in semi-arid plains. In *Journal of Hydrology* (Vol. 610, p. 127984). Elsevier BV. doi:10.1016/j.jhydrol.2022.127984

Sharma, D., Shukla, A., Bhondekar, A., Ghanshyam, C., & Ohja, A. (2016). A technical assessment of IOT for Indian agriculture sector. *IJCA Proceedings on National Symposium on Modern Information & Communication Technologies for Digital India 2016.*

Sharma, K., & Arun, M. R. (2022). Priority Queueing Model-Based IoT Middleware for Load Balancing. *2022 6th International Conference on Intelligent Computing and Control Systems (ICICCS)*, 425–430.

Sharma, S. K., Ghosh, S., & Bhattacharya, B. (2021). A review of recent developments in artificial intelligence for groundwater management. *Journal of Hydrology (Amsterdam)*, *590*, 125–487.

Shaw, K., & Dorea, C. C. (2021). Biodegradation mechanisms and functional microbiology in conventional septic tanks: A systematic review and meta-analysis. *Environmental Science. Water Research & Technology*, *7*(1), 144–155. doi:10.1039/D0EW00795A

Shekhar, S., Evans, M. R., Kang, J. M., & Mohan, P. (2011). Identifying patterns in spatial information: A survey of methods. *Wiley Interdisciplinary Reviews. Data Mining and Knowledge Discovery*, *1*(3), 193–214. doi:10.1002/widm.25

Shekhar, S., Li, W., & Zhang, P. (2011). *Encyclopedia of GIS*. Springer Science & Business Media.

Sheng, S., & Xu, G. (2018). Novel performance prediction model of a biofilm system treating domestic wastewater based on stacked denoising autoencoders deep learning network. *Chemical Engineering Journal*, *347*, 280–290. doi:10.1016/j.cej.2018.04.087

Shihu, S. (2011). Multi-sensor Remote Sensing Technologies in Water System Management. *3rd International Conference on Environmental Science and Information Application Technology*, 152–157. 10.1016/j.proenv.2011.09.027

Shumilova, O., Tockner, K., Sukhodolov, A., Khilchevskyi, V., De Meester, L., Stepanenko, S., Trokhymenko, G., Hernández-Agüero, J. A., & Gleick, P. (2023). Impact of the Russia–Ukraine armed conflict on water resources and water infrastructure. *Nature Sustainability*, *6*(5), 578–586. doi:10.103841893-023-01068-x

Sidwick, J. M. (1991). The preliminary treatment of wastewater. *Journal of Chemical Technology and Biotechnology*, *52*(3), 291–300. doi:10.1002/jctb.280520302

Singh, S., Kaushik, M., Gupta, A., & Malviya, A. K. (2019). Weather Forecasting using Machine Learning Techniques. SSRN *Electronic Journal.*

Singh, K. P., Gupta, S., Ojha, P., & Rai, P. (2012). Predicting adsorptive removal of chlorophenol from aqueous solution using artificial intelligence based modeling approaches. *Environmental Science and Pollution Research International*, *20*(4), 2271–2287. doi:10.100711356-012-1102-y PMID:22851225

Sit, M., Demiray, B. Z., Xiang, Z., Ewing, G. J., Sermet, Y., & Demir, I. (2020). A comprehensive review of deep learning applications in hydrology and water resources. *Water Science and Technology*, *82*(12), 2635–2670. doi:10.2166/wst.2020.369 PMID:33341760

Sivakumar, B. (2011). Water crisis: From conflict to cooperation—an overview. *Hydrological Sciences Journal*, *56*(4), 531–552. doi:10.1080/02626667.2011.580747

Smith, J. E., Brainard, R. E., Carter, G., Grinham, A., de Carvalho, R. L., Petus, C., ... Shaw, E. (2020). Monitoring coral reefs using artificial intelligence: A feasibility assessment. *Remote Sensing in Ecology and Conservation*, *6*(4), 398–411.

Solanki, A., Agrawal, H., & Khare, K. (2015). Predictive Analysis of Water Quality Parameters using Deep Learning. *International Journal of Computer Applications*, *125*(9), 29–34. doi:10.5120/ijca2015905874

Son, T. C., Kim, H. S., & Lee, W. (2008). Prediction of river water quality by artificial neural network model. *Water Science and Technology*, *58*(3), 569–576.

Sonune, A., & Ghate, R. (2004). Developments in wastewater treatment methods. *Desalination*, *167*, 55–63. doi:10.1016/j.desal.2004.06.113

Sridhara, P., Yadav, S. B., Chaudhary, S., Gahlot, K., Arya, A., Dahiya, Y., . . . N. (2023). Internet of Things and Cognitive Radio Networks: Applications, Challenges and Future. In Recent Advances in Metrology (Vol. 906). Springer.

Srinivasan, S., Hoffman, N. G., Morgan, M. T., Matsen, F. A., Fiedler, T. L., Hall, R. W., Ross, F. J., McCoy, C. O., Bumgarner, R., Marrazzo, J. M., & Fredricks, D. N. (2012). Bacterial communities in women with bacterial vaginosis: High-resolution phylogenetic analyses reveal relationships of microbiota to clinical criteria. *PLoS One*, *7*(6), e37818. doi:10.1371/journal.pone.0037818 PMID:22719852

Srivastava, M., Singh, R. K., & Sharma, A. (2018). Privacy-preservation blockchain with edge computing for secure IoT in agriculture. *Procedia Computer Science*, *132*, 668–675.

Sugam, V., Parthiban, P., Ravikumar, K., Das, I. C., & Ashutosh, D. (2023). Steady-state Assessment of Hydraulic Potential at Water Scarce regions of Agniyar River Basin, India using GMS-MODFLOW. *Disaster Advances*, *16*(5), 38–43. doi:10.25303/1605da038043

Sugiura, T., Yamamura, K., Watanabe, Y., Yamakiri, S., & Nakano, N. (2023). Circuits and devices for standalone large-scale integration (LSI) chips and IoT applications: A review. *Chip (Würzburg)*, *2*(3), 100048. doi:10.1016/j.chip.2023.100048

Sugumar, S. J., R, S., Phadke, S., Prasad, S., & R, S. G. (2021). Real Time Water Treatment Plant Monitoring System using IOT and Machine Learning Approach. In *2021 International Conference on Design Innovations for 3Cs Compute Communicate Control (ICDI3C)* (pp. 286–289). 10.1109/ICDI3C53598.2021.00064

Sulea, T., Rohani, N., Baardsnes, J., Corbeil, C. R., Deprez, C., Cepero-Donates, Y., Robert, A., Schrag, J. D., Parat, M., Duchesne, M., Jaramillo, M. L., Purisima, E. O., & Zwaagstra, J. C. (2020, January). Structure-based engineering of pH-dependent antibody binding for selective targeting of solid-tumor microenvironment. *mAbs*, *12*(1), 1682866. doi:10.1080/19420862.2019.1682866 PMID:31777319

Suma, V., Shekar, R. R., & Akshay, K. A. (2019). Gas leakage detection based on IOT. *3rd International Conference on Electronics, Communication and Aerospace Technology (ICECA)*, 1312–1315.

Sun, A. Y., & Scanlon, B. R. (2019). How can Big Data and machine learning benefit environment and water management: A survey of methods, applications, and future directions. *Environmental Research Letters*, *14*(7), 073001. doi:10.1088/1748-9326/ab1b7d

Sun, J., Hu, L., Li, D., Sun, K., & Yang, Z. (2022). Data-driven models for accurate groundwater level prediction and their practical significance in groundwater management. In *Journal of Hydrology* (Vol. 608, p. 127630). Elsevier BV. doi:10.1016/j.jhydrol.2022.127630

Sunku Mohan, V., Sankaran, S., Nanda, P., & Achuthan, K. (2023). Enabling secure lightweight mobile Narrowband IoT (NB-IoT) applications using blockchain. *Journal of Network and Computer Applications*, *219*, 103723. doi:10.1016/j.jnca.2023.103723

Sunny, J. S., Patro, C. P. K., Karnani, K., Pingle, S. C., Lin, F., Anekoji, M., Jones, L. D., Kesari, S., & Ashili, S. (2022). Anomaly detection framework for wearables data: A perspective review on data concepts, data analysis algorithms, and prospects. *Sensors (Basel)*, *22*(3), 756–756. doi:10.339022030756 PMID:35161502

Su, P., Wang, X. X., Lin, Q. D., Peng, J. L., Song, J. X., Fu, J. X., Wang, S. Q., Cheng, D. D., Bai, H. F., & Li, Q. (2019). Variability in macroinvertebrate community structure and its response to ecological factors of the Weihe River Basin, China. *Ecological Engineering*, *140*, 140. doi:10.1016/j.ecoleng.2019.105595

Su, Y., Chen, H., Qi, L., Gao, S., & Shang, J. K. (2013). Strong adsorption of phosphate by amorphous zirconium oxide nanoparticles. *Water Research*, *47*(14), 5018–5026. doi:10.1016/j.watres.2013.05.044 PMID:23850213

Swetha, T. M., Yogitha, T., Hitha, M. K. S., Syamanthika, P., Poorna, S. S., & Anuraj, K. (2021). IOT Based Water Management System For Crops Using Conventional Machine Learning Techniques. In *2021 12th International Conference on Computing Communication and Networking Technologies (ICCCNT)* (pp. 1–4). 10.1109/ICCCNT51525.2021.9579651

Syrmos, E., Sidiropoulos, V., Bechtsis, D., Stergiopoulos, F., Aivazidou, E., Vrakas, D., Vezinias, P., & Vlahavas, I. (2023). An Intelligent Modular Water Monitoring IoT System for Real-Time Quantitative and Qualitative Measurements. *Sustainability (Basel)*, *15*(3), 3. Advance online publication. doi:10.3390u15032127

Tandon, R., Verma, A., & Gupta, P. K. (2022a). Blockchain enabled vehicular networks: a review. *2022 5th International Conference on Multimedia, Signal Processing and Communication Technologies, IMPACT 2022*. 10.1109/IMPACT55510.2022.10029136

Tandon, R., Verma, A., & Gupta, P. K. (2022b). RVTN: Recommender system for vehicle routing in transportation network. *PDGC 2022 - 2022 7th International Conference on Parallel, Distributed and Grid Computing*, 352–356. 10.1109/PDGC56933.2022.10053267

Tandon, R., Verma, A., & Gupta, P. K. (2023). Nature-inspired whale optimization technique for efficient information exchange in vehicular networks. *Proceedings - 2023 12th IEEE International Conference on Communication Systems and Network Technologies, CSNT 2023*, 833–838. 10.1109/CSNT57126.2023.10134671

Taneja, M., Byabazaire, J., Jalodia, N., Davy, A., Olariu, C., & Malone, P. (2020). ML based fog computing assisted data-driven approach for early lameness detection in dairy cattle. *Computers and Electronics in Agriculture*, *171*, 105286. doi:10.1016/j.compag.2020.105286

Tatari, M., Agarwal, P., Alam, M. A., & Ahmed, J. (2022). Review of intelligent building management system. *ICT Systems and Sustainability: Proceedings of ICT4SD 2021, 1*, 167–176.

Taylor, S. J., & Letham, B. (2018). Forecasting at scale. *The American Statistician*, *72*(1), 37–45. doi:10.1080/00031305.2017.1380080

Tchobanoglous, G., Burton, F. L., Eddy, M. &., & Stensel, H. D. (2003). *Wastewater engineering: Treatment and Reuse*. Academic Press.

Thirumalini, S., & Joseph, K. G. (2009). Correlation between Electrical Conductivity and Total Dissolved Solids in Natural Waters. *Malaysian Journal of Science. Series B, Physical & Earth Sciences*, *28*(1), 55–61. doi:10.22452/mjs.vol28no1.7

Thirumalraj, A., & Rajesh, T. (2023). *An Improved ARO Model for Task Offloading in Vehicular Cloud Computing in VANET*. Academic Press.

Thiyagarajan, K., Pappu, S., Vudatha, P., & Niharika, A. V. (2017). Intelligent IoT based water quality monitoring system. *ResearchGate*. https://www.researchgate.net/publication/328802276_Intelligent_IoT_Based_Water_Quality_Monitoring_System

Tirkey, P., Bhattacharya, T., Chakraborty, S., & Baraik, S. (2017). Assessment of groundwater quality and associated health risks: A case study of Ranchi city, Jharkhand, India. *Groundwater for Sustainable Development*, *5*, 85–100. doi:10.1016/j.gsd.2017.05.002

Togneri, R., Kamienski, C., Dantas, R., Prati, R., Toscano, A., Soininen, J.-P., & Cinotti, T. S. (2019). Advancing IoT-Based Smart Irrigation. *IEEE Internet of Things Magazine*, *2*(4), 20–25. doi:10.1109/IOTM.0001.1900046

Tomasso, J. R. (1997). Environmental requirements and noninfectious diseases. *Developments in Aquaculture and Fisheries Science*, *30*, 253–270. doi:10.1016/S0167-9309(97)80012-9

Tomić, A. Š., Antanasijević, D., Ristić, M., Perić-Grujić, A. A., & Pocajt, V. (2018). A linear and nonlinear polynomial neural network modeling of dissolved oxygen content in surface water: Inter- and extrapolation performance with inputs' significance analysis. *The Science of the Total Environment*, *610–611*, 1038–1046. doi:10.1016/j.scitotenv.2017.08.192 PMID:28847097

Topare, N. S., Attar, S. J., & Manfe, M. M. (2011). Sewage/wastewater treatment technologies: A review. *Scientific Reviews and Chemical Communications, 1*(1). https://www.tsijournals.com/abstract/sewagewastewater-treatment-technologies-a-review-11194.html

Torres, A. B. B., da Rocha, A. R., Coelho da Silva, T. L., de Souza, J. N., & Gondim, R. S. (2020). Multilevel data fusion for the Internet of things in smart agriculture. *Computers and Electronics in Agriculture*, *171*, 105309. doi:10.1016/j.compag.2020.105309

Truong, Q. C., El Soueidy, C.-P., Hawchar, L., Li, Y., & Bastidas-Arteaga, E. (2023). Modelling two-dimensional chloride diffusion in repaired RC structures for sustainable maintenance management. *Structures*, *51*, 895–909. doi:10.1016/j.istruc.2023.03.088

Tütmez, B., Hatipoglu, Z., & Kaymak, U. (2006). Modeling electrical conductivity of groundwater using an adaptive neuro-fuzzy inference system. *Computers & Geosciences*, *32*(4), 421–433. doi:10.1016/j.cageo.2005.07.003

Uddin, M. G., Nash, S., Rahman, A., & Olbert, A. I. (2023a). A novel approach for estimating and predicting uncertainty in water quality index model using ML approaches. *Water Research, 229*, 119422. doi:10.1016/j.watres.2022.119422

Uddin, M. G., Nash, S., Rahman, A., & Olbert, A. I. (2023b). Performance analysis of the water quality index model for predicting water state using ML techniques. *Process Safety and Environmental Protection, 169*, 808–828. doi:10.1016/j.psep.2022.11.073

Ullah, R., Abbas, A. W., Ullah, M., Khan, R. U., Khan, I. U., Aslam, N., & Aljameel, S. S. (2021). EEWMP: An IoT-Based Energy-Efficient Water Management Platform for Smart Irrigation. In S. Nazir (Ed.), *Scientific Programming* (Vol. 2021, pp. 1–9). Hindawi Limited. doi:10.1155/2021/5536884

UNESCO. (2021). *Ethics of Artificial Intelligence*. UNESCO. Retrieved from https://en.unesco.org/themes/ethics-artificial-intelligence

UNLV. (2021). *AI algorithm predicts Lake Mead levels*. UNLV News Center. Retrieved from https://www.unlv.edu/news/release/ai-algorithm-predicts-lake-mead-levels

USGS. (2021). *Data Quality Information: U.S. Geological Survey Guidelines and Practices*. United States Geological Survey.

Vallino, E., Ridolfi, L., & Laio, F. (2020). Measuring economic water scarcity in agriculture: a cross-country empirical investigation. In Environmental Science & Policy (Vol. 114, pp. 73–85). Elsevier BV. doi:10.1016/j.envsci.2020.07.017

Verma, A., Tandon, R., & Gupta, P. K. (2022). TrafC-AnTabu: AnTabu routing algorithm for congestion control and traffic lights management using fuzzy model. *Internet Technology Letters*, *5*(2), 1–6. doi:10.1002/itl2.309

Von Haefen, R. H., Van Houtven, G., Naumenko, A., Obenour, D. R., Miller, J. A., Kenney, M. A., Gerst, M. D., & Waters, H. (2023). Estimating the benefits of stream water quality improvements in urbanizing watersheds: An ecological production function approach. *Proceedings of the National Academy of Sciences of the United States of America*, *120*(18), e2120252120. Advance online publication. doi:10.1073/pnas.2120252120 PMID:37094134

Vörösmarty, C. J., Osuna, V. R., Cak, A. D., Bhaduri, A., Bunn, S. E., Corsi, F., ... Uhlenbrook, S. (2018). Ecosystem-based water security and the Sustainable Development Goals (SDGs). *Ecohydrology & Hydrobiology*, *18*(4), 317–333. doi:10.1016/j.ecohyd.2018.07.004

Vulli, A., Srinivasu, P. N., Sashank, M. S. K., Shafi, J., Choi, J., & Ijaz, M. F. (2022). Fine-tuned DenseNet-169 for breast cancer metastasis prediction using FastAI and 1-cycle policy. *Sensors (Basel)*, *22*(8), 2988. doi:10.339022082988 PMID:35458972

Walczak, S. (2019). Artificial neural networks. In Advances in computer and electrical engineering book series (pp. 40–53). doi:10.4018/978-1-5225-7368-5.ch004

Walker, D., Smigaj, M., & Jovanovic, N. (2019). Ephemeral sand river flow detection using satellite optical remote sensing. *Journal of Arid Environments*, *168*, 17–25. doi:10.1016/j.jaridenv.2019.05.006

Water, U. N. (2018). *The United Nations World Water Development Report 2018: Nature-Based Solutions for Water.* UNESCO. Retrieved from https://www.unwater.org/publications/the-united-nations-world-water-development-report-2018-nature-based-solutions-for-water/

Water, U. N. (2021). *Water Scarcity.* United Nations World Water Development Report.

Wee, W. J., Zaini, N. B., Ahmed, A. N., & El-Shafie, A. (2021). A review of models for water level forecasting based on machine learning. In Earth Science Informatics (Vol. 14, Issue 4, pp. 1707–1728). Springer Science and Business Media LLC. doi:10.100712145-021-00664-9

Wen, X., Gong, B., Zhou, J., He, Q., & Qing, X. (2017). Efficient simultaneous partial nitrification, anammox and denitrification (SNAD) system equipped with a real-time dissolved oxygen (DO) intelligent control system and microbial community shifts of different substrate concentrations. *Water Research*, *119*, 201–211. doi:10.1016/j.watres.2017.04.052 PMID:28460292

Whelton, A. J., Dietrich, A. M., Burlingame, G. A., Schechs, M., & Duncan, S. E. (2007). Minerals in drinking water: Impacts on taste and importance to consumer health. *Water Science and Technology*, *55*(5), 283–291. doi:10.2166/wst.2007.190 PMID:17489421

Wissel, B., Boeing, W. J., & Ramcharan, C. W. (2003). Effects of water color on predation regimes and zooplankton assemblages in freshwater lakes. *Limnology and Oceanography*, *48*(5), 1965–1976. doi:10.4319/lo.2003.48.5.1965

World Bank. (2023). *World Bank Group - International Development, Poverty, & Sustainability*. World Bank. https://www.worldbank.org/

Woźnica, A., Absalon, D., Matysik, M., Bąk, M., Cieplok, A., Halabowski, D., Koczorowska, A., Krodkiewska, M., Libera, M., Sierka, E., Spyra, A., Czerniawski, R., Sługocki, Ł., & Łozowski, B. (2023). Analysis of the Salinity of the Vistula River Based on Patrol Monitoring and State Environmental Monitoring. *Water (Basel)*, *15*(5), 838. doi:10.3390/w15050838

Wu, G., Liu, Y., Liu, B., Ren, H., Wang, W., Zhang, X., Yuan, Z., & Yang, M. (2023). Hanjiang River Runoff Change and Its Attribution Analysis Integrating the Inter-Basin Water Transfer. *Water (Basel)*, *15*(16), 2974. doi:10.3390/w15162974

Yadav, D., Pandey, A., Mishra, D., Bagchi, T., Mahapatra, A., Chandrasekhar, P., & Kumar, A. (2022). IoT-enabled smart dustbin with messaging alert system. *International Journal of Information Technology : an Official Journal of Bharati Vidyapeeth's Institute of Computer Applications and Management*, *14*(7), 3601–3609. doi:10.100741870-022-00947-4

Yahya, A. S. A., Ahmed, A. N., Othman, F., Ibrahim, R. K., Afan, H. A., El-Shafie, A., Fai, C. M., Hossain, S., Ehteram, M., & El-Shafie, A. (2019). Water Quality Prediction Model Based Support Vector Machine Model for Ungauged River Catchment under Dual Scenarios. *Water (Basel)*, *11*(6), 1231. doi:10.3390/w11061231

Yang, C.-H., Lee, K.-C., & Li, S.-E. (2020). A mixed activity-based costing and resource constraint optimal decision model for IoT-oriented intelligent building management system portfolios. *Sustainable Cities and Society*, *60*, 102142. doi:10.1016/j.scs.2020.102142

Yang, F., Moayedi, H., & Mosavi, A. (2021). Predicting the degree of dissolved oxygen using three types of Multi-Layer Perceptron-Based artificial neural networks. *Sustainability (Basel)*, *13*(17), 9898. doi:10.3390u13179898

Yaseen, Z. M., Abba, A. H., & Deo, R. C. (2018). Drought prediction: A challenge to meet in Northern Sudan. *The Science of the Total Environment*, *621*, 130–144.

Yeong, D. J., Velasco-Hernandez, G., Barry, J., & Walsh, J. (2021). Sensor and sensor fusion technology in autonomous vehicles: A review. *Sensors (Basel)*, *21*(6), 2140–2140. doi:10.339021062140 PMID:33803889

Yussif, A.-M., Sadeghi, H., & Zayed, T. (2023). Application of Machine Learning for Leak Localization in Water Supply Networks. In Buildings (Vol. 13, Issue 4, p. 849). MDPI AG. doi:10.3390/buildings13040849

Zainab, A., Amina, I., Abdulmuhaimin, M., Sadiku, A. S., & Baballe, M. A. (2023). The water monitoring system's disadvantages. Zenodo *(CERN European Organization for Nuclear Research)*. doi:10.5281/zenodo.8161049

Zambrano-Bigiarini, M., & Baez-Villaneuva, O. M. (2019). Characterizing meteorological droughts in data scare regions using remote sensing estimates of precipitation. In *Extreme Hydroclimatic Events and Multivariate Hazards in a Changing Environment* (pp. 221–246). Elsevier. doi:10.1016/B978-0-12-814899-0.00009-2

Zaresefat, M., & Derakhshani, R. (2023). Revolutionizing Groundwater Management with Hybrid AI Models: A Practical Review. *Water (Basel)*, *15*(9), 1750. doi:10.3390/w15091750

Zehetbauer, T., Plöckinger, A., Emminger, C., & Çakmak, U. D. (2021). Mechanical Design and Performance Analyses of a Rubber-Based Peristaltic Micro-Dosing Pump. *Actuators*, *10*(8), 198. doi:10.3390/act10080198

Zhai, Z., Martínez, J. F., Beltran, V., & Martínez, N. L. (2020). Decision support systems for agriculture 4.0: Survey and challenges. (2020). *Computers and Electronics in Agriculture*, *170*, 105256. doi:10.1016/j.compag.2020.105256

Zhang, C., & Song, Y. (2017). Spatial analysis of water quality in the Haihe River Basin in China. *Scientific Reports*, *7*, 40123.

Zhang, X., Rane, K. P., Kakaravada, I., & Shabaz, M. (2021). Research on vibration monitoring and fault diagnosis of rotating machinery based on IoT technology. *Nonlinear Engineering*, *10*(1), 245–254. doi:10.1515/nleng-2021-0019

Zhao, J., Liu, D., & Huang, R. (2023). A Review of Climate-Smart Agriculture: Recent Advancements, Challenges, and Future Directions. *Sustainability (Basel)*, *15*(4), 1–15. doi:10.3390u15043404

Zheng, W. (2021). Deep Learning for Groundwater Level Prediction in an Arid Region: Case Study in the Middle Reaches of Heihe River Basin, Northwest China. *Journal of Hydrology (Amsterdam)*, *596*, 125758.

Zhu, M., Wang, J., Yang, X., Zhang, Y., Zhang, L., Ren, H., Wu, B., & Ye, L. (2022). *A review of the application of machine learning in water quality evaluation*. Eco-Environment & Health. doi:10.1016/j.eehl.2022.06.001

Zolanvari, M., Yang, Z., Khan, K., Jain, R., & Meskin, N. (2023, February 15). TRUST XAI: Model-Agnostic Explanations for AI With a Case Study on IIoT Security. *IEEE Internet of Things Journal*, *10*(4), 2967–2978. Advance online publication. doi:10.1109/JIOT.2021.3122019

About the Contributors

Abhishek Kumar is an Assistant Director/Associate Professor in the Computer Science & Engineering Department at Chandigarh University, Mohali, Punjab, India. He has a Doctorate in Computer Science from the University of Madras. He is doing Post-Doctoral Fellow in the Ingenium Research Group Lab, Universidad De Castilla-La Mancha, Ciudad Real, and Ciudad Real Spain. He has more than 150 publications in reputed, peer-reviewed National and International Journals, books, and conferences. He is the Series Editor for three books: Quantum Computing with De Gruyter Germany, Intelligent Energy System with Elsevier, & Sustainable Energy with Nova, USA.

Arun Lal Srivastav works as an Associate Professor at Chitkara University, Himachal Pradesh, India. He obtained Ph.D. degree from the Indian Institute of Technology (BHU), Varanasi, India. He has done post-doctoral research at National Chung Hsing University, Taiwan. He is currently involved in teaching Environmental Science, Environmental Engineering, Disaster Management, and Design Thinking to undergraduate engineering students. His research interests include Water Quality Surveillance, Climate Change, Water Treatment, River Ecosystems, Soil Health, Phytoremediation and Waste Management. He has published over 80 research papers in various prestigious journals (Elsevier, Springer Nature, IWA, Wiley, etc.), including book chapters and conference papers. He is the editor of 17 books with Elsevier, Nova, Springer, IGI Global, and Wiley.

Ashutosh Kumar Dubey is an Associate Professor in the Department of Computer Science at Chitkara University in Himachal Pradesh, India. He is a Postdoctoral Fellow at the Ingenium Research Group Lab, Universidad de Castilla-La Mancha, Ciudad Real, Spain. Dr. Dubey completed his BE and M. Tech degrees in Computer Science and Engineering from RGPV, Bhopal, Madhya Pradesh. He completed his PhD in Computer Science and Engineering at JK Lakshmipat University in Jaipur, Rajasthan, India. He is a Senior Member of both IEEE and ACM and possesses more than 16 years of teaching experience. Dr. Dubey has authored and edited fifteen books and has published over 65 articles in peer-reviewed international journals and conference proceedings. He serves as an Editor, Editorial Board Member, and Reviewer for numerous peer-reviewed journals. His research interests encompass Machine Learning, Renewable Energy, Health Informatics, Nature-Inspired Algorithms, Cloud Computing, and Big Data.

Vishal Dutt, a Principal Research Consultant at AVN Innovations, is a renowned freelance trainer in Android development with over seven years of academic teaching experience. He has authored over 50 publications in prestigious national and international journals, SCI and Scopus journals, conferences, and book chapters. Vishal has contributed editorially to two books with Wiley and Eureka publications and is working on three additional publications with Wiley. A sought-after keynote speaker, he has significantly contributed to workshops and webinars across India. He provides peer review services for elite publishers like Elsevier, Springer, and IEEE Access and was a program committee member and reviewer for ICCIPS 2021. His research focuses on Data Science, Data Mining, Machine Learning, Deep Learning, and Remote Sensing, with extensive experience in data analytics using Rapid Miner, Tableau, and WEKA, as well as over six years in Java and Android development.

Narayan Vyas, a Principal Research Consultant at AVN Innovations, is a distinguished academician and expert in advanced technologies. He cleared the NTA UGC NET & JRF in Computer Science & Applications on his first attempt, underscoring his academic excellence. With profound knowledge of the Internet of Things (IoT) and Mobile Application Development, he has trained students worldwide and authored numerous articles in reputable national and international Scopus-indexed conferences and journals. His research spans IoT, Remote Sensing, Machine Learning, Deep Learning, and Computer Vision. A sought-after keynote speaker, Mr. Vyas collaborates with leading publishers like Wiley, IGI Global, and DeGruyter on various book projects, marking his significant contribution to the field.

* * *

Abdullah Alzahrani is a lead researcher and special lecturer in the Electrical and Computer Engineering Department at Oakland University in Rochester, MI, USA. Alzahrani received his Ph.D in the Electrical and Computer Engineering Department at Oakland University. He received his B.S. in electrical engineering in 2014 from Northern Border University located in Arar, KSA, and in 2018 his M.S. in electrical and computer engineering from Rochester Institute of Technology, Rochester, NY, USA. His research interests are in the application of optimization techniques and policy analysis to smart grid technology, electrical power, and control power systems. He is a member of the Saudi Council of Engineering.

Vanshika Ambati is a student of Institute of Pharmaceutical Technology Sri Padmavati Mahila Visvavidyalayam (Women's University), Tirupati, A.P., India.

Ardison is Senior Researcher from Research Center of Public Policy, The Research Organization of Public Governance, Economy and Community Welfare, National Research and Innovation. His expertise is in broadcasting communication and new media.

Manjunatha B. did his BE in Mechanical Engineering, M.Tech in Production Engineering and System Technology and Ph.D in Mechanical Engineering. He is associated with New Horizon college of Engineering, Bengaluru since 2003. He has played a key role in implementing the innovative best practices of teaching-learning and research at NHCE since last 2 decades. He is actively involved in R&D and has published good number of papers in indexed journals and is a recognized research guide under Visveswaraya Technological University. He has guided PG and Ph.D students for their research work.

Neha Bhati is a Research Trainee at AVN Innovations.

Ahmad Budi Setiawan is an Associate Researcher in the field of expertise in Information Systems at the National Research and Innovation Agency (BRIN). Ahmad has an educational background from the Masters Program in Information Technology, University of Indonesia which was completed in 2013. Currently actively conducting research with a research focus on the field of information systems which includes: Information Security Systems, Cybersecurity, IT Governance, electronic business systems and socio-informatics.

Ronak Duggar is a Director & Research Consultant at AVN Innovations.

Dipankar Ghosh is an Associate Professor, Department of Biosciences, JIS University, Kolkata.

Dankan Gowda V. is currently working as an Assistant Professor in the Department of Electronics and Communication Engineering at BMS Institute of Technology and Management in Bangalore. Previously, he worked as a Research Fellow at ADA DRDO and as a Software Engineer at Robert Bosch. With a total experience of 14 years, including teaching and industry, Dr. Gowda has made significant contributions to both academia and research. He has published over 60 research papers in renowned international journals and conferences. In recognition of his innovative work, Dr. Gowda has been granted six patents, including four from Indian authorities and two international patents. His research interests primarily lie in the fields of IoT and Signal Processing, where he has conducted workshops and handled industry projects. Dr. Gowda is passionate about teaching and strives to create an engaging and effective learning environment for his students. He consistently seeks opportunities to enhance his knowledge and stay updated with the latest advancements in his field.

Rajni Gupta received her PhD in computer science from Thapar University, Patiala in 2013. She obtained her Master's degree in mathematics and computing from Thapar University, Patiala in 2009. At present she is working as Assistant Professor in SVKM's Narsee Monjee Institute of Management Studies (NMIMS) University, Mumbai (Maharashtra), India. Previously, she worked as Associate Professor in Computer Science & Engineering Department, Lovely Professional University, Punjab. She has also worked as Assistant Professor, Computer Science & Engineering Department, LNMIIT, Jaipur. Before joining LNMIIT, she was Post-Doctorate Fellow in INRIA, France and Concordia University, Montreal, Canada. She has published her research work in highly reputed scientific citation index journals. She won "Microsoft Azure" educator grand award and got many scholarships to attend international conferences like Grace hopper celebration of women in computing sponsored by Google.

Inzimam Ul Hassan is currently working as an Assistant Professor in computer Science and Engineering Department in Vivekananda Global University, Jaipur, India. He is Pursuing PhD in Computer science and Engineering from Chandigarh university. He has done M Tech from Lovely Professional University, Punjab India and B Tech from Baba Ghulam Shah Badshah University, Rajouri, India. He has about 5 years of experience in teaching.

Mandeep Kaur has completed her PhD in 2021 from lovely professional university, Jalandhar. She has completed her M.Tech in CSE in 2015 from DIET, Kharar under PTU, Jalandhar. She has completed B.Tech in CSE in 2011 from G.G.S.C.M.T. Kharar under PTU, Jalandhar. She is currently working as Assistant Professor in Department of Computer Science and Engineering, Chitkara University, Punjab, India. Previously she has worked in Chandigarh University, Gharuan, Punjab India. She have also worked in SBSSM Govt. College(Girls), Guru Ka, Khuh, Munne. Before that, she was working in GCET, Kahnpur Khuhi, Anandpur Sahib, Punjab. Her area of interest is Fog computing and Cloud computing.

B. Kuber is Faculty of Institute of Pharmaceutical Technology Sri Padmavati Mahila Visvavidyalayam (Women's University), Tirupati, A.P., India.

Zeeshan Lone is working as an Assistant Professor at Vivekananda Global University, Jipur, India.

Poojitha Nalluri is a student of Institute of Pharmaceutical Technology Sri Padmavati Mahila Visvavidyalayam (Women's University), Tirupati, Andhra Pradesh, India.

Htet Ne Oo has completed her Doctoral Degree in Information Technology and having teaching and research experience of more than 13 years. She has been awarded Fellowship by NAM S&T Center and DST GOI. Moreover, she has experience of working in collaborative projects with Busan Digital University, ROK. She is an alumina of Sakura Exchange Program in Science by DST, Japan. Her area of expertise are Computer Networks, Artificial Intelligence, Machine Learning, GIS and Remote Sensing. Currently she is working in AI based precision agriculture, AI based medical diagnosis and AI based smart city development researches.

K.D.V. Prasad works as a Faculty of Research at Symbiosis Institute of Business Management, Hyderabad; Symbiosis International (Deemed University), Pune, India. Dr. Prasad holds a Master's in Computer Applications and a Master's in Software Systems from BITS, Pilani; MBA (Human Resources), IGNOU, New Delhi. Dr. Prasad possesses a Ph.D. in Business Management (Kanpur University), and PhD in Business Administration (RTM Nagpur University. He is AIMA-Certified Management Teacher and Fellow, World HR Board, Carlton Advanced Management Institute, USA; Fellowship of the Management and Business Research Council (FMBRC) by Open Association Research Society, Delaware, USA Dr. Prasad Published over 100 articles in Scopus/WoS indexed journals and 3 books.

Danny Ruhiyat, while studying at South Tangerang Institute of Technology (Institut Teknologi Tangerang Selatan / ITTS), worked at PT. SiCepat Ekspres Indonesia as AIoT Tech Lead. His research activities are related to Artificial Intelligence (AI) and the Internet of Things (IoT).

Abeer Saber was born in Damietta, Dumyat, Egypt, in 1992. She received the B.Sc. Degree in Computer Science and the M.Sc. Degree in Computer Science from Mansoura University, Egypt, in 2013 and 2018, respectively. She is received the Ph.D. in computer science from Menoufia University, Egypt, in 2022. She is currently a Lecturer in information technology with the Faculty of Computers and Artificial intelligence, Damietta University, Egypt. She has published many research articles in prestigious international conferences and reputable journals. She is also a reviewer for many journals. Her current research interests include big data analysis, semantic web, linked open data, Optimization, machine learning, Deep learning, bioinformatics, and IoT.

Ramandeep Sandhu () (57270451000) () is an Assistant Professor at the School of Computer Science and Engineering, Lovely Professional University, Punjab, India. She received a B. Tech, M.Tech in Computer Science Engineering from GNE College Ludhiana, Punjab, India, and Ph.D. in Computer Science Engineering from Lovely Professional University, Punjab, India. She has 14 years of teaching experience of reputed institutes. Her general research areas are Cloud Computing, AI/ML, and Opinion Mining/Sentiment Analysis. She has published 2 books. The first one is named "OOPs Awareness with C++", ISBN: 978-93-5300-939-7. Another one is "A Start to C++ Programming with Object Oriented Concepts", ISBN: 978-93-955819-3-6.

Vaishnavi Srivastava is a Ph.D. research scholar in the School of Life Sciences and Biotechnology, Chhatrapati Shahuji Maharaj University, Kanpur. She has a B.Sc. degree in Biotechnology from Invertis University, Bareilly and M.Sc. degree in Biotechnology from Chhatrapati Shahuji Maharaj University, Kanpur.

Swati is working as an Assistant Professor in Computer Science and Engineering Department in Lovely Professional University, Punjab, India.

Aries Syamsuddin is a Civil Government Employee at Population and Civil Registration Office, Distric Government of Blitar, District Government of Blitar, Indonesia. His expertise is in IoT, cyber security and information system.

Righa Tandon is an accomplished Assistant Professor currently serving at Chitkara University, where she brings with her a wealth of knowledge and experience in the field of Computer Science and Engineering. She completed her Ph.D. in 2021 from Jaypee University of Information Technology after obtaining her B.Tech and M.Tech degrees in the same field. She has shown immense dedication and passion for her area of expertise, which includes information security, fog computing, image processing, blockchain and vehicular networks. Her research work has been published in numerous reputed SCI journals, Scopus indexed publications, and book chapters. In recognition of her innovative contributions, she has also filed four patents. She has served as TPC in various conferences and served as reviewer in various journals/conferences. With a sharp analytical mind and an unwavering commitment to her field, she is a highly regarded and respected member of the academic community.

Arunadevi Thirumalraj did her BE in Computer Science Engineering in Government college of Engineering, Bodinayakanur and she has worked as a Research associate in various industrial and academic research institutes for past 6 years. Now she is pursing ME Computer Science Engineering in K. Ramakrishnan College of Technology, (KRCT), Tiruchirappalli. She is actively involved in the field of innovative teaching and Learning. She has publishing a good number of papers in indexed journals.

Anusuya V. S. did her MSc in Organic Chemistry, Ph D in Analytical Chemistry. She is associated with New Horizon college of Engineering, Bengaluru since 2015. She has played a key role as HOD Chemistry in implementing the innovative best practices of teaching-learning and research at NHCE. She is actively involved in R&D and has published good number of papers in indexed journals and is a recognized research guide under Visveswaraya Technological University. She has guided Ph.D students for their research work.

Heena Wadhwa is currently working as Assistant Professor in Chitkara University, Punjab, India. She received her B.Tech and M.Tech degree from Punjab Technical University, India in 2007 and 2014 respectively. She received her Ph.D. degree in Computer science and Engineering from Lovely Professional University, India in 2022. Her research interests include blockchain, cloud computing, fog computing, IoT, software engineering, and wireless sensor networks. She has published various research papers in reputed peer reviewed SCI and SCOPUS indexed journals and various national/international conferences. She has also served as a reviewer in various journals/conferences. In recognition of her innovation contributions, she has also filed six patents.

Djoko Waluyo is Principal Researcher from Research Center of Public Policy, The Research Organization of Public Governance, Economy and Community Welfare, National Research and Innovation. His expertise is in communication science and new media.

Index

www.ingramcontent.com/pod-product-compliance
Lightning Source LLC
LaVergne TN
LVHW080640011026
849160LV00014B/1272
* 9 7 9 8 3 6 9 3 4 6 8 4 6 *